Generalized Frequency Distributions for Environmental and Water Engineering

A multitude of processes in hydrology and environmental engineering are either random or entail random components that are characterized by random variables. These variables are described by frequency distributions. This book provides an overview of generalized frequency distributions, their properties, and their applications to the fields of water resources and environmental engineering. A variety of generalized distributions are covered, including the Burr–Singh–Maddala distribution; Halphen distributions; three-parameter generalized gamma distribution; generalized beta Lomax distribution; Feller–Pareto distribution; kappa distribution; and four-parameter exponential gamma distribution. The last chapter summarized and compared the above generalized distributions. Using real-world data, this book provides a valuable reference for researchers, graduate students, and professionals interested in frequency analysis.

VIJAY P. SINGH is University Distinguished Professor, Regents Professor, and Caroline and William N. Lehrer Distinguished Chair in Water Engineering at Texas A&M University. He has published more than 1,220 journal articles, 30 books, 70 edited reference books, 113 book chapters, and 314 conference papers in the areas of hydrology, groundwater, hydraulics, irrigation, pollutant transport, copulas, entropy, and water resources. He has received more than 93 national and international awards, including the Arid Lands Hydraulic Engineering Award; the Richard R. Torrens Award; the Norman Medal; the EWRI Lifetime Achievement Award, given by the American Society of Civil Engineers (ASCE); the Ray K. Linsley Award and Founder's Award, given by the American Institute of Hydrology; the Crystal Drop Award and the Ven Te Chow Award, given by the International Water Resources Association; and three honorary doctorates. He is a distinguished member of ASCE and a fellow of EWRI, AWRA, IWRS, ISAE, IASWC, and IE. He has served as the president of the American Institute of Hydrology, serves on the editorial boards of more than 25 journals and 3 book series, and is the president-elect of AAWRE-ASCE.

LAN ZHANG has joined Dewberry in Fairfax, VA since August of 2021. Before that she worked for Texas A & M University and University of Akron. She holds a BS in mechanical engineering, an MS in water resources sciences, and a PhD in civil and environmental engineering. She has published more than 40 articles in the areas of hydrology, copulas, water quality, entropy, and water resources. She has been working on the copula and its applications in hydrology and water resource engineering for more than 10 years.

Generalized Frequency Distributions for Environmental and Water Engineering

VIJAY P. SINGH
Texas A&M University

LAN ZHANG
Dewberry

CAMBRIDGE
UNIVERSITY PRESS

CAMBRIDGE
UNIVERSITY PRESS

University Printing House, Cambridge CB2 8BS, United Kingdom

One Liberty Plaza, 20th Floor, New York, NY 10006, USA

477 Williamstown Road, Port Melbourne, VIC 3207, Australia

314–321, 3rd Floor, Plot 3, Splendor Forum, Jasola District Centre,
New Delhi – 110025, India

103 Penang Road, #05–06/07, Visioncrest Commercial, Singapore 238467

Cambridge University Press is part of the University of Cambridge.

It furthers the University's mission by disseminating knowledge in the pursuit of
education, learning, and research at the highest international levels of excellence.

www.cambridge.org
Information on this title: www.cambridge.org/9781316516843
DOI: 10.1017/9781009025317

First published 2022

A catalogue record for this publication is available from the British Library.

Library of Congress Cataloging-in-Publication Data
Names: Singh, V. P. (Vijay P.) author. | Zhang, Lan, author.
Title: Generalized frequency distributions for environmental and water engineering /
Vijay Singh, Texas A&M University, Lan Zhang, University of Akron.
Description: Cambridge, United Kingdom ; New York, NY : Cambridge University Press,
[2022] | Includes bibliographical references and index.
Identifiers: LCCN 2021037802 (print) | LCCN 2021037803 (ebook) | ISBN 9781316516843
(hardcover) | ISBN 9781009025317 (epub)
Subjects: LCSH: Hydrology–Statistical methods. | Environmental engineering–Statistical
methods. | Distribution (Probability theory)
Classification: LCC GB656.2.S7 S56 2022 (print) | LCC GB656.2.S7 (ebook) |
DDC 628.01/5195–dc23
LC record available at https://lccn.loc.gov/2021037802
LC ebook record available at https://lccn.loc.gov/2021037803

ISBN 978-1-316-51684-3 Hardback

To

VPS: Wife, Anita, who is no more; son, Vinay; daughter, Arti; daughter-in-law, Sonali; son-in-law, Vamsi; and grandsons, Ronin, Kayden, and Davin

LZ: Husband, Bret; and son, Caelan

Contents

Preface

Frequency analysis occupies a prominent place in hydrometeorology, hydrology, hydraulics, and environmental and water resources engineering. One-, two-, and three-parameter distributions are often employed for frequency analysis, but most of these distributions are found to be special cases of generalized distributions. However, these generalized distributions have received limited attention in environmental and water engineering. Further, parameter estimation in frequency analysis is usually done using the methods of moments, cumulative moments, maximum likelihood estimation, and L-moments, but the entropy-based method, despite its obvious advantages, has not yet become common. Understanding generalized distributions will help uncover their underlying hypotheses, which may help estimate their parameters and make informed inferences. Moreover, it will be instructive to describe the estimation of parameters of these distributions using entropy and compare it with other methods. Currently, generalized distributions and their parameter estimation with entropy do not appear to have been discussed under one cover, and this is what constituted the motivation for this book.

The subject matter of the book is divided into 11 chapters. Introducing the theme of the book, Chapter 1 provides a snapshot of the generalized distributions to be discussed in the book and includes a short discussion of the methods of parameter estimation, such as methods of moments, cumulative moments, maximum likelihood estimation, probability weighted moments, L-moments, and least squares; goodness-of-fit statistics; selection of a distribution; and confidence intervals.

The Burr–Singh–Maddala probability distribution, a generalization of the Pareto distribution and the Weibull distribution, is discussed in Chapter 2. The chapter also deals with distribution characteristics; derivation using the entropy theory; parameter estimation with methods of moments, maximum likelihood estimation, entropy, probability weighted moments, L-moments, and cumulative moments; and application to simulated and real-world data.

Chapter 3 discusses the Halphen type A frequency distribution; the derivation of the distribution using entropy theory; the estimation of its parameters using the methods of entropy, moments, and maximum likelihood estimation; and application to synthetic and real-world data.

The Halphen type B frequency distribution is presented in Chapter 4. The chapter discusses the derivation of the distribution using entropy theory; the estimation of its parameters using entropy and methods of moments and maximum likelihood estimation; and application to synthetic and real-world data.

Chapter 5 deals with the Halphen Inverse B distribution. It discusses the derivation of the distribution and its characteristics; the estimation of its parameters using the methods of entropy, moments, and maximum likelihood estimation; and application to synthetic and real-world data.

The three-parameter generalized gamma distribution, a generalization of the two-parameter gamma distribution, exponential distribution, two-parameter gamma distribution, Weibull distribution, and lognormal distribution, is the subject matter of Chapter 6, which includes the derivation using the entropy theory; parameter estimation using the methods of entropy, maximum likelihood estimation, and moments; and application to synthetic and real-world data.

Chapter 7 treats the four-parameter beta Lomax distribution, a generalization of the three-parameter beta distribution, and discusses the derivation using the entropy theory; parameter estimation with the methods of maximum entropy and maximum likelihood estimation; and application to synthetic and real-world data.

The five-parameter Feller–Pareto distribution, as a generalization of the two-parameter beta distribution, is discussed in Chapter 8. The discussion includes the derivation using the entropy theory; parameter estimation with the methods of maximum entropy, maximum likelihood estimation, and moments; and application to synthetic and real-world peak flow data.

Chapter 9 discusses the kappa distribution and its characteristics; parameter estimation using the methods of entropy, maximum likelihood estimation, moments, and L-moments; and application to synthetic and real-world data.

The four-parameter exponential gamma distribution is described in Chapter 10, which is applicable to frequency analysis for a range of random variables, such as floods, drought, wind velocity, and rainfall. This distribution gives rise to a number of distributions useful for frequency analyses in environmental and water engineering. The chapter discusses the derivation of the distribution using the entropy theory; parameter estimation with the methods of entropy, maximum likelihood estimation, and moments; and application to real-world data.

Chapter 11 – the last chapter – summarizes and compares the results of fitting different distributions to three sets of data: peak flow, daily maximum precipitation, and total flow deficit.

The book is meant for graduate students, college faculty, and professionals who are in the fields of hydrology, hydraulics, water quality engineering, hydrometeorology, earth sciences, environmental engineering, and water resources engineering. It can also be used as a reference book for courses on statistical methods in hydrology, hydraulics, watershed science, and environmental and water engineering. It is hoped that the book will help the understanding of frequency distributions and their application.

Acknowledgments

The authors express their gratitude to those who developed generalized frequency distributions, some of which have found a place in the environmental and water resources literature while others have not. They express their profound gratefulness to their families for their support and sacrifice at all times. As a small token of their appreciation the book is affectionately dedicated to them. The authors are also thankful to the Cambridge University Press Editorial Board and production staff for their support and patience.

1

Introduction

1.1 Introduction

A broad range of applications in environmental and water engineering require frequency analysis. Examples of these applications include frequency analyses of flood characteristics, such as peak, volume, duration, and inter-arrival time; flood inundation and damage; drought characteristics, such as severity, areal extent, duration, and inter-arrival time; rainfall amount and duration, number of rainfall events, and day of occurrence; number of frost days; number of low temperature days; number of high temperature days; number of extreme hot days; number of extreme cold days; high wind velocity and duration; number of cloudy days; amount and duration of snowfall and number of snowfall days; sediment concentration, discharge, and yield; water quality parameters due to accidents; oil spill; and time, duration, and spatial extent of bacterial and viral spread. It is implied that these characteristics or variables are random variables. Many distributions have been applied to frequency analysis of each of these variables and there is a vast body of literature on these distributions and their applications. Most of the distributions are one-, two-, or three-parameter distributions and some of them are also four-parameter distributions. Some of the three- and four-parameter distributions are referred to as generalized distributions, because depending on their parameter values they give rise to several simpler distributions.

Some generalized distributions have attracted more attention than others in water engineering. The generalized extreme value (GEV) distribution, which is a three-parameter distribution, has probably been the most popular distribution, partly because it is considered a standard flood frequency distribution in Europe. In the 1970s and 1980s, the Wakeby distribution received quite a bit

1

of attention in the United States and was even labeled as the parent distribution. The Halphen distribution system was presented about two decades ago. Lately, the generalized gamma, the generalized Burr, and the generalized beta distributions have been advocated. There are other generalized distributions that have been applied in other fields but may be of potential interest in environmental and water engineering.

The literature on generalized distributions in water engineering is scattered. It may therefore be useful to discuss these distributions and their characteristics under one cover. This is what this book is aimed at.

1.2 Generalized Distributions

The generalized distributions considered in this book will have at least three parameters and will contain several distributions as special cases, depending on the values of their parameters. The distributions will be: Burr–Singh–Maddala (BSM); generalized gamma; generalized beta Lomax; Halphen types A, B, and IB; kappa; Feller–Pareto; and exponential gamma distributions. Some of these distributions have explicit forms of probability density and cumulative probability distribution functions. The preferred method of parameter estimation depends on the form of the probability density function (PDF) and the cumulative probability distribution function (CDF).

1.3 Distribution Characteristics

It may be interesting to analyze distribution characteristics from two aspects. First, characteristics such as survival function, domain of attraction, non-Debye decay or relaxation, and universality of relaxation provide insights into the distribution characteristics. Second, since a distribution represents a particular stochastic process, it is instructive to discuss the process characteristics, such as memory or memoryless, equilibrium system or non-equilibrium system, interactive or non-interactive system, and stability of family.

1.4 Characterization through Hazard Function

Let the lifetime of an object be characterized by a random variable X with a PDF $f(x)$. Then, the probability of surviving at least up to time x can be defined as

$$S(x) = \int_x^\infty f(x)dx = \int_0^\infty f(x)dx - \int_0^x f(x)dx = 1 - F(x). \qquad (1.1)$$

The probability of death in a small time interval dx is $f(x)$. Since the object has survived up to age x, the instantaneous death rate at age x, $r(x)$, can be defined as

$$r(x) = -\frac{d}{dx}\ln S(x) = \frac{f(x)}{1 - F(x)}. \qquad (1.2a)$$

The quantity $r(x)$ is also called the hazard rate, failure rate, or morbidity rate.

The distribution is characterized by an increasing failure rate (IFR) if $r(x)$ is increasing, $dr(x)/dx > 0$, and by a decreasing failure rate (DFR) if $r(x)$ is decreasing, $dr(x)/dx < 0$. From Equation (1.2a), we further have

$$\frac{dr(x)}{dx} = \frac{f'(x)\big(1 - F(x)\big) + f^2(x)}{\big(1 - F(x)\big)^2}. \qquad (1.2b)$$

Examples of distributions exhibiting IFR are the gamma distribution with $\alpha > 1$, the Weibull distribution with $k > 1$, and the modified extreme value distribution. Examples of distributions characterized as IFR are explained as follows.

1.4.1 Gamma Distribution

The PDF and the CDF of the gamma distribution can be expressed as

$$f(x; \alpha, \beta) = \frac{\beta^\alpha x^{\alpha-1} e^{-\beta x}}{\Gamma(\alpha)}; \quad x \in (0, \infty),\ \alpha, \beta > 0 \qquad (1.3a)$$

$$F(x; \alpha, \beta) = \frac{1}{\Gamma(\alpha)}\gamma(\alpha, \beta x). \qquad (1.3b)$$

In Equation (1.3b), $\gamma(\alpha, \beta x)$ represents the lower incomplete gamma function. The hazard rate can then be given as

$$r(x; \alpha, \beta) = \frac{f(x; \alpha, \beta)}{1 - F(x; \alpha, \beta)} = \frac{\beta^\alpha x^{\alpha-1} e^{-\beta x}}{\Gamma(\alpha) - \gamma(\alpha, \beta x)}. \qquad (1.4)$$

The derivative of Equation (1.4) yields

$$\frac{dr(x; \alpha, \beta)}{dx} = \frac{f'(x)\big(1 - F(x)\big) + f^2(x)}{\big(1 - F(x)\big)^2}. \qquad (1.5)$$

In Equation (1.5) we have

$$f'(x) = \frac{\beta^\alpha}{\Gamma(\alpha)} x^{\alpha-1} e^{-\beta x} \left(\frac{\alpha-1}{x} - \beta\right) \tag{1.6a}$$

$$f'(x)(1 - F(x)) + f^2(x) = f(x)\left[\left(\frac{\alpha-1}{x} - \beta\right) + f(x)\right]. \tag{1.6b}$$

According to Barlow and Proschan (1965), Equation (1.6b) is guaranteed to be greater than zero when $\alpha > 1$, that is, the IFR is obtained. The DFR is obtained when $\alpha < 1$, that is, Equation (1.6b) is guaranteed to be less than zero. For the IFR property of the gamma distribution (i.e., $\alpha > 1$), the failure rate is bounded as

$$\begin{cases} r(x \to 0) = \lim\limits_{x \to 0} \dfrac{f(x; \alpha, \beta)}{1 - F(x; \alpha, \beta)} = 0 \\[3mm] r(x \to \infty) = \lim\limits_{x \to \infty} \dfrac{f(x; \alpha, \beta)}{1 - F(x; \alpha, \beta)} = \lim\limits_{x \to \infty} \dfrac{f'(x)}{-f(x)} = \lim\limits_{x \to \infty} \left(b - \dfrac{a-1}{x}\right) = b > 0. \end{cases}$$

$$\tag{1.7}$$

Figure 1.1 illustrates the IFR and DFR properties of the gamma distribution with different parameters.

1.4.2 Weibull Distribution

The PDF and CDF of the Weibull distribution can be expressed as

$$f(x; k, \lambda) = \frac{k}{\lambda} \left(\frac{x}{\lambda}\right)^{k-1} \exp\left(-\left(\frac{x}{\lambda}\right)^k\right); \quad x \ge 0, \quad k, \lambda > 0 \tag{1.8a}$$

$$F(x) = 1 - \exp\left(-\left(\frac{x}{\lambda}\right)^k\right). \tag{1.8b}$$

The failure rate can be expressed as

$$r(x; k, \lambda) = \frac{f(x)}{1 - F(x)} = \frac{k}{\lambda} \left(\frac{x}{\lambda}\right)^{k-1}. \tag{1.9}$$

The derivative of Equation (1.9) yields

$$\frac{dr}{dx} = \frac{k(k-1)}{\lambda^2} \left(\frac{x}{\lambda}\right)^{k-2}; \quad x > 0. \tag{1.10}$$

Equation (1.10) indicates that the derivative is greater than zero if $k > 1$. In other words, the Weibull distribution has the IFR property if $k > 1$ and has the DFR property if $k < 1$. Figure 1.2 illustrates the IFR and DFR properties for the Weibull distribution.

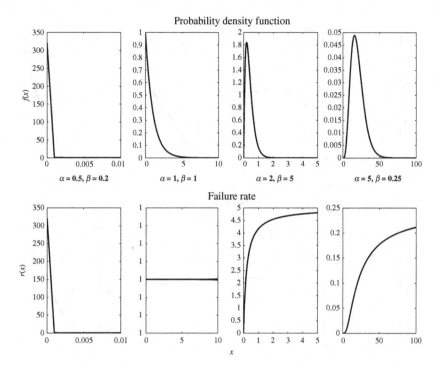

Figure 1.1 PDF and failure rate properties for gamma distribution with different parameters.

1.4.3 Modified Extreme Value Distribution

The PDF and CDF of the modified extreme value distribution can be expressed as

$$f(x; \lambda) = \frac{1}{\lambda} \exp\left(-\frac{e^x - 1}{\lambda} + x\right); \quad x > 0, \ \lambda > 0 \qquad (1.11a)$$

$$F(x; \lambda) = 1 - \exp\left(-\frac{e^x - 1}{\lambda}\right). \qquad (1.11b)$$

Substituting Equations (1.11a) and (1.11b) in Equation (1.2a), the failure rate function can be expressed as

$$r(x; \lambda) = \frac{\exp\left(-\dfrac{e^x - 1}{\lambda} + x\right)}{\lambda \exp\left(-\dfrac{e^x - 1}{\lambda}\right)} = \frac{e^x}{\lambda}. \qquad (1.12)$$

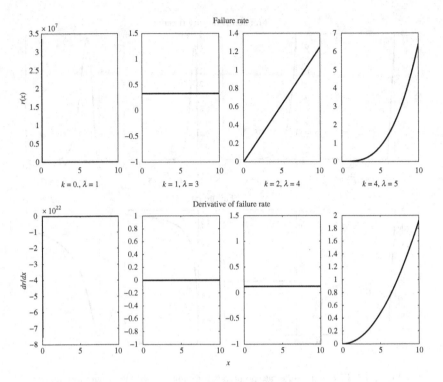

Figure 1.2 Failure rate properties for Weibull distribution.

The derivative of Equation (1.12) yields

$$\frac{dr(x;\lambda)}{dx} = \frac{e^x}{\lambda} > 0; \quad x > 0, \quad \lambda > 0. \tag{1.13}$$

Equation (1.13) indicates that the modified extreme value distribution has the IFR property. Figure 1.3 illustrates the IFR property with parameter $\lambda = 2.3$ as an example.

The Pareto distribution is an example for a distribution with the DFR property. The PDF and CDF of the Pareto distribution are expressed as

$$f(x;c,\alpha) = \frac{\alpha c^\alpha}{x^{\alpha+1}}; \quad x \in [c,\infty), \quad c > 0, \quad \alpha > 0 \tag{1.14a}$$

$$F(x;c,\alpha) = 1 - \left(\frac{c}{x}\right)^\alpha. \tag{1.14b}$$

Substituting Equations (1.14a) and (1.14b) in Equation (1.2a), the failure rate can be expressed as

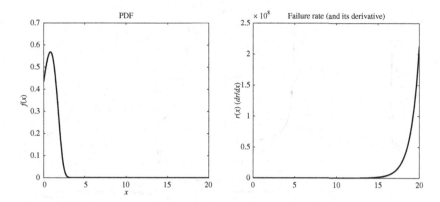

Figure 1.3 IFR property for modified extreme value distribution.

$$r(x; c, \alpha) = \frac{\alpha}{x}. \tag{1.15}$$

The derivative of Equation (1.15) yields

$$\frac{dr(x; c, \alpha)}{dx} = -\frac{\alpha}{x^2} < 0. \tag{1.16}$$

Equation (1.16) indicates that the Pareto distribution has the DFR property. Furthermore, the failure rate of the Pareto distribution is bounded as

$$\lim_{x \to c} r(x; c, \alpha) = \frac{\alpha}{m}; \quad \lim_{x \to \infty} r(x; c, \alpha) = 0. \tag{1.17}$$

Figure 1.4 illustrates the DFR property for the Pareto distribution.

Depending on the problem at hand, $r(x)$ provides a physical explanation for whether a distribution represents the real situation.

In some cases, it is more informative to consider the hazard rate using a transform of X. Let $z = \log x$. Then, the hazard rate is determined for z instead of x. Thus,

$$r(z) = \frac{f(z)}{1 - F(z)} = \frac{dF/dz}{1 - F(z)}. \tag{1.18}$$

If $dr(z)/dz \geq 0$, the distribution is characterized by an increasing proportional failure rate (IPFR), and if $dr(z)/dz \leq 0$, the distribution is characterized by a decreasing proportional failure rate (DPFR). The Pareto distribution has a constant rate, but the lognormal distribution has a monotone IPFR.

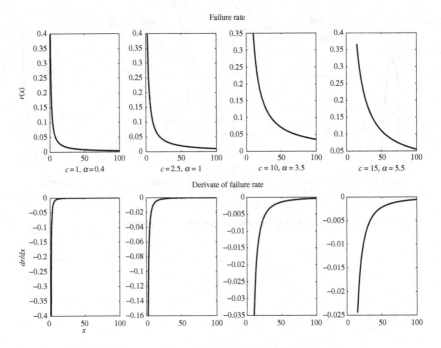

Figure 1.4 DFR property for Pareto distribution with different parameters.

1.4.4 Pareto Distribution

Let $z = t(x) = \ln x$, then we have $dx/dz = e^z$. Then, the density function and distribution function may be expressed as

$$f_Z(z) = f_X\left(t^{-1}(z)\right)\left|\frac{dx}{dz}\right| = \frac{\alpha c^\alpha}{(e^z)^{\alpha+1}}e^z = \alpha c^\alpha e^{-\alpha z} \qquad (1.19a)$$

$$F_Z(z) = 1 - c^\alpha e^{-\alpha z}. \qquad (1.19b)$$

It is seen that Equation (1.19a) belongs to the exponential family. Substituting Equations (1.19a) and (1.19b) in Equation (1.18) we have

$$r(z) = \frac{f(z)}{1 - F(z)} = \alpha. \qquad (1.20)$$

Now we have shown that in the logarithm domain, the exponential type distribution is observed and yields a constant failure rate.

1.4.5 Lognormal Distribution

In the case of the lognormal distribution, let $z = t(x) = \ln x$; its density function may be rewritten as

$$f_Z(z) = f_X\left(t^{-1}(z)\right)\left|\frac{dx}{dz}\right| = \frac{1}{\sqrt{2\pi}\sigma}\exp\left(-\frac{(z-\mu)^2}{2\sigma^2}\right). \tag{1.21}$$

It is seen that Equation (1.21) is the normal distribution. Furthermore, Barlow and Proschan (1965) showed that the truncated normal distribution has the IFR property. It can be concluded that the lognormal distribution has a monotone IPFR property. Figure 1.5 illustrates this property for $\sigma = 1$.

1.4.6 Log-Logistic Distribution

The log-logistic distribution is another example that has the monotone IPFR property. The PDF and CDF of the log-logistic distribution can be expressed as

$$f(x; \alpha, \beta) = \frac{\left(\frac{\beta}{\alpha}\right)\left(\frac{x}{\alpha}\right)^{\beta-1}}{\left(1 + \left(\frac{x}{\alpha}\right)^{\beta}\right)^2}; \quad x \in [0, \infty), \quad \alpha, \beta > 0 \tag{1.22a}$$

$$F(x; \alpha, \beta) = \frac{\left(\frac{x}{\alpha}\right)^{\beta}}{1 + \left(\frac{x}{\alpha}\right)^{\beta}}. \tag{1.22b}$$

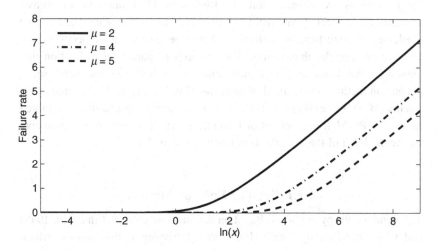

Figure 1.5 IPFR property for lognormal distribution.

Let $z = \ln(x/\alpha)$; we have $x = \alpha \exp(z)$ and $dx/dz = \alpha \exp(z)$. Then Equations (1.22a) and (1.22b) can be rewritten as

$$f(z; \alpha, \beta) = \frac{\beta \exp(\beta z)}{1 + \exp(\beta z)} \tag{1.23a}$$

$$F(z; \alpha, \beta) = \frac{\exp(\beta z)}{1 + \exp(\beta z)} \tag{1.23b}$$

$$r(z; \alpha, \beta) = \frac{\beta \exp(\beta z)}{1 + \exp(\beta z)} \tag{1.24}$$

$$\frac{dr}{dz} = \frac{\beta^2 \exp(\beta z)\left(1 + \exp(\beta z)\right) - \beta^2 \exp(\beta z)\exp(\beta z)}{\left(1 + \exp(\beta z)\right)^2} = \frac{\beta^2 \exp(\beta z)}{\left(1 + \exp(\beta z)\right)^2} > 0. \tag{1.25}$$

It is seen that different distributions exhibit different hazard rate characteristics and these characteristics can serve as a basis for selecting a distribution that closely mimics real-world characteristics.

1.5 Methods of Parameter Estimation

Fundamental to applying a distribution to frequency analysis is the estimation of its parameters. It would be ideal to estimate these parameters from physically measurable system characteristics, but that is not plausible at this time and the parameters are instead estimated from data. Of course, as the dataset changes, the estimated parameter values may change. Many methods can be used to estimate distribution parameters but some are considered better and are hence more popular than others. The methods of parameter estimation discussed in this book are (1) regular entropy method, (2) parameter space expansion method, (3) method of moments (MOM), (4) method of maximum likelihood (ML) estimation (MLE), (5) method of probability weighted moments (PWMs), (6) method of L-moments, and (7) method of cumulative moments. Each of these methods is briefly presented here.

1.5.1 Regular Entropy Method

The regular entropy method is based on the Shannon entropy (Shannon, 1948) and been described by Singh (1998) for 21 frequency distributions, which include several three-parameter distributions. For a continuous random variable X with a PDF $f(x)$ and CDF $F(x)$, the Shannon entropy $H(X)$ or $H[f(x)]$ can be defined as

$$H(x) = -\int f(x)\ln f(x)dx. \tag{1.26}$$

Fundamental to the entropy method for parameter estimation is the specification of constraints that the probability distribution must satisfy. These constraints encode the information about the random variable. Let $g_i(x)$, $i = 1, \ldots, m$, be the functions of the random variable X describing the characteristics of X. Then, the i-th constraint, C_i, can be expressed as

$$C_i = \int g_i(x)f(x)dx = E[g_i(x)], \quad i = 1, 2, \ldots, m, \tag{1.27}$$

where m is the number of constraints. If $i = 0$ and $g_0(x) = 1$, Equation (1.27) leads to the total probability as

$$C_0 = \int f(x)dx = 1. \tag{1.28}$$

The next step is to maximize entropy given by Equation (1.26) in accordance with the principle of maximum entropy (Jaynes, 1957), subject to Equation (1.27). Entropy maximizing can be done using the method of Lagrange multipliers for which the Lagrangian function L can be written as

$$L = -\int f(x)\ln f(x)dx - (\lambda_0 - 1)\left(\int f(x)dx - 1\right) - \sum_{i=1}^{m}\lambda_i\left(\int g_i(x)f(x)dx - C_i\right), \tag{1.29}$$

where $\lambda_i, i = 0, 1, \ldots, m$, are the unknown Lagrange multipliers. Now, differentiating Equation (1.29) with respect to $f(x)$ in accordance with the Lagrange–Euler calculus of variation, one obtains

$$\frac{\partial L}{\partial f(x)} = -\int \ln f(x)dx - \int dx - (\lambda_0 - 1)\int dx - \sum_{i=1}^{m}\lambda_i\left(\int g_i(x)dx\right). \tag{1.30}$$

Equating Equation (1.15a) to zero leads to the maximum entropy distribution

$$f(x) = \exp\left(-\sum_{i=0}^{m}\lambda_i g_i(x)\right). \tag{1.31}$$

Substituting Equation (1.31) in Equation (1.27), we obtain the constraints as

$$C_i = \int g_i(x) \exp\left(-\sum_{i=0}^{m} \lambda_i g_i(x)\right) dx = \frac{\int g_i(x) \exp\left(-\sum_{i=1}^{m} \lambda_i g_i(x)\right) dx}{\exp(\lambda_0)}, \quad i = 1, 2, \ldots, m.$$

(1.32)

Now the unknown Lagrange multipliers are determined from the known constraints. Substituting Equation (1.31) in Equation (1.28), we get λ_0 as a function of $\lambda_1, \ldots, \lambda_m$ as $\lambda_0 = \lambda_0(\lambda_1, \ldots, \lambda_m)$:

$$\exp(\lambda_0) = Z = \int \exp\left(-\sum_{i=1}^{m} \lambda_i g_i(x)\right) dx, \quad \text{or}$$

$$\lambda_0 = \ln Z = \ln\left[\int \exp\left(-\sum_{i=1}^{m} \lambda_i g_i(x)\right) dx\right].$$

(1.33)

In Equation (1.33), Z is called the partition function and it is a convex function.

Substituting Equation (1.33) in Equation (1.32), one can express the constraints as

$$C_i = E[g_i(x)] = \frac{\int g_i(x) \exp\left(-\sum_{i=1}^{m} \lambda_i g_i(x)\right) dx}{\int \exp\left(-\sum_{i=1}^{m} \lambda_i g_i(x)\right) dx}, \quad i = 1, 2, \ldots, m. \quad (1.34)$$

Differentiating λ_0 with respect to $\lambda_1, \lambda_2, \ldots, \lambda_m$ individually, we get the relation between Lagrange multipliers and constraints as

$$\frac{\partial \lambda_0}{\partial \lambda_i} = -\frac{\int g_i(x) \exp\left(-\sum_{i=1}^{m} \lambda_i g_i(x)\right) dx}{\int \exp\left(-\sum_{i=1}^{m} \lambda_i g_i(x)\right) dx} = -C_i, \quad i = 1, 2, \ldots, m. \quad (1.35)$$

Similarly, it can be shown that

$$\frac{\partial^2 \lambda_0}{\partial \lambda_i^2} = E[g_i^2(x)] - \{E[g_i(x)]\}^2, \quad i = 1, 2, \ldots, m \quad (1.36)$$

and

$$\frac{\partial^2 \lambda_0}{\partial \lambda_i \partial \lambda_j} = E[g_i(x)g_j(x)] - E[g_i(x)]E[g_j(x)] = \text{cov}[g_i(x), g_j(x)]. \quad (1.37)$$

To this end, parameters can be obtained by using the relations of Equations (1.34) and (1.35), which lead to distribution parameters in terms of given constraints.

Example 1.1 Estimate the parameters of gamma distribution using the maximum entropy method.

Solution: The gamma distribution can be given as

$$f(x) = \frac{\beta^\alpha}{\Gamma(\alpha)} x^{\alpha-1} e^{-\beta x}. \tag{1.38}$$

For the gamma distribution, the constraints needed are the mean and the geometric mean, which are expressed as

$$C_0 = \int_0^\infty f(x) = 1, \quad C_1 = \int_0^\infty x \, f(x) = E(x), \quad C_2 = \int_0^\infty \ln x \, f(x) dx = E(\ln x). \tag{1.39}$$

Then, the maximum entropy-based distribution following Equation (1.31) can be written as

$$f(x) = \exp(-\lambda_0 - \lambda_1 x - \lambda_2 \ln x). \tag{1.40}$$

From Equation (1.40), one can express λ_0 as a function of λ_1 and λ_2 as

$$\lambda_0 = \ln \int_0^\infty x^{-\lambda_2} \exp(-\lambda_1 x) dx = (\lambda_2 - 1)\ln \lambda_1 + \ln \Gamma(1 - \lambda_2). \tag{1.41}$$

Taking the partial derivative of λ_0 with respect to λ_1 and λ_2 we getas

$$\frac{\partial \lambda_0}{\partial \lambda_1} = \frac{(\lambda_2 - 1)}{\lambda_1}; \quad \frac{\partial \lambda_0}{\partial \lambda_2} = \ln \lambda_1 - \Psi(1 - \lambda_2); \quad \lambda_1 > 0, \quad \lambda_2 < 1. \tag{1.42}$$

From Equations (1.34) and (1.35) we have for Equation (1.42)

$$\frac{\partial \lambda_0}{\partial \lambda_1} = \frac{\lambda_2 - 1}{\lambda_1} = -E(x); \quad \frac{\partial \lambda_0}{\partial \lambda_2} = \ln \lambda_1 - \Psi(1 - \lambda_2) = -E(\ln x). \tag{1.43}$$

Setting $E(x)$ and $E(\ln x)$ to the sample $\bar{x}$ and $\overline{\ln x}$, we can solve Equation (1.43) numerically to obtain the estimated $\hat{\lambda}_1$ and $\hat{\lambda}_2$, and Equation (1.40) may be rewritten using the estimated Lagrange multipliers as

$$f(x) = \frac{\hat{\lambda}_1^{1-\hat{\lambda}_2}}{\Gamma(1 - \hat{\lambda}_2)} x^{-\hat{\lambda}_2} \exp(-\hat{\lambda}_1 x). \tag{1.44}$$

Comparing Equation (1.44) with the gamma distribution equation (1.38), we have

$$\beta = \lambda_1, \quad \alpha = 1 - \lambda_2. \tag{1.45}$$

1.5.2 Parameter Space Expansion Method

The parameter space expansion method is similar to the regular entropy method but is based on the expansion of parameter space and hence its name. Singh and Rajagopal (1986) developed this method and Singh (1998) presented its application to 21 distributions. In this method, the probability distribution whose parameters are to be estimated is expressed using the entropy theory in terms of Lagrange multipliers, say $\lambda_i, i = 0, 1, 2, \ldots, m$, and distribution parameters $a_i, i = 1, 2, \ldots, n$. Then, $\lambda_0 = \lambda_0(\lambda_1, \lambda_2, \ldots, \lambda_m)$ is expressed and is inserted in the entropy expression. Thereafter, the point where entropy will be maximum, H_{max}, can be determined by differentiating maximum entropy with respect to Lagrange multipliers and distribution parameters and equating each derivative to zero as follows:

$$\frac{\partial H}{\partial \lambda_i} = 0, \quad i = 1, 2, \ldots, m - 1 \tag{1.46}$$

$$\frac{\partial H}{\partial a_i} = 0, \quad i = 1, 2, \ldots, n. \tag{1.47}$$

The solution of Equations (1.46) and (1.47) leads to the estimates of distribution parameters.

Example 1.2 Estimate the parameters of the PDF of the gamma distribution given as Equation (1.38) using the parameter space expansion method.

Solution: To apply the parameter space expansion method, the constraints of Equation (1.39) can be reformulated as

$$C_0 = \int_0^\infty f(x)dx = 1, \quad C_1 = \int_0^\infty \beta x f(x)dx = E(\beta x),$$

$$C_2 = \int_0^\infty \ln(\beta x)^{\alpha-1} f(x)dx = E\left(\ln(\beta x)^{\alpha-1}\right). \tag{1.48}$$

Now with the constraints in Equation (1.48), the PDF can be written as

$$f(x) = \exp\left(-\lambda_0 - \lambda_1\beta x - \lambda_2\ln(\beta x)^{\alpha-1}\right). \tag{1.49}$$

Then, the partition function, $\exp(\lambda_0)$, can be solved as

$$\exp(\lambda_0) = \int_0^\infty \exp\left(-\lambda_1\beta x - \lambda_2\ln(\beta x)^{\alpha-1}\right)dx$$
$$= \frac{\lambda_1^{\lambda_2(\alpha-1)-1}}{\beta}\Gamma\left(1 - \lambda_2(\alpha-1)\right) \tag{1.50a}$$

or

$$\lambda_0 = \left(\lambda_2(\alpha-1) - 1\right)\ln\lambda_1 - \ln\beta + \ln[\Gamma\left(1 - \lambda_2(\alpha-1)\right)]. \tag{1.50b}$$

In addition, the entropy for the PDF, Equation (1.49), can be given as

$$H = -\int_0^\infty f(x)\ln f(x)dx = \int_0^\infty \left(\lambda_0 + \lambda_1\beta x + \lambda_2\ln\left((\beta x)^{\alpha-1}\right)\right)f(x)dx$$

$$= \lambda_0 + \lambda_1 E(\beta x) + \lambda_2 E\left(\ln(\beta x)^{\alpha-1}\right)$$

$$= \left(\lambda_2(\alpha-1) - 1\right)\ln\lambda_1 - \ln\beta + \ln\Gamma\left(1 - \lambda_2(\alpha-1)\right)$$

$$+ \lambda_1 E(\beta x) + \lambda_2 E\left(\ln(\beta x)^{\alpha-1}\right). \tag{1.51}$$

Differentiating Equation (1.51) with respect to parameters $\lambda_1, \lambda_2, \alpha$, and β, one has

$$\frac{\partial H}{\partial \lambda_1} = \frac{\lambda_2(\alpha-1) - 1}{\lambda_1} + E[\beta x] = 0 \tag{1.52}$$

$$\frac{\partial H}{\partial \lambda_2} = (\alpha-1)\ln\lambda_1 - (\alpha-1)\Psi\left(1 - \lambda_2(\alpha-1)\right) + E\left(\ln(\beta x)^{\alpha-1}\right) = 0 \tag{1.53}$$

$$\frac{\partial H}{\partial \alpha} = \lambda_2\ln\lambda_1 - \lambda_2\Psi\left(1 - \lambda_2(\alpha-1)\right) + E(\ln\beta x) = 0 \tag{1.54}$$

$$\frac{\partial H}{\partial \beta} = -\frac{1}{\beta} + \frac{\lambda_1}{\beta} E(\beta x) + \lambda_2 E\left(\frac{\alpha - 1}{\beta}\right) = 0. \qquad (1.55)$$

The maximum entropy-based distribution is then obtained by setting the partial derivative to zero in Equations (1.52)–(1.55), and we obtain

$$\frac{\partial H}{\partial \lambda_1} = 0 \Rightarrow -\frac{\lambda_2(\alpha - 1) - 1}{\lambda_1} = E[\beta x] \qquad (1.56)$$

$$\frac{\partial H}{\partial \lambda_2} = 0 \Rightarrow (\alpha - 1)\Psi\left(1 - \lambda_2(\alpha - 1)\right) - (\alpha - 1)\ln\lambda_1 = E\left[\ln\left((\beta x)^{\alpha - 1}\right)\right]$$

$$(1.57)$$

$$\frac{\partial H}{\partial \alpha} = 0 \Rightarrow \lambda_2\Psi\left(1 - \lambda_2(\alpha - 1)\right) - \lambda_2\ln\lambda_1 = E[\ln\beta x] \qquad (1.58)$$

$$\frac{\partial H}{\partial \beta} = 0 \Rightarrow \frac{1 - \lambda_2(\alpha - 1)}{\lambda_1} = E[\beta x]. \qquad (1.59)$$

From Equations (1.57) and (1.59), we have $E[\beta x] = (1 - \lambda_2(\alpha - 1))/\lambda_1$, that is, $\partial H/\partial\lambda_1 = \partial H/\partial\beta$. And by solving the system of equations, we can estimate the parameters.

1.5.3 MOM

The MOM is one of the most popular methods because moments are descriptors of distribution characteristics and have geometric meaning. Moreover, for linear hydrologic systems there is the theorem of moments that relates moments of system output or response to system input and system impulse response. There are also variations of the MOM, such as partial or incomplete moments and geometric moments. A detailed description of the MOM is given by Singh (1988). Ashkar et al. (1988) developed a generalized MOM and applied it to the generalized gamma distribution.

Let a probability distribution have m parameters. Then, the MOM entails expressing m moments of the probability distribution. This gives rise to m equations, which are then solved for parameters in terms of moments that are determined from data. Moments can be determined about the origin or about the centroid of the distribution or about any arbitrarily chosen point. The rth moment of $f(x)$ about the origin can be expressed as

$$M_r = \int x^r f(x)dx, \quad r = 1, 2, \ldots, m. \qquad (1.60)$$

If the moment is taken about an arbitrary point b, then Equation (1.60) can be modified as

$$M_r^b = \int (x - b)^r f(x) dx, \quad r = 1, 2, \ldots, m. \tag{1.61}$$

The first four moments are able to capture some of the most important properties of the probability distribution. These moments have the following meaning:

M_0 represents the area under the curve. For the PDF, the area is the total probability and equals one.

M_1 represents the mean or the centroid and is often denoted by μ. In surface water hydrology, this is also referred to as the lag or residence time.

M_2^μ represents the variance, that is, the measure of spread or dispersion around the mean. It is often denoted as σ^2 and its square root σ is called the standard deviation. This measure is used in uncertainty and reliability analysis. The coefficient of variation is defined by the ratio of the standard deviation to the mean:

$$CV = \frac{\sigma}{\mu}. \tag{1.62}$$

M_3^μ measures the skewness or asymmetry. It is viewed with respect to the skewness of the normal distribution for which it has a value of zero. It is employed for extreme value analysis.

M_4^μ represents kurtosis, which is a measure of peakedness or flatness. It is measured with respect to the normal distribution for which it has a value of 3.

In many practical applications, moment ratios, defined below, are employed:

$$\alpha_i = \frac{M_i^\mu}{(M_2^\mu)^{0.5i}} = \frac{M_i^\mu}{\sigma^i}, \quad i = 1, 2, 3, 4. \tag{1.63}$$

These ratios, which are dimensionless, are also called shape factors. Equation (1.63) shows that $\alpha_1 = 0$ and $\alpha_2 = 1$, so α_3 and α_4 are useful. Therefore,

$$\alpha_3 = \frac{M_3^\mu}{\sigma^3} \tag{1.64a}$$

$$\alpha_4 = \frac{M_4^\mu}{\sigma^4}. \tag{1.64b}$$

When moment ratios are plotted, the resulting diagram is called the moment diagram. Bobée et al. (1993) presented two kinds of moment diagrams and discussed their applications in hydrology.

Equation (1.64b) defines the coefficient of skewness, C_s or γ_1, whose unbiased estimate, especially for a small sample of size n, is obtained as

$$C_s = \frac{n^2}{(n-1)(n-2)} \frac{M_3^{\bar{x}}}{(m_2^{\bar{x}})^{\frac{3}{2}}} = \frac{n^2}{(n-1)(n-2)} \frac{M_3^{\bar{x}}}{s^3}, \qquad (1.65)$$

where $\bar{x}$ is the sample arithmetic mean, $m_2^{\bar{x}} = s^2$ is the sample variance, and $m_3^{\bar{x}}$ is the third sample central moment. Sometimes an unbiased estimate is also obtained as

$$\gamma_1 = C_s = \frac{n}{n-1} \frac{M_3^{\bar{x}}}{(m_2^{\bar{x}})^{\frac{3}{2}}} = \frac{n}{n-1} \frac{M_3^{\bar{x}}}{s^3}. \qquad (1.66)$$

For large values of n, $n^2/((n-1)(n-2))$ and $n/(n-1)$ are nearly equal.

Likewise, the coefficient of flatness, γ_2, is given by Equation (1.66), and its less biased estimate for small samples is given as

$$\gamma_2 = \frac{n^3}{(n-1)(n-2)(n-3)} \frac{M_4^{\bar{x}}}{s^4}. \qquad (1.67)$$

Figure 1.6 illustrates the moment diagram for some distributions.

It can be shown that moments about the origin and those about the centroid are related with the use of binomial theorem:

$$(a+b)^n = \sum_{i=0}^{n} \binom{n}{i} a^i b^{n-i}. \qquad (1.68)$$

With the binomial theorem, the rth moment about the origin (M_r) may be expressed using the moments about the centroid $(M_j^{\mu}; j = 0, 1, \ldots, r)$ as

$$M_r = E(X^r) = E[(X - \mu) + \mu]^r = E\left[\sum_{i=0}^{r} \binom{r}{i} \mu^i (X - \mu)^{r-i}\right]$$
$$= \sum_{i=0}^{r} \binom{r}{i} \mu_i E(X - \mu)^{r-i} = \sum_{i=0}^{r} \binom{r}{i} \mu_i M_{r-i}^{\mu} \qquad (1.69)$$

Similarly, the rth moment about the centroid (M_r^{μ}) may be expressed using the moments about the origin $(M_j; j = 0, 1, \ldots, r)$ as

$$M_r^{\mu} = E(X - \mu)^r = E\left[\sum_{i=0}^{r} \binom{r}{i} (-\mu)^i X^{r-i}\right]$$
$$= \sum_{i=0}^{r} (-1)^i \binom{r}{i} \mu^i E(X^{r-i}) = \sum_{i=0}^{r} (-1)^i \binom{r}{i} \mu^i M_{r-i}. \qquad (1.70)$$

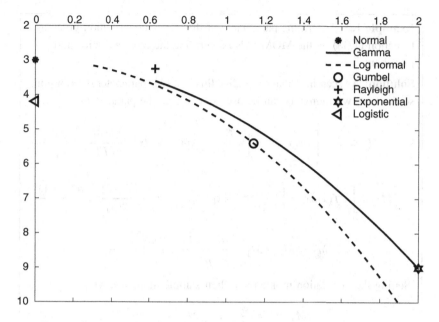

Figure 1.6 Moment diagram of some distributions.

In particular,

$$M_0 = M_0^\mu = 1 \tag{1.71}$$

$$M_1 = M_1^\mu + \mu M_0^\mu = \mu \tag{1.72}$$

$$M_2 = M_2^\mu + \mu^2 \tag{1.73}$$

$$M_3 = M_3^\mu + 3\mu M_2^\mu + \mu^3 \tag{1.74}$$

$$M_4 = M_4^\mu + 4\mu M_3^\mu + 6\mu^2 M_2^\mu + \mu^3. \tag{1.75}$$

Likewise,

$$M_0^\mu = M_0 = 1 \tag{1.76}$$

$$M_1^\mu = 0 \tag{1.77}$$

$$M_2^\mu = M_2 - \mu^2 \tag{1.78}$$

$$M_3^\mu = M_3 - 3\mu M_2 + 2\mu^3 \tag{1.79}$$

$$M_4^\mu = M_4 - 4\mu M_3 + 6\mu^2 M_2 - 3\mu^4. \tag{1.80}$$

Example 1.3 Determine parameters of the gamma distribution given as Equation (1.38) by the MOM. Also determine the moment ratios and plot them.

Solution: As seen in Equation (1.28), this is a two-parameter distribution, so the first two moments can be used to estimate the parameters.

$$M_1 = \mu = \int_0^\infty x f(x)dx = \int_0^\infty \frac{\beta^\alpha}{\Gamma(\alpha)} x^\alpha \exp(-\beta x)dx = \frac{\Gamma(\alpha+1)}{\beta\Gamma(\alpha)} = \frac{\alpha}{\beta}$$

$$M_2 = \int_0^\infty x^2 f(x)dx = \int_0^\infty \frac{\beta^\alpha}{\Gamma(\alpha)} x^{\alpha+1}\exp(-\beta x)dx = \frac{\Gamma(\alpha+2)}{\beta^2\Gamma(\alpha)} = \frac{\alpha(\alpha+1)}{\beta^2}$$

$$M_2^\mu = M_2 - M_1^2 = \frac{\alpha(\alpha+1)}{\beta^2} - \left(\frac{\alpha}{\beta}\right)^2 = \frac{\alpha}{\beta^2}.$$

Setting the population moments to their sample moments, we get

$$M_1 = \mu = \frac{\alpha}{\beta} = \bar{x}, \quad M_2^\mu = \frac{\alpha}{\beta^2} = s^2.$$

The distribution parameter can then be solved as

$$\hat{\alpha} = \frac{\bar{x}^2}{s^2}, \quad \hat{\beta} = \frac{\bar{x}}{s^2}.$$

As discussed earlier, the third and fourth central moments are needed to compute the moment ratios.

$$M_3^\mu = \int_0^\infty (x-\mu)^3 f(x)dx = \int_0^\infty (x^3 - 3\mu x^2 + 3\mu^2 x - \mu^3)\frac{\beta^\alpha}{\Gamma(\alpha)} x^{\alpha-1}\exp(-\beta x)dx = \frac{2\alpha}{\beta^3}$$

$$M_4^\mu = \int_0^\infty (x-\mu)^4 f(x)dx = \int_0^\infty (x^4 - 4\mu x^3 + 6\mu^2 x^2 - 4\mu^3 x + \mu^4)f(x)dx$$

$$= \frac{\alpha(\alpha+1)(\alpha+2)(\alpha+3)}{\beta^4} - 4\left(\frac{\alpha}{\beta}\right)\left(\frac{\alpha(\alpha+1)(\alpha+2)}{\beta^3}\right)$$

$$+ 6\left(\frac{\alpha}{\beta}\right)^2\left(\frac{\alpha(\alpha+1)}{\beta^2}\right) - 3\left(\frac{\alpha}{\beta}\right)^4 = \frac{3\alpha(\alpha+2)}{\beta^4}.$$

Then, we have

$$\gamma_1 = \frac{M_3^\mu}{(M_2^\mu)^{1.5}} = \frac{(2\alpha/\beta^3)}{(\alpha/\beta^2)^{\frac{3}{2}}} = \frac{2}{\sqrt{\alpha}}$$

$$\gamma_2 = \frac{M_4^\mu}{\left(M_2^\mu\right)^2} = \frac{\left(3\alpha(\alpha+2)\right)/\beta^4}{\alpha^2/\beta^4} = \frac{3\alpha(\alpha+2)}{\alpha^2} = 3 + \frac{6}{\alpha}.$$

From the computed coefficient of skewness and kurtosis, it is seen that the moment ratio of the gamma distribution is only dependent on the shape parameter α.

1.5.4 Method of PWMs

The method of PWMs is useful for estimating parameters of those distributions whose inverse form can be explicitly stated as $x = x(F)$, such as Wakeby and kappa distributions. Greenwood et al. (1979) developed the PWM method and derived relations between PWMs and parameters of six distributions, namely, generalized lambda, Wakeby, Weibull, Gumbel, logistic, and kappa. Hosking (1986) developed the theory of PWMs and used it for estimating parameters of several distributions. Landwehr et al. (1979a, b) discussed inference procedures with PWMs, Haktamir (1996) presented a modified PWM method, and Wang (1997) defined partial PWMs.

The PWM of a CDF $F(x)$, denoted as $M_{i,j,k}$, of order (i, j, k) where i, j, and k are real numbers, can be expressed as

$$M_{i,j,k} = \int_0^1 \left(x(F)\right)^i F^j (1 - F)^k dF. \tag{1.81}$$

Equation (1.81) is the expectation of the quantity inside the integral

$$M_{i,j,k} = E\left(x^i F^j (1 - F)^k\right). \tag{1.82}$$

Equation (1.82) shows that if $j = k = 0$, then the PWM reduces to the ordinary moment; that is,

$$M_{i,0,0} = M_i = \int x^i f(x) dx. \tag{1.83}$$

PWMs exist for all nonnegative j and k if ordinary moments exist and can be expressed as

$$M_{i,0,k} = \sum_{j=0}^k \binom{k}{j} M_{i,j,0} \tag{1.84}$$

$$M_{i,j,0} = \sum_{k=0}^{j} \binom{j}{k} (-1)^j M_{i,0,k}. \tag{1.85}$$

PWMs can be related to the distribution parameters. For practical applications, we use PWMs in which x enters linearly as

$$M_{1,0,k} = a_k = E\left(x(1-F)^k\right), \quad k = 0,1,2,\ldots \tag{1.86}$$

$$M_{1,j,0} = b_j = E(xF^j), \quad j = 0,1,2,\ldots \tag{1.87}$$

Consider a sample of size $n = k + j + 1$ and let $E(x_{j+1,k+1}^i)$ denote the i-th moment about the origin of the $(j+1)$-th order statistic for the sample. Then, one can write

$$M_{i,j,k} = B(j+1,k+1)E(x_{j+1,k+1}^i), \tag{1.88}$$

where $B(\cdot,\ \cdot)$ denotes the beta function defined in terms of the gamma function as

$$B(a,b) = \frac{\Gamma(a)\Gamma(b)}{\Gamma(a+b)}. \tag{1.89}$$

If $j = 0$, then Equation (1.88) reduces to

$$M_{i,0,k} = B(1,k+1)E(x_{1,k+1}^i) = \frac{E(x_{1,k+1}^i)}{k+1}. \tag{1.90}$$

Equation (1.90) states that $(k+1)\,M_{i,0,k}$ is the i-th moment about the origin of the first-order statistic for a sample of size $k+1$. In a similar manner, if $k = 0$, then Equation (1.88) reduces to

$$M_{i,j,0} = B(j+1,1)E(x_{j+1,1}^i) = \frac{E(x_{j+1,1}^i)}{j+1}. \tag{1.91}$$

Equation (1.91) states that $(j+1)\,M_{i,j,0}$ is the i-th moment about the origin of the $(j+1)$-order statistic for a sample of size $j+1$. The expected value of the range of X for the sample can be written as

$$E(x_{n,n} - x_{1,n}) = n(M_{1,n-1,0} - M_{1,0,n-1}). \tag{1.92}$$

Now from Equations (1.91) and (1.92), it can be seen that $a_{k-1} = E(x_{1:k})$ and $b_k = E(x_{k:k})$ are the expected values of extreme order statistics. Further, a_r and b_r are related to each other as

$$a_r = \sum_{k=0}^{r} (-1)^r \binom{r}{k} b_k \tag{1.93}$$

$$b_r = \sum_{k=0}^{r} (-1)^r \binom{r}{k} a_k. \tag{1.94}$$

In particular,

$$a_0 = b_0 \tag{1.95}$$

$$a_1 = b_0 - b_1 \quad b_1 = a_0 - a_1 \tag{1.96}$$

$$a_2 = b_0 - 2b_1 + b_2 \quad b_2 = a_0 - 2a_1 + a_2 \tag{1.97}$$

$$a_3 = b_0 - 3b_1 + 3b_2 - b_3 \quad b_3 = a_0 - 3a_1 + 3a_2 - a_3. \tag{1.98}$$

Furthermore, for the ordered sample $x_{1:n} \leq x_{2:n} \leq x_{n:n}$, the unbiased estimate for the PWM can be given as

$$a_r^{\text{sample}} = \frac{1}{n} \sum_{i=1}^{n-r} \frac{\binom{n-i}{r}}{\binom{n-1}{r}} x_{i:n} = \frac{1}{n} \sum_{i=1}^{n} \frac{(n-i)(n-i-1)\cdots(n-i+1)}{(n-1)(n-2)\cdots(n-r)} x_{i:n}.$$

$$\tag{1.99}$$

More specifically, the first several moments can be given as

$$a_0^{\text{sample}} = \frac{1}{n} \sum_{i=1}^{n} x_{i:n} = \bar{x} \tag{1.100}$$

$$a_1^{\text{sample}} = \frac{1}{n(n-1)} \sum_{i=1}^{n-1} \binom{n-i}{1} x_{i:n} = \frac{1}{n(n-1)} \sum_{i=1}^{n-1} (n-i) x_{i:n} \tag{1.101}$$

$$a_2^{\text{sample}} = \frac{1}{n\binom{n-1}{2}} \sum_{i=1}^{n-2} \binom{n-i}{2} x_{i:n} = \frac{1}{n(n-1)(n-2)} \sum_{i=1}^{n-2} (n-i)(n-i-1) x_{i:n}$$

$$\tag{1.102}$$

$$a_3^{\text{sample}} = \frac{1}{n\binom{n-1}{3}} \sum_{i=1}^{n-3} \binom{n-i}{3} x_{i:n}$$

$$= \frac{1}{n(n-1)(n-2)(n-3)} \sum_{i=1}^{n-3} (n-i)(n-i-1)(n-i-2) x_{i:n} \tag{1.103}$$

and

$$b_r^{\text{sample}} = \frac{1}{n} \sum_{i=r+1}^{n} \frac{\binom{i-1}{r}}{\binom{n-1}{r}} x_{i:n} = \frac{1}{n} \sum_{i=1}^{n} \frac{(i-1)(i-2)\cdots(i-r)}{(n-1)(n-2)\cdots(n-r)} x_{i:n}.$$

$$\tag{1.104}$$

More specifically, the first several moments can be given as

$$b_0^{\text{sample}} = \frac{1}{n}\sum_{i=1}^{n} x_{i:n} = \bar{x} = a_0^{\text{sample}} \tag{1.105}$$

$$b_1^{\text{sample}} = \frac{1}{n(n-1)}\sum_{i=2}^{n}(i-1)x_{i:n} \tag{1.106}$$

$$b_2^{\text{sample}} = \frac{1}{n\binom{n-1}{2}}\sum_{i=3}^{n}\binom{i-1}{2}x_{i:n} = \frac{1}{n(n-1)(n-2)}\sum_{i=3}^{n}(i-1)(i-2)x_{i:n}$$

$$\tag{1.107}$$

$$b_3^{\text{sample}} = \frac{1}{n\binom{n-1}{3}}\sum_{i=4}^{n}\binom{i-1}{3}x_{i:n}$$

$$= \frac{1}{n(n-1)(n-2)(n-3)}\sum_{i=4}^{n}(i-1)(i-2)(i-3)x_{i:n}. \tag{1.108}$$

Example 1.4 Estimate the parameters of the GEV distribution using the method of PWMs.

Solution: The PDF and CDF of the GEV distribution can be given as

$$f(x;\mu,\sigma,\xi) = \left(1 + \frac{\xi(x-\mu)}{\sigma}\right)^{-\frac{1}{\xi}-1} \exp\left(-\left(1 + \frac{\xi(x-\mu)}{\sigma}\right)^{-\frac{1}{\xi}}\right)$$

$$\tag{1.109a}$$

$$F(x;\mu,\sigma,\xi) = \exp\left(-\left(1 + \frac{\xi(x-\mu)}{\sigma}\right)^{-\frac{1}{\xi}}\right). \tag{1.109b}$$

From Equation (1.109b), we can express x as a function of $F(x)$ as

$$x = \frac{\sigma\left((-\ln F)^{-\xi} - 1\right)}{\xi} + \mu. \tag{1.110}$$

For the three-parameter GEV distribution, we apply Equation (1.91) to compute $M_{1,0,0}$, $M_{1,1,0}$, $M_{1,2,0}$ and apply Equation (1.60) to compute the corresponding sample estimates as

$$M_{1,0,0} = \int_0^1 x\, dF = \int_0^1 \left(\frac{\sigma\left((-\ln F)^{-\xi} - 1\right)}{\xi} + \mu \right) dF = \mu - \frac{\sigma}{\xi} + \frac{\sigma}{\xi} \int_0^1 (-\ln F)^{-\xi}\, dF$$

$$= \mu - \frac{\sigma}{\xi} + \frac{\sigma}{\xi}\Gamma(1 - \xi) = b_0^{\text{sample}} = \bar{x}$$

$$(1.111)$$

$$M_{1,1,0} = \int_0^1 xF\, dF = \int_0^1 \left(\frac{\sigma\left((-\ln F)^{-\xi} - 1\right)}{\xi} + \mu \right) F\, dF$$

$$= \frac{1}{2}\left(\mu - \frac{\sigma}{\xi}\right) + \frac{\sigma}{\xi}\int_0^1 (-\ln F)^{-\xi} F\; dF = \frac{1}{2}\left(\mu - \frac{\sigma}{\xi}\right) + \frac{\sigma}{\xi}\left(\frac{1}{2}\right)^{-\xi+1}$$

$$\times \Gamma(1 - \xi) = b_1^{\text{sample}}$$

$$(1.112)$$

$$M_{1,2,0} = \int_0^1 xF^2\, dF = \int_0^1 \left(\frac{\left(\sigma(-\ln F)^{-\xi} - 1\right)}{\xi} + \mu \right) F^2\, dF$$

$$= \frac{1}{3}\left(\mu - \frac{\sigma}{\xi}\right) + \frac{\sigma}{\xi}\int_0^1 \frac{\sigma(-\ln F)^{-\xi}}{\xi} F^2\, dF = \frac{1}{3}\left(\mu - \frac{\sigma}{\xi}\right) + \frac{\sigma}{\xi}\left(\frac{1}{3}\right)^{1-\xi}$$

$$\times \Gamma(1 - \xi) = b_2^{\text{sample}}.$$

$$(1.113)$$

To this point, we have three equations and three unknowns, and the parameters μ, σ, and ξ may be estimated numerically by solving the system of equations.

Example 1.5 Estimate the parameters of the Pareto distribution using the method of PWMs.

Solution: The PDF and CDF of the Pareto distribution can be expressed as

$$f(x; c, a) = \frac{ac^a}{x^{a+1}}; \quad a > 0, \quad x \in [c, \infty) \tag{1.114a}$$

$$F(x; c, a) = 1 - \left(\frac{c}{x}\right)^a. \tag{1.114b}$$

From Equation (1.114b), we can express variable x as a function of F:

$$x = c(1 - F)^{-\frac{1}{\alpha}}.$$ (1.115)

For the two-parameter Pareto distribution, we apply Equation (1.86) to compute $M_{1,0,0}, M_{1,0,1}$ and apply Equation (1.59) to compute the corresponding sample estimates as

$$M_{1,0,0} = \int_0^1 x \, dF = \int_0^1 c(1 - F)^{-\frac{1}{\alpha}} dF = \begin{cases} \infty, & \alpha \leq 1 \\ \dfrac{c\alpha}{\alpha - 1}, & \alpha > 1 \end{cases} = a_0^{\text{sample}}$$ (1.116a)

$$M_{1,0,1} = \int_0^1 x(1 - F) \, dF = \int_0^1 c(1 - F)^{-\frac{1}{\alpha}+1} dF = \begin{cases} \infty, & \alpha \leq \dfrac{1}{2} \\ \dfrac{c\alpha}{2\alpha - 1}, & \alpha > \dfrac{1}{2} \end{cases} = a_1^{\text{sample}}.$$ (1.116b)

To this point, we have two equations and two unknowns, and we can estimate the parameters c and α under the condition that $\alpha > 1$.

1.5.5 Method of L-Moments

Hosking (1986, 1990) developed the theory of L-moments and the method of L-moments. This method has since become one of the most popular methods of parameter estimation. Using combinations of higher order statistics, Wang (1997) developed a generalization of L-moments. The L-moments can be expressed as functions of PWMs, which descriptively summarize the distribution characteristics, such as location, scale, and shape. They are less biased than ordinary moments, because they are always linear combinations of ranked or order statistics.

Let x be a real-valued ordered random variate from a sample of size n such that $x_{1:n} \leq x_{2:n} \leq \cdots \leq x_{n:n}$ with CDF $F(x)$ and quantile $x(F)$. Then, the rth L-moment, L_r, is defined as a linear function of expected order statistics as

$$L_r \equiv \frac{1}{r} \sum_{j=0}^{r-1} (-1)^j \binom{r-1}{j} E(x_{r-k:r}), \quad r = 1, 2, \ldots,$$ (1.117)

where

$$E(x_{j:r}) = \frac{r!}{(r-j)!(j-1)!} \int_0^1 xF^{j-1}(1-F)^{r-j}dF. \tag{1.118}$$

Applying the binomial expansion in $F(x)$, Hosking (1990) rewrote Equation (1.117) as

$$L_r = \int_0^1 xP^*_{r-1}(F)dF, \tag{1.119}$$

where

$$P^*_{r-1}(F) = \sum_{k=0}^{r-1} p^*_{r-1,k}F^k; \quad p^*_{r-1} = (-1)^{r-1-k}\binom{r-1}{k}\binom{r-1+k}{k}. \tag{1.120}$$

Hence, the most important first four L-moments are given as

$$L_1 = E(x) = \int_0^1 x\,dF \tag{1.121}$$

$$L_2 = \frac{1}{2}E(x_{2:2} - x_{1:2}) = \int_0^1 x(2F-1)dF \tag{1.122}$$

$$L_3 = \frac{1}{3}E(x_{3:3} - 2x_{2:3} + x_{1:3}) = \int_0^1 x(6F^2 - 6F + 1)dF \tag{1.123}$$

$$L_4 = \frac{1}{4}E(x_{4:4} - 3x_{3:4} + 3x_{2:4} - x_{1:4}) = \int_0^1 x(20F^3 - 30F^2 + 12F - 1)dF. \tag{1.124}$$

The corresponding sample statistics of the ordered sample can then be given as

$$l_1 = \frac{1}{n}\sum_{i=1}^n x_i \tag{1.125}$$

$$l_2 = \frac{1}{2}\binom{n}{2}^{-1} \sum\sum_{i>J}(x_{i:n} - x_{j:n}) \tag{1.126}$$

$$l_3 = \frac{1}{3}\binom{n}{3}^{-1} \sum\sum\sum_{i>j>k}(x_{i:n} - 2x_{j:n} - x_{k:n}) \tag{1.127}$$

$$l_4 = \frac{1}{4}\binom{n}{4}^{-1} \sum_{i>j>k>m} \sum \sum \sum (x_{i:n} - 3x_{j:n} + 3x_{k:n} - x_{m:n}). \qquad (1.128)$$

L-moments may also be estimated using a plotting-position formula (Hosking, 1990) as

$$l_r = \sum_{i=1}^{n} P_{r-1}^*(p_{i:n})x_{i:n}; \quad p_{i:n} = \frac{(i+\gamma)}{(n+\delta)}. \qquad (1.129)$$

In Equation (1.129), if $\gamma = -0.35$ and $\delta = 0$, we have $p_{i:n} = (i - 0.35)/n$, that is, the plotting-position formula proposed by Cunnane (1989). Hosking (1990) noted that L-moments are linear functions of PWMs. Hence, L-moments can be expressed in terms of a_r and b_r of PWMs as

$$L_r = (-1)^r \sum_{k=0}^{r-1} p_{r-1,k}^* a_k = \sum_{k=0}^{r-1} p_{r-1,k}^* b_k; \quad r = 1, 2, \ldots \qquad (1.130)$$

Equation (1.130) yields

$$L_1 = a_0 = b_0 \qquad (1.131)$$

$$L_2 = a_0 - 2a_1 = 2b_1 - b_0 \qquad (1.132)$$

$$L_3 = a_0 - 6a_1 + 6a_2 = 6b_2 - 6b_1 + b_0 \qquad (1.133)$$

$$L_4 = a_0 - 12a_1 + 30a_2 - 20a_3 = 20b_3 - 30b_2 + 12b_1 - b_0. \qquad (1.134)$$

Furthermore, similar to the conventional moment ratios, Hosking (1990) defined L-moment ratios τ_r as

$$L\text{-}C_v: \quad \tau_2 = \frac{L_2}{L_1} \qquad (1.135)$$

$$\tau_r = \frac{L_r}{L_2}, \quad r \geq 3. \qquad (1.136)$$

More specifically, the L-skewness and L-kurtosis are given as

$$L\text{-skewness:} \quad \tau_3 = \frac{L_3}{L_2} \qquad (1.137)$$

$$L\text{-kurtosis:} \quad \tau_4 = \frac{L_4}{L_2}. \qquad (1.138)$$

In the above equations, τ_2, τ_3, and τ_4 represent the measurements of scale, skewness, and kurtosis, respectively. The first L-moment is the mean. Vogel and Fennessey (1993) argued that L-moment diagrams, (i.e., τ_3 vs. τ_4) should replace product moment diagrams.

Example 1.6 Rework **Example 1.4** using the method of L-moments.

Solution: To estimate the parameters using the method of L-moments for the GEV distribution whose PDF and CDF are given as Equations (1.109a) and (1.109b), we will again evaluate the first three population and sample L-moments. The first three population L-moments of the GEV distribution may be expressed with the use of Equations (1.121)–(1.123) as

$$L_1 = M_{1,0,0} = \int_0^1 x\, dF = \mu - \frac{\sigma}{\xi} + \frac{\sigma}{\xi}\Gamma(1-\xi) \qquad (1.139)$$

$$L_2 = \int_0^1 x(2F-1)dF = 2M_{1,1,0} - M_{1,0,0} = \frac{\sigma}{\xi}\Gamma(1-\xi)(2^\xi - 1) \qquad (1.140)$$

$$L_3 = \int_0^1 x(6F^2 - 6F + 1)dF$$
$$= 6M_{1,2,0} - 6M_{1,1,0} + M_{1,0,0} = \frac{\sigma}{\xi}\Gamma(1-\xi)\left(2(3^{-\xi}) - 3(2^{-\xi}) + 1\right). \qquad (1.141)$$

Setting Equations (1.139)–(1.141) equal to the corresponding sample estimates, we get

$$\mu - \frac{\sigma}{\xi} + \frac{\sigma}{\xi}\Gamma(1-\xi) = b_0^{\text{sample}} = \bar{x} \qquad (1.142)$$

$$\frac{\sigma}{\xi}\Gamma(1-\xi)(2^\xi - 1) = 2b_1^{\text{sample}} - b_0^{\text{sample}} \qquad (1.143)$$

$$\frac{\sigma}{\xi}\Gamma(1-\xi)\left(2(3^{-\xi}) - 3(2^{-\xi}) + 1\right) = 6b_2^{\text{sample}} - 6b_1^{\text{sample}} + b_0^{\text{sample}}. \qquad (1.144)$$

We can then estimate the parameters by solving the system of equations (1.142)–(1.144).

Additionally, we can estimate the parameters by introducing the L-skewness. From Equations (1.143) and (1.144), we can compute the L-skewness as

$$\tau_3 = \frac{L_3}{L_2} = \frac{2(3^{-\xi}) - 3(2^{-\xi}) + 1}{2^{-\xi} - 1} = \frac{2(3^{-\xi} - 1)}{2^{-\xi} - 1} - 3. \qquad (1.145)$$

Setting $\tau_3 = t_3$ in which t_3 is the sample estimate, we can estimate
parameter $\hat{\xi}$. Substituting $\hat{\xi}$ in Equation (1.143) we can estimate $\hat{\sigma}$.
Finally, we can estimate parameter $\hat{\mu}$ by substituting $\hat{\xi}$ and $\hat{\sigma}$ in
Equation (1.142).

1.5.6 MLE Method

The method of MLE is one of the most widely used methods of parameter
estimation. Hosking (1985) suggested a correction for the bias in the ML
estimators of the Gumbel distribution parameters. Koch (1991) studied bias
error in the ML estimators. Clarke (1996) developed residual ML methods for
analyzing hydrologic data. The MLE method has also been widely used in
rainfall-runoff modeling (Sorooshian et al., 1983).

Let $f(x; \alpha_i)$, $i = 1, 2, \ldots, m$, be the PDF with parameters α_i. Then, for an
independent and identically distributed (IID) random sample of data, x_i, $i = 1$,
$2, \ldots, n$, the joint PDF can be defined as

$$f(x_1, x_2, \ldots, x_n; \alpha_1, \alpha_2, \ldots, \alpha_m) = \prod_{i=1}^{n} f(x_i; \alpha_1, \ldots, \alpha_m). \qquad (1.146)$$

The joint PDF is regarded as the likelihood function, L, because the likelihood
of obtaining the random sample x_i, $i = 1, 2, \ldots, n$, from the population of X is
proportional to the product of individual PDFs or equivalently joint PDF.
Therefore, the likelihood function is given as

$$l = \prod_{i=1}^{n} f(x_i; \alpha_1, \ldots, \alpha_m). \qquad (1.147)$$

The unknown parameters should be determined so that the likelihood that the
sample drawn is the one from X. This can be accomplished by maximizing the
likelihood function L. However, it is mathematically more convenient to take
the logarithm of L as

$$LL = \ln L = \ln \prod_{i=1}^{n} f(x_i; \alpha_1, \ldots, \alpha_m) = \sum_{i=1}^{n} \ln f(x_i; \alpha_1, \ldots, \alpha_m). \qquad (1.148)$$

For obtaining the maximum LL, Equation (1.148) is differentiated with respect
to the parameters as

$$\frac{\partial LL}{\partial a_i} = 0; \quad i = 1, 2, \ldots, m. \tag{1.149}$$

Equation (1.149) is a system of m equations whose solution leads to a_i, $i = 1$, $2, \ldots, m$.

Example 1.7 Rework **Example 1.1** to estimate the parameters using the method of MLE.

Solution: The PDF of the gamma distribution is given as Equation (1.38). Then, the log-likelihood function for the random sample of size n may be written as

$$LL = n\big(\alpha \ln \beta - \ln \Gamma(\alpha)\big) + (\alpha - 1)\sum_{i=1}^{n}\ln x_i - \beta\sum_{i=1}^{n} x_i. \tag{1.150}$$

Differentiating Equation (1.150) with respect to parameters α and β and setting the derivative as zero, we have

$$\frac{\partial LL}{\partial \alpha} = n\big(\ln \beta - \Psi(\alpha)\big) + \sum_{i=1}^{n}\ln x_i = 0 \tag{1.151}$$

$$\frac{\partial LL}{\partial \beta} = \frac{n\alpha}{\beta} - \sum_{i=1}^{n} x_i = 0. \tag{1.152}$$

In Equation (1.151), $\Psi(\alpha) = d\ln(\Gamma(\alpha))/d\alpha$.

Now the parameters can be estimated by solving the system of equations given as Equations (1.151) and (1.152) numerically.

1.5.7 Method of Cumulative Moments

The usual practice of frequency analysis involves determining a PDF, estimating its parameters, determining the probabilities or quantiles or CDF, which requires the integration of PDF, and then constructing confidence bands. Sometimes integration can be difficult. To avoid this problem, Burr (1942) developed the theory of cumulative moments, which is utilized here. In this theory, $F(x)$ is fitted by moments, and parameters are determined by mean, standard deviation, and moment ratios α_3 and α_4. A cumulative moment for $F(x)$ about a point a can be defined as

$$M_j(a) = \int_a^\infty (x-a)^j (1-F(x)) dx - \int_{-\infty}^a (x-a)^j F(x) dx; \quad x \in (-\infty, +\infty).$$

(1.153)

If $a = 0$ and $0 \le x < \infty$, then the cumulative moment can be defined as

$$M_j(0) = \int_0^\infty x^j (1-F(x)) dx.$$

(1.154)

It is straightforward to derive relations between cumulative moments about an arbitrary point a and those about the origin ($a = 0$).

From Equation (1.153) we can express the j-th-order cumulative moment about point a as

$$M_j(a) = \int_0^\infty (x-a)^j (1-F(x) dx - \int_{-\infty}^0 (x-a)^j F(x) dx$$
$$- \left(\int_0^a (x-a)^j (1-F(x)) dx + \int_0^a (x-a)^j F(x) dx \right)$$
$$= \int_0^\infty (x-a)^j (1-F(x)) dx - \int_{-\infty}^0 (x-a)^j F(x) dx - \int_0^a (x-a)^j dx.$$

(1.155)

Applying the binomial theorem, Equation (1.153) can be rewritten as

$$M_j(a) = \int_0^\infty \sum_{i=0}^j \binom{j}{i} (-a)^i x^{j-i} (1-F(x)) dx - \int_{-\infty}^0 \sum_{i=0}^j \binom{j}{i} (-a)^i x^{j-i} F(x) dx + \frac{(-a)^{j+1}}{j+1}$$
$$= \sum_{i=0}^j \binom{j}{i} (-a)^i \left(\int_0^\infty x^{j-i} (1-F(x) dx - \int_{-\infty}^0 x^{j-i} F(x) dx \right) + \frac{(-a)^{j+1}}{j+1}$$
$$= \sum_{i=0}^j \binom{j}{i} (-a)^i M_{j-i} + \frac{(-a)^{j+1}}{j+1}.$$

(1.156)

Similarly, we can express the cumulative moment about the origin as a function of the cumulative moments about point a as follows:

$$M_j = \int_0^\infty x^j (1 - F(x)) dx - \int_{-\infty}^0 x^j F(x) dx$$

$$= \int_a^\infty ((x - a) + a)^j (1 - F(x)) dx + \int_0^a x^j (1 - F(x)) dx$$

$$- \int_{-\infty}^a ((x - a) + a)^j F(x) dx + \int_0^a x^j F(x) dx$$

$$= \int_a^\infty ((x - a) + a)^j (1 - F(x)) dx - \int_{-\infty}^a ((x - a) + a)^j F(x) dx + \frac{a^{j+1}}{j+1}.$$

$$(1.157)$$

Applying binomial theorem to Equation (1.157), we have

$$M_j = \int_a^\infty \sum_{i=0}^j \binom{j}{i} a^i (x - a)^{j-i} (1 - F(x)) dx - \int_{-\infty}^a \sum_{i=0}^j \binom{j}{i} a^i (x - a)^{j-i} F(x) dx + \frac{a^{j+1}}{j+1}$$

$$= \sum_{i=0}^j \binom{j}{i} a^i \left(\int_a^\infty (x - a)^{j-i} (1 - F(x)) dx - \int_{-\infty}^a (x - a)^{j-i} F(x) dx \right) + \frac{a^{j+1}}{j+1}$$

$$= \sum_{i=0}^j \binom{j}{i} M_{j-i}(a) + \frac{a^{j+1}}{j+1}.$$

$$(1.158)$$

It is also possible to write cumulative moments about point a in terms of ordinary moments by integrating Equation (1.153) as

$$M_j(a) = \int_a^\infty (x - a)^j (1 - F(x)) dx - \int_{-\infty}^a (x - a)^j F(x) dx$$

$$= \frac{1}{j+1} (x - a)^{j+1} (1 - F(x)) \Big|_a^\infty + \frac{1}{j+1} \int_a^\infty (x - a)^{j+1} f(x) dx$$

$$- \frac{1}{j+1} (x - a)^{j+1} F(x) \Big|_{-\infty}^a + \frac{1}{j+1} \int_a^\infty (x - a)^{j+1} f(x) dx$$

$$(1.159)$$

$$= \frac{1}{j+1} \int_{-\infty}^{+\infty} (x - a)^{j+1} f(x) dx = \frac{1}{j+1} \mu_{j+1}(a).$$

In Equation (1.159), $\mu_{j+1}(a)$ denotes the $(j + 1)$-th moment about a.

Furthermore, applying binomial theorem, Equation (1.159) is rewritten as

$$M_j(a) = \frac{1}{j+1} \int_{-\infty}^{\infty} \sum_{i=0}^{j+1} \binom{j+1}{i}(-a)^i x^{j+1-i} f(x) dx$$

$$= \frac{1}{j+1} \sum_{i=0}^{j+1} (-a)^i \binom{j+1}{i} \mu'_{j+1-i}. \qquad (1.160)$$

In Equation (1.160), μ'_{j+1-i} denotes the $(j+1-i)$-th moment about zero.

If we expand $(x-a)^{j+1}$ about the first moment (i.e., mean: $\mu_1 = \mu'_1$), we have $(x-a)^{j+1} = [(x-\mu'_1) + (\mu'_1 - a)]^{j+1}$ and Equation (1.160) can be rewritten as

$$M_j(a) = \frac{1}{j+1} \int_{-\infty}^{\infty} \sum_{i=0}^{j+1} \binom{j+1}{i}(\mu'_1 - a)^i (x-\mu'_1)^{j+1-i} f(x) dx$$

$$= \frac{1}{j+1} \sum_{i=0}^{j+1} \binom{j+1}{i}(\mu'_1 - a)^i \mu_{j+1-i}. \qquad (1.161)$$

More specifically, we have

$$M_1(a) = \frac{1}{2} \sum_{i=0}^{2} \binom{2}{i}(\mu_1 - a)^i \mu_{2-i} = \frac{1}{2}\left(\mu_2 + 2(\mu_1 - a)\mu_1 + (\mu_1 - a)^2\right)$$

$$\qquad (1.162)$$

$$M_2(a) = \frac{1}{3} \sum_{i=0}^{3} \binom{3}{i}(\mu_1 - a)^i \mu_{3-i}$$
$$= \frac{1}{3}\left(\mu_3 + 3(\mu_1 - a)\mu_2 + 3(\mu_1 - a)^2\mu_1 + (\mu_1 - a)^3\right) \qquad (1.163)$$

$$M_3(a) = \frac{1}{4} \sum_{i=0}^{4} \binom{4}{i}(\mu_1 - a)^i \mu_{4-i}$$
$$= \frac{1}{4}\left(\mu_4 + 4(\mu_1 - a)\mu_3 + 6(\mu_1 - a)^2\mu_2 + 4(\mu_1 - a)^3\mu_1 + (\mu_1 - a)^4\right).$$

$$\qquad (1.164)$$

In Equation (1.161), μ_{j+1-i} denotes the $(j+1-i)$-th central moment (i.e., the moment about the mean).

Setting $a = 0$, Equations (1.160) and (1.161) can be rewritten, respectively, as

$$M_j = \frac{1}{j+1} \mu'_{j+1} \qquad (1.165)$$

$$M_j = \frac{1}{j+1} \sum_{i=0}^{j+1} \binom{j+1}{i} \mu_1'^i \mu_{j+1-i}'. \tag{1.166}$$

Furthermore, the central moments may be expressed using the cumulative moments about a as

$$\mu_j = j \sum_{i=0}^{j-1} \binom{j-1}{i} (a - \mu_1)^i M_{j-1-i}(a) + (a - \mu_1)^j, \quad j > 1. \tag{1.167}$$

In Equation (1.167), $\mu_1 - a = M_0(a)$.

More specifically, we have

$$\mu_2 = 2 \sum_{i=0}^{1} \binom{1}{i} (a - \mu_1)^i M_{1-i}(a) + (a - \mu_1)^2$$
$$= 2[M_1(a) + (a - \mu_1)M_0(a)] + (a - \mu_1)^2 = 2M_1(a) - M_0(a)^2 \tag{1.168}$$

$$\mu_3 = 3 \sum_{i=0}^{2} \binom{2}{i} (a - \mu_1)^i M_{2-i}(a) + (a - \mu_1)^j$$
$$= 3\left(M_2(a) - 2M_0(a)M_1(a) + M_0^3(a)\right) - M_0^3(a) \tag{1.169}$$
$$= 3M_2(a) - 6M_0(a)M_1(a) + 2M_0^3(a)$$

$$\mu_4 = 4 \sum_{i=0}^{3} \binom{3}{i} (a - \mu_1)^i M_{3-i}(a) + (a - \mu_1)^4$$
$$= 4\left(M_3(a) - 3M_0^3(a)M_2(a) + 3M_0^2(a)M_1(a) - M_0^4(a)\right) + M_0^4(a)$$
$$= 4M_3(a) - 12M_0^3(a)M_2(a) + 12M_0^2(a)M_1(a) - 3M_0^4(a). \tag{1.170}$$

Then, the moment ratio may be computed from the cumulative moments about a as

$$\alpha_3 = \frac{\mu_3}{\mu_2^{\frac{3}{2}}} = \frac{3M_2(a) - 6M_0(a)M_1(a) + 2M_0^3(a)}{\left(2M_1(a) - M_0(a)^2\right)^{\frac{3}{2}}} \tag{1.171}$$

$$\alpha_4 = \frac{\mu_4}{\mu_2^2} = \frac{4M_3(a) - 12M_2(a)M_0(a) + 12M_1(a)M_0^2(a) - 3M_0^4(a)}{\left(2M_1(a) - M_0^2(a)\right)^2} \tag{1.172}$$

$$\alpha_j = \frac{\mu_j}{\mu_2^{\frac{j}{2}}} = \frac{j \sum_{i=0}^{j-1} \binom{j-1}{i} \left(-M_0(a)\right)^i M_{j-1-i}(a) + \left(-M_0(a)\right)^j}{\left(2M_1(a) - M_0^2(a)\right)^{\frac{j}{2}}}. \tag{1.173}$$

Example 1.8 Consider a CDF $F(x) = 1 - (1/(1 + x^c)^k)$, $x \in [0, \infty)$; $c, k \geq 1$. Determine parameters c and k using cumulative moments. Also, derive the relation between the cumulative moments and ordinary moments.

Solution: The solution of this example is given in Burr (1942). Following Burr (1942), we outline the solution here. The PDF of the distribution is given as

$$f(x) = \frac{dF(x)}{dx} = \frac{kcx^{c-1}}{(1 + x^c)^{k+1}}. \tag{1.174}$$

The j-th cumulative moment of the distribution is then evaluated with the use of Equation (1.154) as

$$M_j = \int_0^\infty x^j \left(1 - F(x)\right) dx = \int_0^\infty \frac{x^j}{(1 + x^c)^k} dx. \tag{1.175}$$

Equation (1.175) is a special integral (De Haan, 1867) and can be solved as

$$M_j = \begin{cases} \dfrac{\pi}{c \sin\left(\dfrac{j+1}{c}\right)} \left(1 - \dfrac{j+1}{c}\right)\left(1 - \dfrac{j+1}{2c}\right) \cdots \left(1 - \dfrac{j+1}{(k-1)c}\right), & \text{if } j < c - 1 \\[4mm] \dfrac{\Gamma\left(\dfrac{j+1}{c}\right)\Gamma\left(k - \dfrac{j+1}{c}\right)}{c\Gamma(k)}, & \text{if } j \in [c-1, ck-1] \end{cases}$$

$$\tag{1.176}$$

Now we can relate the cumulative moments to the ordinary moments. For this two-parameter distribution, we write the first two moments:

$$M_0 = \frac{\pi}{c \sin\left(\dfrac{1}{c}\right)} \left(1 - \dfrac{1}{c}\right)\left(1 - \dfrac{1}{2c}\right) \cdots \left(1 - \dfrac{1}{(k-1)c}\right) \tag{1.177}$$

$$M_1 = \begin{cases} \dfrac{\pi}{c \sin\left(\dfrac{2}{c}\right)} \left(1 - \dfrac{2}{c}\right)\left(1 - \dfrac{2}{2c}\right) \cdots \left(1 - \dfrac{2}{(k-1)c}\right), & \text{if } c < 2 \\[4mm] \dfrac{\Gamma\left(\dfrac{2}{c}\right)\Gamma\left(k - \dfrac{2}{c}\right)}{c\Gamma(k)}, & \text{if } c \geq 2 \end{cases}$$

$$\tag{1.178}$$

From Equation (1.157), we have

$$M_0 = \mu'_1; \quad M_1 = \frac{1}{2}\left(\mu_2 + \mu'^2_1\right). \tag{1.179}$$

Equating $M_0 = \bar{x}$; $M_1 = (1/2)(S^2 + \bar{x}^2)$ where S^2 denotes the sample variance, we can estimate parameters c and k by solving the system of equations.

1.5.8 Least Squares Method

The least squares method has been commonly applied for regression analysis. It may also be applied to estimate the parameters of a PDF fitted to the observed random variables. Let f_{em} represent the empirical density for the random sample $\mathbf{x} = \{x_i: \ i = 1, 2, \ldots, n\}$ and the parameters of the parametric density function candidate $f(x; \boldsymbol{\alpha})$; $\boldsymbol{\alpha} = \{\alpha_1, \alpha_2, \ldots, \alpha_m\}$. The least squares is computed as

$$E(\boldsymbol{\alpha}) = \sum_{i=1}^{n} \left(f_{em}(x_i) - f(x_i; \boldsymbol{\alpha})\right)^2. \tag{1.180}$$

To this end, the error can be minimized by taking the derivative with respect to the parameters as

$$\frac{\partial E}{\partial \alpha_1} = 0, \ \frac{\partial E}{\partial \alpha_2} = 0, \ \ldots, \ \frac{\partial E}{\partial \alpha_m} = 0. \tag{1.181}$$

As a result, the parameters may be estimated by solving Equation (1.181) simultaneously.

1.6 Selection of a Distribution

Let sample $\mathbf{x} = \{x_i: i = 1, 2, \ldots, n\}$ be a random variable. We now address the question of how to select the distribution candidates for further frequency analysis with the following steps:

(1) **Graph the nonparametric histogram or kernel density for the sample data:** This step allows to obtain a rough idea about the kind of distribution that may be applied. Figure 1.7 graphs the histograms from the synthetic data of random variables simulated from different populations. The

Table 1.1. *Sample statistics of synthetic data of random variables.*

Variable	Mean	Standard deviation	Skewness	Kurtosis
A	1.999	0.896	0.886	1.074
B	9.668	10.292	−1.902	7.861
C	9.981	1.985	0.009	−0.036

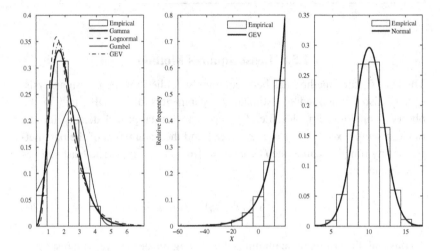

Figure 1.7 Relative frequency histograms and fitted distributions for the synthetic data of random variables simulated from different populations.

histograms graphed clearly show different shapes. The histogram in (A) indicates that the random variable data are bell-shaped and right-skewed. The histogram in (B) indicates that the random variable data are J-shaped and left-skewed. And the histogram in (C) indicates the random variable data are bell-shaped with no obvious skewness.

(2) **Evaluate the essential sample statistics:** Besides the visual evidence, the statistics including sample mean, standard deviation, skewness, and kurtosis are very essential for selecting the distribution candidates. Using the synthetic random data, Table 1.1 lists the first four sample statistics that dominate the shape of a frequency distribution. The sample statistics listed in Table 1.1 further confirm the conclusion from the visual evidence.

(3) **Select the distribution candidates based on the histogram and sample statistics:** Based on the non-empirical histogram (or kernel density) graph

and the sample statistics, we can choose several different frequency distribution candidates for the target random variables. For the right-skewed and tailed random variable shown in (A), we may choose gamma, lognormal, Gumbel, and GEV distributions as candidate distributions. For the left-skewed and tailed random variable shown in (B), we may choose Weibull and GEV distributions. For the bell-shaped random variables without obvious skewness and excess kurtosis shown in (C), we may choose normal distribution as the candidate distribution. Using MLE, Table 1.2 lists the estimated parameters and the log-likelihood for each candidate. The frequencies computed from the fitted distributions are also shown in Figure 1.7. For the variable shown as (A), it is seen that there exist minimal differences among gamma, lognormal, and GEV distributions. In addition, the gamma distribution yields the largest log-likelihood. Thus, we can select gamma, GEV, and normal distributions for variables A, B, and C, respectively.

1.7 Goodness-of-Fit Measures

Let $X = \{x_1, \ldots, x_n\}$ be the IID random variable following the true probability distribution function F. For a fitted distribution $\{\hat{F}(x; \hat{\alpha}), \hat{\alpha}: \text{fitted parameters}\}$ to random variable X, its goodness-of-fit may be expressed by testing the null hypothesis of $H_0: F = \hat{F}$ versus the alternative $H_1: F \neq \hat{F}$.

For testing, there are a number of formal goodness-of-fit statistics based on the measurement of the distance between the empirical CDF $[F_n(x)]$ and the fitted parametric CDF $[\hat{F}(x; \hat{\alpha})]$. These include Kolmogorov–Smirnov (KS) statistic D_N (Kolmogorov, 1933; Smirnov, 1948), Cramér–von Mises (CM) statistic W_N^2 (Cramér, 1928; von Mises, 1928), Anderson–Darling (AD) statistic A_N^2 (Anderson and Darling, 1952), modified weighted Watson statistic U_N^2 (Stock and Watson, 1989), and Liao and Shimokawa statistic L_N (Liao and Shimokawa, 1999). The chi-square goodness-of-fit test is also commonly applied, which measures the difference between the empirical frequency and the frequency computed from the fitted parametric distribution. In what follows, we first briefly discuss each goodness-of-fit statistic followed by the evaluation with the use of the P-value.

Kolmogorov–Smirnov (KS) statistic D_N: The KS test statistic can be expressed theoretically as

$$D_N = \sup_{x \in \mathbb{R}} |F_n(x) - F(x)|, \tag{1.182}$$

Table 1.2. Estimated parameters and corresponding log-likelihood.

Variable	Gamma	Lognormal[a]	Gumbel[b]	GEV[b]	Normal[a]
A	$\alpha = 5.00$ $\beta = 0.40$ $LL = -12,368$	$\mu = 0.59$ $\sigma = 0.47$ $LL = -12,517$	$\mu = 2.48$ $\sigma = 1.10$ $LL = -15,304$	$\xi = -0.02$ $\mu = 1.60$ $\sigma = 0.72$ $LL = -12,399$	
B				$\xi = -1.00$ $\mu = 10.33$ $\sigma = 9.66$ $LL = -33,352$	
C					$\mu = 9.98$ $\sigma = 1.99$ $LL = -21,045$

[a] μ and σ represent the mean and standard deviation.
[b] μ and σ represent the location and shape parameters.

where $F_n(x)$ is the empirical distribution estimated as n/N, where n is the cumulative number of sample events at class limit n. Applying the fitted distribution function $\hat{F}(x; \hat{a})$, Equation (1.182) can be rewritten as

$$D_N = \max(\hat{\delta}_i), \quad \hat{\delta}_i = \max\left[\frac{i}{N} - \hat{F}(x_i; \hat{a}), \hat{F}(x_i; \hat{a}) - \frac{i-1}{N}\right], \quad i \in [1, N].$$

(1.183)

Cramer–von Mises (CM) statistic W_N^2: The CM test statistic can be expressed theoretically as

$$W^2 = \int_{-\infty}^{\infty} [F_n(x) - F(x)]^2 dF(x).$$

(1.184)

Applying the fitted probability distribution $\hat{F}(x; \hat{a})$, Equation (1.184) can be rewritten as

$$W_N^2 = \frac{1}{12N} + \sum_{i=1}^{N}\left[\hat{F}(x_i; \hat{a}) - \frac{2i-1}{2N}\right]^2.$$

(1.185)

Anderson–Darling (AD) statistic A_N^2: The AD test statistic can be expressed theoretically as

$$A = n \int_{-\infty}^{\infty} \frac{\left(F_n(x) - F(x)\right)^2}{F(x)\left(1 - F(x)\right)} dF(x).$$

(1.186)

Applying the fitted probability distribution $\hat{F}(x; \hat{a})$, Equation (1.186) can be rewritten as

$$A_N^2 = -N - \frac{1}{N}\sum_{i=1}^{N}(2i - 1)\ln\left(\hat{F}(x_i; \hat{a})\left(1 - \hat{F}(x_{n+1-i}; \hat{a})\right)\right).$$

(1.187)

Modified weighted Watson statistic U_N^2: Applying the fitted probability distribution $\hat{F}(x; \hat{a})$, the U_N^2 test statistic can be expressed as

$$U_N^2 = N^2 \sum_{i=1}^{N} d_i^2 - N\left(\sum_{i=1}^{N} d_i\right)^2; \quad d_i = \frac{\hat{F}(x_i; \hat{a}) - \frac{i}{N+1}}{\sqrt{i(N-i+1)}}.$$

(1.188)

Liao and Shimokawa statistic L_N: Applying the fitted probability distribution $\hat{F}(x;\hat{a})$, the L_N test statistic can be expressed as

$$L_N = \frac{1}{\sqrt{N}} + \sum_{i=1}^{N} \frac{\max\left[\frac{i}{N} - \hat{F}(x_i;\hat{a}),\ \hat{F}(x_i;\hat{a}) - \frac{i-1}{N}\right]}{\sqrt{\hat{F}(x_i;\hat{a})\left(1 - \hat{F}(x_i;\hat{a})\right)}}. \tag{1.189}$$

In Equations (1.188) and (1.189), N is the sample size.

Conventionally, the P-value of the above statistics is computed using the limiting probability distribution for each specific test statistic. To avoid misidentification of the limiting probability distribution, the parametric bootstrap simulation method is widely applied to estimate the P-value using the following procedure:

(1) Estimate the parameter vector $\hat{a}$ of the probability distribution $\hat{F}(x_i;a)$.
(2) Compute the test statistics of D_N, W_N^2, A_N^2, or U_N^2, L_N.
(3) With a larger number of M, for $k = 1 : M$ to proceed:
   (a) Generate random variable $x^{(k)}$ with sample size N from the fitted probability distribution $\hat{F}(x_i;\hat{a})$.
   (b) Re-estimate the parameter vector $\hat{a}^*$ from the hypothesized distribution using the random sample generated from step (a).
   (c) Compute the test statistics of D_N^*, W_N^{2*}, A_N^{2*}, U_N^{2*}, and L_N^* using steps (a) and (b).
   (d) Repeat steps (a)–(c) M times.
(4) Compute the P-value using

$$P\text{-value} = \frac{\sum_{i=1}^{M} 1\left(D_N^*(i) > D_N\right)}{M}. \tag{1.190}$$

Replacing (D_N^*, D_N) by (W_N^{2*}, W_N^2), (A_N^{2*}, A_N^2), (U_N^{2*}, U_N^2), and (L_N^*, L_N) in Equation (1.190), one can simulate the P-values for other statistics.

From common practice, we may set $\alpha_{\text{level}} = 0.05$, which means the hypothesized parametric univariate distribution cannot be rejected if P-value $\geq 0.05 = \alpha_{\text{level}}$. Furthermore, the larger the M, the closer the simulated P-value to its true P-value.

Additionally, the root mean square error (RMSE), relative error (RE), and bias have also been commonly applied to evaluate the model performance. Let f_{em} and f_{fit} represent the empirical and fitted frequency for the random variable **x**, then the RMSE, RE, and bias may be expressed as

$$\text{RMSE} = \sqrt{\frac{\sum_{i=1}^{m}\left(f_{em}(i) - f_{fit}(i)\right)^2}{m}} \qquad (1.191)$$

$$\text{RE} = \frac{1}{m}\sum_{i=1}^{m}\frac{|f_{em}(i) - f_{fit}(i)|}{f_{em}(i)} \qquad (1.192)$$

$$\text{Bias} = E(f_{em}) - E(f_{fit}) = \frac{1}{m}\sum_{i=1}^{m}\left(f_{em}(i) - f_{fit}(i)\right). \qquad (1.193)$$

In Equations (1.191)–(1.193), m represents the number of bins applied to compute empirical frequency.

Example 1.9 Evaluate the goodness-of-fit of the gamma distribution to synthetic data of variable (A) using the KS test as an example.

Solution: The gamma distribution can be given as

$$f(x; \alpha, \beta) = \frac{1}{\beta^{\alpha}\Gamma(\alpha)}x^{\alpha-1}\exp\left(-\frac{x}{\beta}\right). \qquad (1.194)$$

Step 1: Order the variable A values in the ascending order. The parameters estimated using MLE are listed in Table 1.2.

Step 2: Compute the test statistics. Using the first sample as an example, we illustrate the computation procedure as follows:

$$x_{(1)} = 0.1449, \quad N = 10{,}000,$$

$$F(x_{(1)}) = gamcdf(x_{(1)}; 5.00, 0.40) = 3.86 \times 10^{-5};$$

$$\hat{\delta}_1 = \max\left(\frac{1}{N} - F(x_{(1)}), F(x_{(1)}) - \frac{1-1}{N}\right)$$

$$= \max(10^{-4} - 3.86 \times 10^{-5}, 3.86 \times 10^{-5} - 0) = 6.14 \times 10^{-5}.$$

Similarly, we can compute $\hat{\delta}_2, \ldots, \hat{\delta}_{10{,}000}$.

Applying Equation (1.101a), we can compute the KS statistic:

$D_n = \max(\hat{\delta}) = 0.0058.$

Step 3: Apply parametric bootstrap method N times to approximate the P-value with the given significance level α. Here we choose $N = 5{,}000$, $\alpha = 0.05$. Again, we will use one parametric bootstrap example to illustrate the procedure.

- Generate IID random variable data from the fitted gamma distribution with parameters $\alpha = \hat{5}.00$ and $\beta = \hat{0}.40$, and sample size of 10,000.
- Arrange the generated random variable data in ascending order.
- Re-estimate the parameters for the generated random variable, and we have $\hat{\alpha}_1 = 4.90$ and $\hat{\beta}_1 = 0.41$.
- Compute the KS statistic with the generated random variable data, and we have $D_n^1 = 0.0059$.

Repeat Step 3 for $N = 5,000$ times, and we can finally compute the P-value of KS test; we have

$$P\text{-value} = \frac{\sum\limits_{i=1}^{5,000} (D_n^i > D_n)}{5,000} = 0.61 > 0.05.$$

The P-value of the KS statistic shows that the fitted gamma distribution is appropriate to model variable A.

Chi-square goodness-of-fit test: Rather than measuring the difference between the empirical CDF and the fitted parametric CDF, the chi-square goodness-of-fit test deals with the frequency directly. As per its name, the limiting distribution is the chi-square distribution with its statistic expressed as

$$\chi_{K-m-1}^2 = \sum_{i=0}^{K} \frac{(o_i - e_i)^2}{e_i}. \tag{1.195}$$

In Equation (1.195), o_i is the observed frequency count for the level i of a variable; e_i is the corresponding expected frequency count from the fitted probability distribution; K is the number of levels of the random variable; m is the number of parameters of the fitted probability distribution; and $K - m - 1$ is the degree of freedom of the limiting chi-square distribution. In other words, Equation (1.195) actually compares the relative frequency computed from a histogram with K-bins to the fitted parametric distribution; that is, (1) level i is equivalent to bin i of the histogram and (2) the number of level K is equivalent to the total number of bins (K) of the histogram.

The simplest rule of thumb to determine the number of bins for a histogram is given as

$$K = [1 + \log_2 n].\tag{1.196}$$

Example 1.10 Evaluate the goodness-of-fit of the gamma distribution to synthetic data of variable (A) using the chi-square test.

Solution: To apply the chi-square goodness-of-fit test, we will need the frequency histogram shown in Table 1.3.

The number of bins is computed as $K = [1 + \log_2 10{,}000] = 15$. The degrees of freedom for the limiting chi-squared distribution are computed as $d.f. = K -$ number of parameters $- 1 = 15 - 2 - 1 = 12$.

Table 1.3 lists the observed frequency (O_i), the corresponding interval, and the expected frequency from the fitted distribution (E_i).

Applying Equation (1.195) using the observed and expected frequencies listed in Table 1.3, we obtain:

$$\text{Test statistic} = 9.27, \quad \text{Critical value} = \chi_2^{-1}(0.95, 12) = 21.03$$

$$P\text{-value} = 1 - \chi_2(9, 27, 12) = 0.68 > 0.05.$$

Again, the chi-square test shows that the gamma distribution can be applied to model variable A. In fact, synthetic variable A is simulated from the gamma population $(\alpha = 5, \ \beta = 0.4)$.

Example 1.11 Evaluate the model performance using RMSE, RE, and bias for synthetic data set A.

Solution: Table 1.3 lists the empirical and expected frequencies computed for variable A for gamma distribution. Using Table 1.3, the model performances are computed for the gamma distribution as

$$\text{RMSE}^{\text{gamma}} = \sqrt{\frac{\sum_{i=1}^{15}\left(OF(i) - EF(i)\right)^2}{15}} = 17.09;$$

$$\text{RE}^{\text{gamma}} = \left[\frac{1}{m} \sum_{i=1}^{m} \frac{|f_{em}(i) - f_{fit}(i)|}{f_{em}(i)} \right] = \frac{1}{15} \left[\sum_{i=1}^{15} \frac{|OF(i) - EF(i)|}{OF(i)} \right]$$

$$= 0.0998 = 9.98\%$$

$$\text{Bias}^{\text{gamma}} = E(f_{em}) - E(f_{fit}) = \frac{1}{15} \sum_{i=1}^{15} \left(OF(i) - EF(i) \right) = 0.11.$$

Additionally, Table 1.4 lists the expected frequencies for variable A using the fitted Gumbel, lognormal, and GEV distributions with the same interval as that for the gamma distribution. Now using the observed frequencies listed in Table 1.3 and expected frequencies listed in Table 1.4, RMSE, RE, and bias may then be computed as listed in Table 1.5. The model performance measures listed in Table 1.5 and those computed for the gamma distribution agree with the model selection based on the MLE method, that is, the gamma distribution best fits variable A.

Table 1.3. *Empirical (observed) and expected frequencies for variable* A.

Interval	X value	Observed frequency	Expected frequency
1	[0.14, 0.60]	177	184.88
2	[0.60, 1.06]	1,127	1,091.29
3	[1.06, 1.51]	1,968	1,991.06
4	[1.51, 1.96]	2,182	2,166.14
5	[1.96, 2.42]	1,780	1,776.19
6	[2.42, 2.87]	1,177	1,219.98
7	[2.87, 3.33]	750	742.35
8	[3.33, 3.78]	408	413.67
9	[3.78, 4.23]	221	215.63
10	[4.23, 4.69]	123	106.66
11	[4.69, 5.14]	49	50.58
12	[5.14, 5.60]	21	23.17
13	[5.60, 6.05]	10	10.31
14	[6.05, 6.51]	3	4.47
15	[6.51, 6.96]	4	1.90

Table 1.4. *Expected frequencies for variable A (Gumbel, lognormal, and GEV distributions).*

X value	Gumbel	Lognormal	GEV
[0.14, 0.60]	527.91	94.69	192.19
[0.60, 1.06]	738.90	1,167.86	1,013.11
[1.06, 1.51]	994.51	2,256.24	2,022.53
[1.51, 1.96]	1,261.88	2,199.23	2,258.98
[1.96, 2.42]	1,465.22	1,621.35	1,801.65
[2.42, 2.87]	1,489.40	1,054.34	1,186.42
[2.87, 3.33]	1,241.11	646.73	700.88
[3.33, 3.78]	769.87	386.56	388.95
[3.78, 4.23]	309.14	229.01	207.91
[4.23, 4.69]	65.82	135.72	108.50
[4.69, 5.14]	5.59	80.88	55.66
[5.14, 5.60]	0.13	48.61	28.17
[5.60, 6.05]	0.00	29.50	14.09
[6.05, 6.51]	0.00	18.10	6.97
[6.51, 6.96]	0.00	11.23	3.41

Table 1.5. *Model performances for Gumbel, lognormal, and GEV distributions.*

Distributions	RMSE	RE	Bias
Gumbel	420.35	0.73	75.37
Lognormal	98.60	0.79	1.33
GEV	41.59	0.19	0.7

References

Anderson, T.W. and Darling, D.A. (1952). Asymptotic theory of certain "goodness-of-fit" criteria based on stochastic processes. *Annals of Mathematical Statistics*, Vol. 23, pp. 193–212.

Ashkar, F., Bobée, B., Leroux, D., and Morisette, D. (1988). The generalized method of moments as applied to the generalized gamma distribution. *Stochastic Hydrology and Hydraulics*, Vol. 2, pp. 161–174.

Barlow, R.E. and Proschan, F. (1965). *Mathematical Theory of Reliability*. John Wiley & Sons, Inc, New York.

Bobée, B., Perreault, L., and Ashkar, F. (1993). Two kinds of moment diagrams and their applications in hydrology. *Stochastic Hydrology and Hydraulics*, Vol. 7, pp. 41–65.

Burr, I.W. (1942). Cumulative frequency functions. *The Annals of Mathematical Statistics*, Vol. 13, No. 2, pp. 215–232.

Clarke, R.T. (1996). Residual maximum likelihood (REML) methods for analyzing hydrological data series. *Journal of Hydrology*, Vol. 182, pp. 277–295.

Cramér, H. (1928). On the composition of elementary errors. *Scandinavian Actuarial Journal*, Vol. 1, pp. 13–74. doi: 10.1080/03461238.1928.10416862.

Cunnane, C. (1989). *Statistical Distribution for Flood Frequency Analysis*. WMO Operational Hydrology Report No. 33, WMO-No. 718, Geneva, Switzerland.

De Haan, D.B. (1867). *Nouvelles tables d'Integrales definies*. G. E. Stechert, New York.

Greenwood, J.A., Landwehr, J.M., Matalas, N.C., and Wallis, J.R. (1979). Probability-weighted moments: Definition and relation to parameters of several distributions expressible in inverse form. Water Resources research, Vol. 15, pp. 1049–1054.

Haktamir, T. (1996). Probability-weighted moments without plotting position formula. *Journal of Hydrologic Engineering*, Vol. 1, No. 2, pp. 89–91.

Hosking, J.R.M. (1985). A correction for the bias of maximum likelihood estimators of Gumbel parameters – Comment. *Journal of Hydrology*, Vol. 78, pp. 393–396.

Hosking, J.R.M. (1986). *The Theory of Probability-Weighted Moments*. Technical Report RC 12210, Mathematics, 160 pp., IBM Thomas Watson Research Center, Yorktown Heights, New York.

Hosking, J.R.M. (1990). L-moments: Analysis and estimation of distribution using linear combination of order statistics. *Journal of Royal Statistical Society, Series B*, Vol. 52, No. 1, pp. 105–124.

Jaynes, E.T. (1957). Information theory and statistical mechanics, I. *Physical Review*, Vol. 106, pp. 620–630.

Koch, S.P. (1991). Bias error in maximum likelihood estimation. *Journal of Hydrology*, Vol. 122, pp. 289–300.

Kolmogorov, A. (1933). Sulla determinazione empirica di una legge di distribuzione. *Giornale dell'Istituto Italiano degli Attuari*, Vol. 4, pp. 83–91.

Landwehr, J.M., Matalas, N.C., and Wallis, J.R. (1979a). Probability-weighted moments compared with some traditional techniques in estimating Gumbel parameters and quantiles. *Water Resources Research*, Vol. 15, pp. 1055–1064.

Landwehr, J.M., Matalas, N.C., and Wallis, J.R. (1979b). Estimation of parameters and quantiles of Wakeby distribution. *Water Resources Research*, Vol. 15, pp. 1361–1379.

Liao, M. and Shimokawa, T. (1999). A new goodness-of-fit for type I extreme value and 2-parameter Weibull distributions with estimated parameters. *Optimization*, Vol. 64, No. 1, pp. 23–48.

von Mises, R.E. (1928). *Wahrscheinlichkeit, Statistik und wahreit*. Julius Springer, Vienna.

Shannon, C.E. (1948). The mathematical theory of communication, I and II. *Bell System Technical Journal*, Vol. 27, pp. 379–423.

Singh, V.P. (1988). *Hydrologic Systems*, Vol. 1: Rainfall-Runoff Modeling. Prentice Hall, Engelwood Cliffs, NJ.

Singh, V.P. (1998). *Entropy-Based Parameter Estimation in Hydrology*. Kluwer Academic Publishers (now Springer), Dordrecht, the Netherlands.

Singh, V.P. and Rajagopal, A.K. (1986). A new method of parameter estimation for hydrologic frequency analysis. *Hydrological Science and Technology*, Vol. 2, No.3, pp. 33–40.

Smirnov, N. (1948). Table for estimating the goodness-of-fit of empirical distributions. *Annals of Mathematical Statistics*, Vol. 19, pp. 279–281. doi: 10.1214/aoms/1177730256.

Sorooshian, S., Gupta, V.K., and Fulton, J.L. (1983). Evaluation of maximum likelihood parameter estimation techniques for conceptual rainfall-runoff models: Influence of calibration data variability and length on model credibility. *Water Resources Research*, Vol. 19, No. 1, pp. 251–259.

Stock, J.H. and Watson, M.W. (1989). Interpreting the evidence on money-income casualty. *Journal of Econometrics*, Vol. 40, pp. 161–181.

Vogel, R.M. and Fennessey, N.M. (1993). L-moment diagrams should replace product moment diagrams. *Water Resources research*, Vol. 29, No. 6, pp. 1745–1752.

Wang, Q.J. (1997). LH moments for statistical analysis of extreme events. *Water Resources Research*, Vol. 33, No. 12, pp. 2841–2848.

2
Burr–Singh–Maddala Distribution

2.1 Introduction

Burr (1942) derived 12 cumulative probability distribution functions (CDFs) that constitute what is often referred to as the Burr family. The 12th member of this family, frequently referred to as the Burr XII distribution, is one of the most frequently used distributions in environmental and water resources engineering. Using a somewhat different approach, Singh and Maddala (1976) derived a similar distribution for describing the size distribution of incomes, which consists of the Burr XII distribution as a special case. This distribution is now referred to as the Burr–Singh–Maddala (BSM) distribution. Brouers (2015) provided a comprehensive exposition of the BSM distribution and its properties. This chapter draws significantly from his exposition.

The BSM distribution has been employed in a variety of fields, such as hydrology (Papalexiou and Koutsoyiannis, 2012), forestry (Dubey and Gove, 2015), sorption theories (Brouers, 2014a, b; Brouers and Al-Musawi, 2015), fractal kinetics (Brouers and Sotolongo-Costa, 2005), and relaxation and reaction phenomena (Weron and Kotulski, 1997; Brouers and Sotolongo-Costa, 2005). This distribution possesses a number of interesting characteristics that are useful in environmental and water engineering. The distribution and its parameters can be derived using the entropy theory. Therefore, the objective of this chapter is to (1) discuss the characteristics of the BSM distribution, (2) derive the distribution using the differential equation approach and the entropy theory, (3) estimate the distribution parameters using the entropy theory as well as other methods, and (4) illustrate the application of the distribution using annual peak flow and daily maximum precipitation.

2.2 Characteristics of BSM Distribution

The probability density function (PDF), $f(x)$, of the BSM distribution can be expressed as

$$f(x) = \frac{a}{b}\left(\frac{x}{b}\right)^{a-1}\left(1 + c\left(\frac{x}{b}\right)^{a}\right)^{-\frac{1}{c}-1}, \quad a, b, c > 0, x \in [0, \infty), \qquad (2.1)$$

where a and c are the shape or form parameters, b is the scale parameter, and x is the value of random variable X. For different values of parameters, the BSM distribution is shown in Figure 2.1. The CDF, $F(x)$, can be expressed by integrating Equation (2.1) as follows. Let

$$w = 1 + c\left(\frac{x}{b}\right)^{a}. \qquad (2.2)$$

Then, taking the derivative of Equation (2.2), we get

$$dw = \frac{ac}{b}\left(\frac{x}{b}\right)^{a-1}dx. \qquad (2.3)$$

With the intermediate variable w defined in Equation (2.2), the density function $f(x)$ can be expressed as

$$f(x) = \frac{dF}{dw}\frac{dw}{dx} \Rightarrow \frac{dF}{dw} = f(x)\left(\frac{dw}{dx}\right)^{-1} = \left(\frac{a}{b}\right)\left(\frac{x}{b}\right)^{a-1}w^{-\frac{1}{c}-1}\left[\frac{ac}{b}\left(\frac{x}{b}\right)^{a-1}\right]^{-1} = \frac{1}{c}w^{-\frac{1}{c}-1}. \qquad (2.4)$$

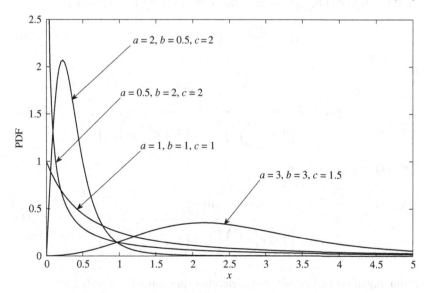

Figure 2.1 PDF of BSM distribution with different parameters.

Equation (2.4) integrates to

$$F(w) = \int_{1}^{w} \frac{1}{c} w^{-\frac{1}{c}-1} dw = 1 - w^{-\frac{1}{c}}, \quad F(w) = 0 \text{ if } w = 1. \quad (2.5)$$

Substituting Equation (2.2) into Equation (2.5), we obtain

$$F(x) = 1 - \left(1 + c\left(\frac{x}{b}\right)^a\right)^{-1/c}, \quad (2.6)$$

which is the CDF of the BSM distribution. Its survival function, $SF(x)$, can be written as

$$SF(x) = 1 - F(x) = \left(1 + c\left(\frac{x}{b}\right)^a\right)^{-1/c}. \quad (2.7)$$

If $a = 1$, $c = q - 1$, where q is the Tsallis entropy index (Tsallis, 1988), then Equation (2.7) becomes

$$SF(x) = 1 - F(x) = \left(1 + (q-1)\left(\frac{x}{b}\right)\right)^{\frac{1}{1-q}}. \quad (2.8)$$

Equation (2.8) has the same form as the PDF of the Tsallis distribution.

Figure 2.1 illustrates the shape of the BSM distribution with different parameters.

From the PDF, CDF, and survival function of the BSM distribution, the PDF function can be expressed in the form stated by Burr (1942) as

$$f(x) = \frac{dF(x)}{dx} = g(x)F(x)\left(1 - F(x)\right) \quad (2.9)$$

and

$$g(x) = \frac{a\left(\frac{x}{b}\right)^a}{x\left(1 + c\left(\frac{x}{b}\right)^a\right)\left(1 - \left(1 + c\left(\frac{x}{b}\right)^a\right)^{-\frac{1}{c}}\right)}. \quad (2.10)$$

Let $g(x) = g_1(x)/x$. Then we have

$$g_1(x) = \frac{a\left(\frac{x}{b}\right)^a}{\left(1 + c\left(\frac{x}{b}\right)^a\right)\left(1 - \left(1 + c\left(\frac{x}{b}\right)^a\right)^{-\frac{1}{c}}\right)}. \quad (2.11)$$

From Equation (2.11), we may determine the limit for $x \to 0$ and $x \to \infty$. When $x \to 0$, it is easy to show that

$$\lim_{x \to 0} g_1(x) = \lim_{x \to 0} \frac{a\left(\frac{x}{b}\right)^a}{\left(1 + c\left(\frac{x}{b}\right)^a\right)\left(1 - \left(1 + c\left(\frac{x}{b}\right)^a\right)^{-\frac{1}{c}}\right)}$$

$$= a \lim_{x \to 0} \left(\frac{1}{c} + \frac{c - 1}{c\left(c\left(\left(c\left(\frac{x}{b}\right)^a + 1\right)^{\frac{1}{c}} - 1\right) + 1\right)}\right) = a \tag{2.12}$$

$$\lim_{x \to \infty} g_1(x) = \lim_{x \to \infty} \frac{a\left(\frac{x}{b}\right)^a}{\left(1 + c\left(\frac{x}{b}\right)^a\right)\left(1 - \left(1 + c\left(\frac{x}{b}\right)^a\right)^{-\frac{1}{c}}\right)}$$

$$= \lim_{x \to \infty} \frac{a}{\left(\left(\frac{x}{b}\right)^{-a} + c\right)\left(1 - \left(1 + c\left(\frac{x}{b}\right)^a\right)^{-\frac{1}{c}}\right)} = \frac{a}{c}. \tag{2.13}$$

Now, according to Equations (2.12) and (2.13), we can show the CDF of the BSM distribution asymptotically exhibits power laws as

$$F(x) \to x^a, \text{ if } x \to 0; \quad F(x) \to x^{-\frac{a}{c}}, \text{ if } x \to \infty. \tag{2.14}$$

That is, in the BSM distribution, power law exponents a and a/c correspond to the asymptotic behavior of X. In other words, the BSM distribution and some of the distributions obtained as its special cases can asymptotically lead to power laws for large and/or small values of X. For some values of parameters, these distributions can be shown to belong to the family of Levy heavy tail distributions (Sornette, 2003; Brouers, 2015). The limiting distributions (special cases) of BSM distributions are discussed below (as shown in Figure 2.2).

Case I: If $c = 1$, then the PDF of the BSM distribution (Equation (2.1)) reduces to

$$f(x; a, b) = \frac{a}{b}\left(\frac{x}{b}\right)^{a-1}\left(1 + \left(\frac{x}{b}\right)^a\right)^{-2}. \tag{2.15}$$

Equation (2.15) is the PDF of the log-logistic Hill–Fisk distribution (Brouers, 2015).

Case II: If $a \to \infty$, then the following limiting distributions are obtained.

• The two-parameter generalized logistic distribution is obtained if $c > 0$:

$$f(x; b, c) = \frac{e^{\frac{x}{b}}}{cb(1 + e^{\frac{x}{b}})^{\frac{1}{c}+1}} \tag{2.16}$$

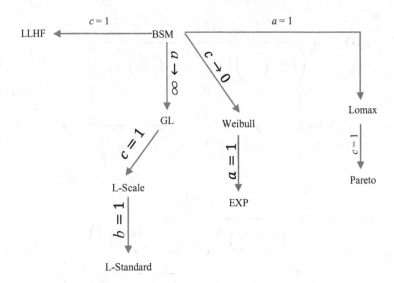

Figure 2.2 Tree diagram for BSM and its special distributions.
LLHF: log-logistic-Hill-Fisk; GL: generalized logistic; L-scale: logistic with scale
parameter L-standard: standard logistic; EXP: exponential

- The logistic distribution with scale parameter b is obtained if $c = 1$:

$$f(x;b) = \frac{e^{\frac{x}{b}}}{b\left(1 + e^{\frac{x}{b}}\right)^2} \tag{2.17}$$

- The standard logistic distribution is obtained if $b = 1$:

$$f(x) = \frac{e^x}{(1 + e^x)^2} \tag{2.18}$$

Case III: If $c \to 0, a, b > 0$ the Weibull distribution is obtained as the limiting
distribution of the BSM distribution:

$$f(x;a,b) = \frac{a}{b}\left(\frac{x}{b}\right)^{a-1} \exp\left(-\left(\frac{x}{b}\right)^a\right). \tag{2.19}$$

In Equation (2.19), if $a = 1$, the Weibull PDF reduces to the exponential PDF:

$$f(x;b) = \frac{1}{b} \exp\left(-\frac{x}{b}\right). \tag{2.20}$$

Case IV: If $a = 1$; $b, c > 0$, the Lomax distribution is obtained:

$$f(x; b, c) = \frac{1}{b} \left(1 + \frac{cx}{b} \right)^{-\left(\frac{1}{c} + 1 \right)}. \tag{2.21}$$

In Equation (2.21), if $c = 1$, the Pareto distribution with scale parameter b is obtained:

$$f(x; b) = \frac{1}{b \left(1 + \frac{x}{b} \right)^2}. \tag{2.22}$$

The standard Pareto distribution is obtained if $b = 1$ in Equation (2.22):

$$f(x) = (1 + x)^{-2}. \tag{2.23}$$

2.3 Characterization through Hazard Function

The hazard rate, $r(x)$, can be defined as

$$r(x) = \frac{f(x)}{1 - F(x)}. \tag{2.24}$$

The quantity $r(x)$ is also called the failure rate or morbidity rate. For the BSM distribution,

$$r(x) = \frac{\frac{a}{b} \left(\frac{x}{b} \right)^{a-1}}{1 + c \left(\frac{x}{b} \right)^a}. \tag{2.25}$$

Differentiating Equation (2.25), we get

$$\frac{dr}{dx} = -\frac{a \left(\frac{x}{b} \right)^a \left(c \left(\frac{x}{b} \right)^a - a + 1 \right)}{x^2 \left(1 + c \left(\frac{x}{b} \right)^a \right)^2}. \tag{2.26}$$

The distribution is characterized by an increasing failure rate (IFR) if $r(x)$ is increasing, $dr(x)/dx \geq 0$, and by a decreasing failure rate (DFR) if $r(x)$ is decreasing, $dr(x)/dx \leq 0$. The probability distributions can be characterized by IFR or DFR. Depending on the problem at hand, $r(x)$ provides a physical explanation as to whether this distribution represents the real situation.

Let $z = \log x$. Then, the hazard rate is determined for z instead of x. Thus,

$$r(z) = \frac{f(z)}{1 - F(z)} = \frac{dF/dz}{1 - F(z)}. \tag{2.27}$$

For the BSM distribution, Equation (2.27) may be rewritten as follows:

$$f(z) = f(x(z))\frac{dx}{dz} = \frac{a\exp(az)}{b^a}\left(1 + \frac{c}{b^a}\exp(az)\right)^{-\frac{1}{c}-1} \tag{2.28}$$

$$1 - F(z) = \left(1 + \frac{c}{b^a}\exp(az)\right)^{-\frac{1}{c}} \tag{2.29}$$

$$r(z) = \frac{f(z)}{1 - F(z)} = \frac{a\exp(az)}{b^a + c\exp(az)}. \tag{2.30}$$

If $dr(z)/dz \geq 0$, the distribution is characterized by increasing proportional failure rate (IPFR), and if $dr(z)/dz \leq 0$, the distribution is characterized by decreasing proportional failure rate (DPFR). In the case of the BSM distribution, the first derivative of Equation (2.30) is given as

$$\frac{dr}{dz} = \frac{a^2 b^a \exp(az)^2}{b^a + c\exp(az)}; \quad a, b, c > 0. \tag{2.31}$$

With the given constraints on the parameters, it is clear that $dr/dz \geq 0$ for the BSM distribution. As a result, the BSM distribution has the property of IPFR.

2.4 Derivation of BSM Distribution

The BSM distribution can be derived in different ways, which are discussed here.

2.4.1 Singh–Maddala Method

Consider the negative of the Pareto transform and let it be denoted by y. Then, we have $y = -\log(1 - F), z = \log(x), y = f(z), y' > 0, y'' > 0$. It is assumed that $r(z)$ asymptotically reaches a constant; it first increases with an increasing rate and then with a decreasing rate; and the rate of increase is zero where $r(z)$ is zero. This assumption leads to the hypothesis

$$y'' = ay'(a_0 - y'), \tag{2.32}$$

where y' is the proportional failure rate and a and a_0 are constants. This differential equation can be rearranged as

$$\frac{y''}{y'} + \frac{y''}{a_0 - y'} = aa_0. \tag{2.33}$$

Integration of Equation (2.33) yields

$$\log y' - \log(a_0 - y') = aa_0 z + c_0, \tag{2.34}$$

where c_0 is the constant of integration. Equation (2.34) can be solved as

$$\frac{y'}{a_0 - y'} = \exp(aa_0 z + c_0) \Rightarrow y' = \frac{a_0 \exp(aa_0 z + c_0)}{1 + \exp(aa_0 z + c_0)}. \tag{2.35}$$

Equation (2.35) is the three-parameter log-logistic distribution. Integration of Equation (2.35) yields

$$y = \frac{1}{a} \log\left(1 + \exp(aa_0 z + c_0)\right) + c_1, \tag{2.36}$$

where c_1 is the constant of integration. Now substituting $-\log(1 - F)$ for y and $\log x$ for z, Equation (2.36), after some algebraic manipulation, becomes

$$\log(1 - F) = -c_1 - \frac{1}{a} \log\left(1 + \exp(c_0) x^{aa_0}\right) \tag{2.37}$$

$$F = 1 - \exp(-c_1)\left(1 + \exp(c_0) x^{aa_0}\right)^{-\frac{1}{a}}$$

$$= 1 - \exp\left(-c_1 - \frac{c_0}{a}\right)\left(\exp(-c_0) + x^{aa_0}\right)^{-\frac{1}{a}}. \tag{2.38}$$

In Equation (2.38), let $c = \exp\left(-c_1 + \frac{c_0}{a}\right)$ and $b = \exp(-c_0)$. Then, Equation (2.38) may be rewritten as

$$F = 1 - \frac{c}{(b + x^{aa_0})^{\frac{1}{a}}}; \quad x \in [0, \infty). \tag{2.39}$$

Substituting $F = 0$ if $x = 0$ into Equation (2.39), we have $c = b^{\frac{1}{a}}$. Then, Equation (2.39) may be rewritten as

$$F = 1 - \frac{b^{\frac{1}{a}}}{(b + x^{aa_0})^{\frac{1}{a}}}. \tag{2.40}$$

Rearranging Equation (2.40), we have

$$F = 1 - \frac{1}{(1 + a_1 x^{a_2})^{a_3}}; \quad a_1 = \frac{1}{b}, a_2 = aa_0, a_3 = \frac{1}{a}. \tag{2.41}$$

Equation (2.41) is the same as Equation (2.6) and is characterized by IPFR.

2.4.2 Another Singh–Maddala Method

The BSM distribution can be derived using a model of decay. Since the CDF, $F(x)$, is a certain mass at a point x, the rate of decay is dF/dx and decays to zero as $x \to \infty$. If the initial mass is standardized to be one, and dF/dx depends only on the left-out mass $(1 - F)$, then the process can be said to be memory-less. The Poisson process is memoryless and follows

$$\frac{dF}{dx} = a(1 - F).$$

(2.42)

As another memoryless process, the Pareto process follows

$$\frac{dF}{dx} = a(1 - F)^{1+\frac{1}{a}}.$$

(2.43)

In Equations (2.42) and (2.43), a is a parameter.

The Weibull process, which leads to the Weibull distribution, follows

$$\frac{dF}{dx} = ax^b(1 - F).$$

(2.44)

In Equation (2.44), a and b are parameters. Equation (2.44) introduces memory and hence the Weibull process has memory. Elements of Equation (2.43) and (2.44) can be combined to obtain a generalization:

$$\frac{dF}{dx} = ax^b(1 - F)^c,$$

(2.45)

where a, b, and c are parameters.

Integration of Equation (2.45) yields

$$F = 1 - \left(1 + \frac{a(c-1)}{b+1}x^{b+1}\right)^{\frac{1}{1-c}}.$$

(2.46)

The solution presented in Equation (2.46) reproduces Equation (2.41) if $a_1 = \frac{a(c-1)}{b+1}, a_2 = b + 1; a_3\frac{1}{c-1}.$

The Fisk (1961) distribution, also called the log-logistic Hill–Fisk distribution, can be derived from

$$\frac{dF}{dw} = \frac{\exp(w)}{(1 + \exp(w))^2}; \quad \exp(w) = \left(\frac{x}{x_0}\right)^a.$$

(2.47)

It can be shown that

$$f(x) = \frac{dF}{dx} = \frac{dF}{dw}\frac{dw}{dx} = \frac{a\left(\frac{1}{x_0^a}\right)x^{a-1}}{\left(1 + \frac{1}{x_0^a}x^a\right)^2}.$$

(2.48)

In Equation (2.48), let $a_1 = 1/x_0^a$. Then, Equation (2.48) may be rewritten as

$$f(x) = \frac{a a_1 x^{a-1}}{(1 + a_1 x^a)^2}.$$

(2.49)

2.4.3 Differential Equation Method

Following Brouers (2015), the differential equation for deriving the BSM distribution can be written as

$$\frac{dF}{dx} = g(x) F(x) (1 - F(x)),$$

(2.50)

where $g(x)$ can be expressed as Equation (2.10) and the corresponding $g_1(x) = x g(x)$ is expressed as Equation (2.11).

As discussed through Equations (2.12) and (2.13), function $g_1(x)$ varies slowly from a to a/c. It tends to a if x tends to zero, and it tends to a/c if x tends to infinity. Solution of Equation (2.50) yields Equation (2.6) as follows:

$$\int_0^F \frac{dF}{F(1-F)} = \int_0^x \frac{a \left(\frac{x}{b}\right)^a}{x \left(1 + c \left(\frac{x}{b}\right)^a\right) \left(1 - \left(1 + c \left(\frac{x}{b}\right)^a\right)^{-\frac{1}{c}}\right)} dx.$$

(2.51)

The left side of Equation (2.51) can be integrated as

$$\int_0^F \frac{dF}{F(1-F)} = \log\left(\frac{F}{1-F}\right).$$

(2.52)

The right side of Equation (2.51) can be integrated as

$$\int_0^x g(x) dx = \int_0^x \frac{a \left(\frac{x}{b}\right)^a}{x \left(1 + c \left(\frac{x}{b}\right)^a\right) \left(1 - \left(1 + c \left(\frac{x}{b}\right)^a\right)^{-\frac{1}{c}}\right)} dx$$

$$= \frac{1}{c} \int_0^x \frac{1}{\left(1 + c \left(\frac{x}{b}\right)^a\right) \left(1 - \left(1 + c \left(\frac{x}{b}\right)^a\right)^{-\frac{1}{c}}\right)} d\left(1 + c\left(\frac{x}{b}\right)^a\right).$$

(2.53)

Using the transformation given by Equation (2.2), Equation (2.53) may be evaluated using the intermediate variable w as:

$$\int_0^x g(x)dx = \frac{1}{c}\int_1^{b\left(\frac{w-1}{c}\right)^{\frac{1}{a}}} \frac{1}{w\left(1-w^{-\frac{1}{c}}\right)}dw = \frac{1}{c}\int_1^{b\left(\frac{w-1}{c}\right)^{\frac{1}{a}}} \left(\frac{1}{w}+\frac{w^{-\frac{1}{c}-1}}{1-w^{-\frac{1}{c}}}\right)dw = \log\left(w^{\frac{1}{c}}\left(1-w^{-\frac{1}{c}}\right)\right).$$

$$(2.54)$$

Equating Equations (2.52) and (2.54), we have

$$\frac{F}{1-F} = w^{\frac{1}{c}}-1 \Rightarrow F = 1-w^{-\frac{1}{c}}.$$

$$(2.55)$$

Resubstituting Equation (2.2) into Equation (2.55), we have

$$F(x) = 1 - \left(1+c\left(\frac{x}{b}\right)^a\right)^{-\frac{1}{c}}.$$

$$(2.56)$$

Now, we have shown that Equation (2.56) so derived is the CDF of the BSM distribution.

2.4.4 Entropy Method

The Shannon entropy $H(x)$ for random variable X can be expressed as

$$H(X) = -\int_0^\infty f(x)\ln f(x)dx.$$

$$(2.57)$$

In order to derive the BSM distribution, we first determine the constraints that the distribution must satisfy. Substituting Equation (2.1) in Equation (2.57), we obtain

$$H(X) = -\int_0^\infty f(x)\ln\left(\frac{a}{b}\left(\frac{x}{b}\right)^{a-1}\left(1+c\left(\frac{x}{b}\right)^a\right)^{-\frac{1}{c}-1}\right)dx.$$

$$(2.58)$$

Equation (2.58) yields

$$H(X) = -\ln\left(\frac{a}{b}\right) - (a-1)E\left(\ln\left(\frac{x}{b}\right)\right) + \left(1+\frac{1}{c}\right)E\left(\ln\left(1+c\left(\frac{x}{b}\right)^a\right)\right),$$

$$(2.59)$$

where $E(\cdot)$ is the expectation of $(\cdot)$. Thus, the constraints can be obtained from Equation (2.59) as

$$C_0 = \int_0^\infty f(x)dx = 1$$

$$(2.60)$$

$$C_1 = \int_0^\infty f(x)\ln\left(\frac{x}{b}\right)dx = E\left(\ln\left(\frac{x}{b}\right)\right)$$

$$(2.61)$$

$$C_2 = \int_0^\infty f(x)\ln\left(1 + c\left(\frac{x}{b}\right)^a\right)dx = E\left(\ln\left(1 + c\left(\frac{x}{b}\right)^a\right)\right). \tag{2.62}$$

The next step is to construct the Lagrangian function L as

$$L = -\int_0^\infty f(x)\ln f(x)dx - (\lambda_0 - 1)\left[\int_0^\infty f(x)dx - C_0\right] - \lambda_1\left[\int_0^\infty f(x)\ln\left(\frac{x}{b}\right)dx - C_1\right]$$
$$- \lambda_2\left[\int_0^\infty f(x)\ln\left(1 + c\left(\frac{x}{b}\right)^a\right)dx\right], \tag{2.63}$$

where λ_0, λ_1, and λ_2 are the unknown Lagrange multipliers that will be determined in terms of constraints given by Equations (2.60)–(2.62). Differentiating Equation (2.63) with respect to $f(x)$ and equating the derivative to zero, we obtain

$$\frac{\partial L}{\partial f(x)} = -\ln f(x) - 1 - (\lambda_0 - 1) - \lambda_1\ln\left(\frac{x}{b}\right) - \lambda_2\ln\left(1 + c\left(\frac{x}{b}\right)^a\right). \tag{2.64}$$

Equation (2.64) yields

$$f(x) = \exp\left\{-\lambda_0 - \lambda_1\ln\left(\frac{x}{b}\right) - \lambda_2\ln\left(1 + c\left(\frac{x}{b}\right)^a\right)\right\}. \tag{2.65}$$

Equation (2.65) can be written as

$$f(x) = \exp(-\lambda_0)\left(\frac{x}{b}\right)^{-\lambda_1}\left(1 + c\left(\frac{x}{b}\right)^a\right)^{-\lambda_2}. \tag{2.66}$$

Comparing Equation (2.66) with Equation (2.1), it is observed that

$$\lambda_0 = -\ln\left(\frac{a}{b}\right), \quad \lambda_1 = 1 - a, \quad \lambda_2 = 1 + \frac{1}{c}. \tag{2.67}$$

2.5 Parameter Estimation

The BSM distribution parameters can be estimated in different ways, which are now discussed.

2.5.1 Regular Entropy Method

Using the entropy theory, Equation (2.66) can be cast in terms of the partition function Z as

$$f(x) = \frac{1}{Z(\lambda_1, \lambda_2)} \left(\frac{x}{b}\right)^{-\lambda_1} \left(1 + c\left(\frac{x}{b}\right)^a\right)^{-\lambda_2}. \tag{2.68}$$

In Equation (2.68), the partition function Z is defined as

$$Z(\lambda_0) = Z(\lambda_1, \lambda_2) = \exp(\lambda_0), \quad \lambda_0 = \lambda_0(\lambda_1, \lambda_2). \tag{2.69}$$

From Equation (2.60), the partition function Z may be written as

$$Z(\lambda_1, \lambda_2) = \int_0^\infty \left(\frac{x}{b}\right)^{-\lambda_1} \left(1 + c\left(\frac{x}{b}\right)^a\right)^{-\lambda_2} dx. \tag{2.70}$$

The Lagrange multipliers are determined in terms of constraints. To that end, it can be shown that

$$\frac{\partial \ln Z(\lambda_1, \lambda_2)}{\partial \lambda_1} = -C_1 \tag{2.71}$$

$$\frac{\partial \ln Z(\lambda_1, \lambda_2)}{\partial \lambda_2} = -C_2. \tag{2.72}$$

Further, Equation (2.70) can be explicitly solved as follows.

Equation (2.70) can be written as

$$Z(\lambda_1, \lambda_2) = \left(\frac{1}{b}\right)^{-\lambda_1} \int_0^\infty x^{-\lambda_1} \left(1 + \frac{c}{b^a} x^a\right)^{-\lambda_2} dx. \tag{2.73}$$

Let $w = \frac{c}{b^a} x^a$; we have $dw = \frac{ac}{b^a} x^{a-1} dx$. Therefore, Equation (2.73) can be rearranged as

$$Z(\lambda_1, \lambda_2) = \frac{b}{a} c^{\frac{\lambda_1 - 1}{a}} \int_0^\infty w^{\frac{-\lambda_1 - a + 1}{a}} (1 + w)^{-\lambda_2} dw. \tag{2.74}$$

Applying the beta function of the second kind, Equation (2.74) can be evaluated as

$$Z(\lambda_1, \lambda_2) = \frac{b}{a} c^{\frac{\lambda_1 - 1}{a}} B\left(\frac{1 - \lambda_1}{a}, \lambda_2 + \frac{\lambda_1 - 1}{a}\right) = \frac{b}{a} c^{\frac{\lambda_1 - 1}{a}} \frac{\Gamma\left(\frac{(1 - \lambda_1)}{a}\right) \Gamma\left(\lambda_2 + \frac{\lambda_1 - 1}{a}\right)}{\Gamma(\lambda_2)}. \tag{2.75}$$

In Equation (2.75), we obtain the constraints on the Lagrange multipliers as $\lambda_1 \langle 1, \lambda_2 \rangle \frac{1 - \lambda_1}{a} > 0$. Taking the logarithm of Equation (2.75), we have

$$\lambda_0 = \ln Z(\lambda_1, \lambda_2)$$
$$= \ln b - \ln a + \frac{\lambda_1 - 1}{a} \ln c + \ln \Gamma\left(\frac{1 - \lambda_1}{a}\right) + \ln \Gamma\left(\lambda_2 + \frac{\lambda_1 - 1}{a}\right)$$
$$- \ln \Gamma(\lambda_2). \tag{2.76}$$

Differentiating Equation (2.76) with respect to Lagrange multipliers, we obtain

$$\frac{\partial \lambda_0}{\partial \lambda_1} = \frac{\ln c}{a} - \frac{1}{a}\psi\left(\frac{(1 - \lambda_1)}{a}\right) + \frac{1}{a}\psi\left(\lambda_2 + \frac{\lambda_1 - 1}{a}\right) = -E\left[\ln\left(\frac{x}{b}\right)\right] \tag{2.77}$$

$$\frac{\partial \lambda_0}{\partial \lambda_2} = \psi\left(\lambda_2 + \frac{\lambda_1 - 1}{a}\right) - \psi(\lambda_2) = -E\left[\ln\left(1 + c\left(\frac{x}{b}\right)^a\right)\right]. \tag{2.78}$$

In Equations (2.77) and (2.78), $\psi(x) = (d \ln \Gamma(x))/dx$. Equating Equation (2.77) to Equation (2.71) and Equation (2.78) to Equation (2.72), we get the expressions for Lagrange multipliers in terms of constraints and then distribution parameters in terms of constraints. Observing Equations (2.71) and (2.72) and Equations (2.77) and (2.78), it can be noted that the distribution parameters of the BSM distribution are also embedded in the constraint equations. Thus, we need three more equations for parameter estimation using the regular entropy method. These three equations are as follows:

$$\frac{\partial^2 \lambda_0}{\partial \lambda_1^2} = \frac{1}{a^2}\psi_1\left(\frac{1 - \lambda_1}{a}\right) + \frac{1}{a^2}\psi_1\left(\lambda_2 + \frac{\lambda_1 - 1}{a}\right) = \mathrm{var}\left[\ln\left(\frac{x}{b}\right)\right] \tag{2.79}$$

$$\frac{\partial^2 \lambda_0}{\partial \lambda_2^2} = \psi_1\left(\lambda_2 + \frac{\lambda_1 - 1}{a}\right) - \psi_1(\lambda_2) = \mathrm{var}\left[\ln\left(1 + c\left(\frac{x}{b}\right)^a\right)\right] \tag{2.80}$$

$$\frac{\partial^2 \lambda_0}{\partial \lambda_1 \partial \lambda_2} = \frac{1}{a}\psi_1\left(\lambda_2 + \frac{\lambda_1 - 1}{a}\right) = \mathrm{cov}\left[\ln\left(\frac{x}{b}\right), \ln\left(1 + c\left(\frac{x}{b}\right)^a\right)\right]. \tag{2.81}$$

Now using Equations (2.77)–(2.81), the parameters may be estimated numerically by solving the system of equations with the constraints of the parameters: $a, b, c > 0$; $\lambda_1 < 1$; $\lambda_2 > 0$.

2.5.2 Parameter Space Expansion Method

The BSM distribution parameters can be estimated using another entropy-based method, called the parameter space expansion method (Singh and Rajagopal, 1986).

2.5.2.1 Derivation of Entropy Function

Following the regular procedure, the entropy-based PDF of the BSM distribution with the constraints of Equations (2.60)–(2.62) is given as

$$f(x) = \exp\left(-\ln\left(\frac{b}{a}c^{\frac{\lambda_1-1}{a}}B\left(\frac{1-\lambda_1}{a}, \lambda_2-\frac{1-\lambda_1}{a}\right)\right) - \lambda_1\ln\left(\frac{x}{b}\right) - \lambda_2\ln\left(1+c\left(\frac{x}{b}\right)^a\right)\right).$$

(2.82)

Substituting Equation (2.82) into Equation (2.57) we get

$$H(x) = -\int[\ln f(x)]f(x)dx$$

$$= \ln a - \ln b + \frac{1-\lambda_1}{a}\ln c - \ln\Gamma\left(\frac{1-\lambda_1}{a}\right) - \ln\Gamma\left(\lambda_2-\frac{1-\lambda_1}{a}\right) + \ln\Gamma(\lambda_2)$$

$$+ \lambda_1 E\left(\ln\left(\frac{x}{b}\right)\right) + \lambda_2 E\left(\ln\left(1+c\left(\frac{x}{b}\right)^a\right)\right).$$

(2.83)

2.5.2.2 Relation between Parameters and Constraints
Differentiating Equation (2.83) with respect to the Lagrange multipliers and
parameters a, b, and c and equating each derivative to zero, we get

$$\frac{\partial H(x)}{\partial\lambda_1} = -\frac{1}{a}\ln c + \frac{1}{a}\psi\left(\frac{1-\lambda_1}{a}\right) - \frac{1}{a}\psi\left(\lambda_2-\frac{1-\lambda_1}{a}\right) + E\left(\ln\left(\frac{x}{b}\right)\right) = 0 \quad (2.84)$$

$$\frac{\partial H(x)}{\partial\lambda_2} = -\psi\left(\lambda_2-\frac{1-\lambda_1}{a}\right) + \psi(\lambda_2) + E\left(\ln\left(1+c\left(\frac{x}{b}\right)^a\right)\right) = 0 \quad (2.85)$$

$$\frac{\partial H(x)}{\partial a} = \frac{1}{a} - \frac{1-\lambda_1}{a^2}\ln c + \frac{1}{a^2}\psi\left(\frac{1-\lambda_1}{a}\right) - \frac{1}{a^2}\psi\left(\lambda_2-\frac{1-\lambda_1}{a}\right) + \frac{\lambda_2}{b}E\left(\frac{c\left(\frac{x}{b}\right)^a\ln\left(\frac{x}{b}\right)}{1+c\left(\frac{x}{b}\right)^a}\right) = 0$$

(2.86)

$$\frac{\partial H(x)}{\partial b} = -\frac{1}{b} - \frac{\lambda_1}{b} - \frac{a\lambda_2}{b}E\left(\frac{c\left(\frac{x}{b}\right)^a}{1+c\left(\frac{x}{b}\right)^a}\right) = 0 \quad (2.87)$$

$$\frac{\partial H(x)}{\partial c} = \frac{1-\lambda_1}{ac} + \lambda_2 E\left(\frac{\left(\frac{x}{b}\right)^a}{1+c\left(\frac{x}{b}\right)^a}\right). \quad (2.88)$$

From Equation (2.67) we have $d\lambda_1 = -da$; $d\lambda_2 = -(1/c^2)dc$. Now Equation
(2.86) can be rewritten as

$$\frac{\partial H}{\partial a} = \frac{\partial H}{\partial\lambda_1}\frac{d\lambda_1}{da} = \frac{\partial H}{\partial\lambda_1} = -\frac{1}{a}\ln c - \frac{1}{a}\psi\left(\frac{1-\lambda_1}{a}\right) + \frac{1}{a}\psi\left(\lambda_2-\frac{1-\lambda_1}{a}\right) - E\left(\ln\left(\frac{x}{b}\right)\right).$$

(2.89)

Equating (2.89) with Equation (2.86) and substituting Equation (2.77), we have

$$\frac{1}{a} + \frac{1-a}{a^2}\left(\psi(1) - \psi\left(\frac{1}{c}\right)\right) = E\left(\ln\left(\frac{x}{b}\right)\right) - \frac{1+\frac{1}{c}}{b}E\left(\frac{c\left(\frac{x}{b}\right)^a \ln\left(\frac{x}{b}\right)}{1+c\left(\frac{x}{b}\right)^a}\right).$$

(2.90)

Similarly, Equation (2.88) may be rewritten as

$$\frac{\partial H}{\partial c} = \frac{\partial H}{\partial \lambda_2}\frac{d\lambda_2}{dc} = -\frac{1}{c^2}\left(-\psi\left(\lambda_2 - \frac{1-\lambda_1}{a}\right) + \psi(\lambda_2) + E\left(\ln\left(1+c\left(\frac{x}{b}\right)^a\right)\right)\right).$$

(2.91)

Equating Equation (2.91) with Equation (2.88) and substituting Equation (2.77), we have

$$(c+1)E\left(\frac{\left(\frac{x}{b}\right)^a}{1+c\left(\frac{x}{b}\right)^a}\right) = \frac{1}{c}E\left(\ln\left(1+c\left(\frac{x}{b}\right)^a\right)\right).$$

(2.92)

Substituting Equation (2.77) into Equation (2.87), we have

$$a\left(1+\frac{1}{c}\right)E\left(\frac{c\left(\frac{x}{b}\right)^a}{1+c\left(\frac{x}{b}\right)^a}\right) = a - 2.$$

(2.93)

To this end, Equations (2.90), (2.92), and (2.93) may be solved simultaneously to estimate the parameters.

2.5.3 Method of Moments

Parameters a, b, and c of the BSM distribution can be estimated using the first three moments. The k-th moment of $f(x)$, $M_k(x)$, about the origin can be written as

$$M_k(x) = \int_0^\infty x^k \left(\frac{a}{b}\right)\left(\frac{x}{b}\right)^{a-1}\left(1+c\left(\frac{x}{b}\right)^a\right)^{-\frac{1}{c}-1} dx.$$

(2.94)

Again, by setting $c(x/b)^a = w$, Equation (2.94) can be solved as

$$M_k(x) = \int_0^\infty c^{-\left(\frac{k}{a}+1\right)}b^k w^{\frac{k}{a}}(1+w)^{-\left(\frac{k}{a}+1\right)}dw = c^{-\left(\frac{k}{a}+1\right)}b^k B\left(\frac{k}{a}+1, \frac{1}{c}-\frac{k}{a}\right); \quad k < \frac{a}{c}.$$

(2.95)

Applying the beta function, Equation (2.95) can be rewritten as

$$M_k(x) = \frac{c^{-\left(\frac{k}{a}+1\right)} b^k \Gamma\left(\frac{k}{a}+1\right) \Gamma\left(\frac{1}{c}-\frac{k}{a}\right)}{\Gamma\left(\frac{1}{c}+1\right)} = b^k c^{-\left(1+\frac{k}{a}\right)} B\left(1+\frac{k}{a}, \frac{1}{c}-\frac{k}{a}\right).$$

(2.96)

Furthermore, the first four noncentral moments are given as

$$M_1 = \frac{b\Gamma\left(1+\frac{1}{a}\right)\Gamma\left(\frac{1}{c}-\frac{1}{a}\right)}{c^{1+\frac{1}{a}}\Gamma\left(\frac{1}{c}+1\right)} = bc^{-\left(1+\frac{1}{a}\right)} B\left(1+\frac{1}{a},\frac{1}{c}-\frac{1}{a}\right) = \frac{b}{a} c^{-\frac{1}{a}} B\left(\frac{1}{a},\frac{1}{c}-\frac{1}{a}\right)$$

(2.97)

$$M_2 = \frac{b^2\Gamma\left(\frac{2}{a}+1\right)\Gamma\left(\frac{1}{c}-\frac{2}{a}\right)}{c^{\frac{2}{a}+1}\Gamma\left(\frac{1}{c}+1\right)} = b^2 c^{-\left(1+\frac{2}{a}\right)} B\left(1+\frac{2}{a},\frac{1}{c}-\frac{2}{a}\right) = \frac{2b^2}{a} c^{-\frac{2}{a}} B\left(\frac{2}{a},\frac{1}{c}-\frac{2}{a}\right)$$

(2.98)

$$M_3 = \frac{b^3\Gamma\left(\frac{3}{a}+1\right)\Gamma\left(\frac{1}{c}-\frac{3}{a}\right)}{C^{\frac{3}{a}+1}\Gamma\left(1+\frac{1}{c}\right)} = b^3 c^{-\left(1+\frac{3}{a}\right)} B\left(1+\frac{3}{a},\frac{1}{c}-\frac{3}{a}\right) = \frac{3b^3}{a} c^{-\frac{3}{a}} B\left(\frac{3}{a},\frac{1}{c}-\frac{3}{a}\right)$$

(2.99)

$$M_4 = \frac{b^4\Gamma\left(\frac{4}{a}+1\right)\Gamma\left(\frac{1}{c}-\frac{4}{a}\right)}{C^{\frac{4}{a}+1}\Gamma\left(1+\frac{1}{c}\right)} = b^4 c^{-\left(1+\frac{4}{a}\right)} B\left(1+\frac{4}{a},\frac{1}{c}-\frac{4}{a}\right) = \frac{4b^4}{a} c^{-\frac{4}{a}} B\left(\frac{4}{a},\frac{1}{c}-\frac{4}{a}\right).$$

(2.100)

Equations (2.97)–(2.99) can be solved for parameters a, b, and c.

2.5.4 Maximum Likelihood Estimation Method

For the maximum likelihood estimation (MLE) method, the log-likelihood function of a sample of size drawn from the BSM distribution, the log-likelihood function can be expressed as

$$\ln L = \sum_{i=1}^{n} \ln\left(\frac{a}{b}\left(\frac{x_i}{b}\right)^{a-1}\left(1+c\left(\frac{x_i}{b}\right)^a\right)^{-\frac{1}{c}-1}\right)$$
$$= n\ln a - na\ln b + (a-1)\sum_{i=1}^{n}\ln(x_i) - \left(\frac{1}{c}+1\right)\sum_{i=1}^{n}\ln\left(1+c\left(\frac{x_i}{b}\right)^a\right).$$

(2.101)

Differentiating Equation (2.101) with respect to parameters a, b, and c, separately, and equating each derivative to zero, we obtain

$$\frac{\partial \ln L}{\partial a} = \frac{n}{a} - n \ln b + \sum_{i=1}^{n} \ln(x_i) - \left(\frac{1}{c} + 1\right) \sum_{i=1}^{n} \frac{c \left(\frac{x_i}{b}\right)^a \ln\left(\frac{x_i}{b}\right)}{1 + c\left(\frac{x_i}{b}\right)^a} \qquad (2.102)$$

$$\frac{\partial \ln L}{\partial b} = -\frac{na}{b} + \left(1 + \frac{1}{c}\right) \sum_{i=1}^{n} \frac{ab^{-a-1} c x_i^a}{1 + c\left(\frac{x_i}{b}\right)^a} \qquad (2.103)$$

$$\frac{\partial \ln L}{\partial c} = \frac{1}{c^2} \sum_{i=1}^{n} \ln\left(1 + c\left(\frac{x_i}{b}\right)^a\right) - \left(1 + \frac{1}{c}\right) \sum_{i=1}^{n} \frac{\left(\frac{x_i}{b}\right)^a}{1 + c\left(\frac{x_i}{b}\right)^a}. \qquad (2.104)$$

Finally, the parameters can be estimated by solving the system of equations numerically simultaneously. For the parameters estimated using the MLE method, the confidence interval may be constructed from the Hessian matrix as follows.

From Equations (2.101) to (2.104), the elements of the observed Hessian matrix are given as

$$H_{11} = \frac{\partial^2 LL}{\partial a^2} = \sum_{i=1}^{n} \left(\frac{c \ln^2\left(\frac{x_i}{b}\right) \left(\frac{x_i}{b}\right)^{2a} (c+1)}{\left(c\left(\frac{x_i}{b}\right)^a + 1\right)^2} - \frac{\ln^2\left(\frac{x_i}{b}\right) \left(\frac{x_i}{b}\right)^a (c+1)}{c\left(\frac{x_i}{b}\right)^a + 1} - \frac{1}{a^2} \right)$$

$$(2.105)$$

$$H_{12} = H_{21} = \frac{\partial^2 LL}{\partial a \partial b} = \sum_{i=1}^{n} \left(\frac{(1-c)\left(\frac{x_i}{b}\right)^a + c\left(\frac{x_i}{b}\right)^{2a} + a(1+c)\ln\left(\frac{x_i}{b}\right)\left(\frac{x_i}{b}\right)^a - 1}{b\left(c\left(\frac{x_i}{b}\right)^a + 1\right)^2} \right)$$

$$(2.106)$$

$$H_{13} = H_{31} = \frac{\partial^2 LL}{\partial a \partial c} = \sum_{i=1}^{n} \left(\frac{\ln\left(\frac{x_i}{b}\right)\left(\frac{x_i}{b}\right)^a \left(\left(\frac{x_i}{b}\right)^a - 1\right)}{\left(c\left(\frac{x_i}{b}\right)^a + 1\right)^2} \right) \qquad (2.107)$$

$$H_{22} = \frac{\partial^2 LL}{\partial b^2} = \sum_{i=1}^{n} \left(-\frac{a\left(a\left(\frac{x}{b}\right)^a - c\left(\frac{x}{b}\right)^a + \left(\frac{x}{b}\right)^a + c\left(\frac{x}{b}\right)^{2a} + ac\left(\frac{x}{b}\right)^a - 1\right)}{b^2\left(c\left(\frac{x}{b}\right)^a + 1\right)^2} \right)$$

$$(2.108)$$

$$H_{23} = H_{32} = \frac{\partial^2 LL}{\partial b \partial c} = \sum_{i=1}^{n} \left(-\frac{a\left(\frac{x}{b}\right)^a \left(\left(\frac{x}{b}\right)^a - 1\right)}{b\left(c\left(\frac{x}{b}\right)^a + 1\right)^2} \right) \qquad (2.109)$$

$$H_{33} = \frac{\partial^2 LL}{\partial c^2} = \sum_{i=1}^{n} \left(\frac{2\left(\frac{x}{b}\right)^a}{c^2\left(c\left(\frac{x}{b}\right)^a + 1\right)} - \frac{2\ln\left(c\left(\frac{x}{b}\right)^a + 1\right)}{c^3} + \frac{\left(\frac{x}{b}\right)^{2a}(c+1)}{c\left(c\left(\frac{x}{b}\right)^a + 1\right)^2} \right).$$

(2.110)

Finally, the observed information matrix $\left(\text{i.e., Fisher information: } I = \left(-H\right)^{-1}\right)$ can be computed by taking the inverse of the negative Hessian matrix. Considering a 95% confidence interval, we have

$$a = \hat{a} \pm 1.96(I_{11})^{0.5}; \quad b = \hat{b} \pm 1.96(I_{22})^{0.5}; \quad c = \hat{c} \pm 1.96(I_{33})^{0.5}.$$

(2.111)

2.5.5 Probability Weighted Moments Method

The probability weighted moments (PWMs) method is useful for estimating parameters of those distributions that can be expressed explicitly in inverse form, such as Wakeby, Gumbel, kappa, Weibull, logistic, and BSM, among others (Greenwood et al. 1979). Following Greenwood et al. (1979), the PWM is given as

$$M_{i,j,k} = E\left(X^i F^j (1-F)^k\right) = \int_0^1 x(F)^i F^j (1-F)^k dF.$$

(2.112)

From Equation (2.6), we can write random variable x as a function of F:

$$x = bc^{-\frac{1}{a}}\left((1-F)^{-c} - 1\right)^{\frac{1}{a}}.$$

(2.113)

Now setting $i = 1$ and $j = 0$, Equation (2.113) is rewritten for the BSM distribution as

$$M_{1,0,k} = \int_0^1 bc^{-\frac{1}{a}}\left((1-F)^{-c} - 1\right)^{\frac{1}{a}}(1-F)^k dF.$$

(2.114)

Let $t = 1 - F$. Equation (2.114) is rewritten as

$$M_{1,0,k} = \int_0^1 bc^{-\frac{1}{a}}(t^{-c} - 1)^{\frac{1}{a}} t^k dt.$$

(2.115)

Let $y = t^{-c} - 1$. Then Equation (2.115) can be solved as

$$M_{1,0,k} = bc^{-\left(1+\frac{1}{a}\right)} \int_0^\infty y^{\frac{1}{a}}(1+y)^{-\left(1+\frac{1+k}{c}\right)}dy. \tag{2.116}$$

Again, the integral of Equation (2.116) can be written through the beta function of the second kind as

$$M_{1,0,k} = bc^{-\left(1+\frac{1}{a}\right)}B\left(\frac{1}{a}+1, \frac{1+k}{c}-\frac{1}{a}\right). \tag{2.117}$$

To this end, by equating the PWM to its sample estimate of $\alpha_k = \frac{1}{n}\sum_{i=1}^n \frac{(n-i)(n-i-1)...(n-i+1)}{(n-1)(n-2)...(n-k)} x_{i:n}$, we can estimate the parameters by solving the system of equations:

$$M_{1,0,0} = bc^{-\left(1+\frac{1}{a}\right)}B\left(1+\frac{1}{a}, \frac{1}{c}-\frac{1}{a}\right) = \alpha_0 = \frac{1}{n}\sum_{i=1}^n x_{(i)} \tag{2.118}$$

$$M_{1,0,1} = bc^{-\left(1+\frac{1}{a}\right)}B\left(1+\frac{1}{a}, \frac{2}{c}-\frac{1}{a}\right) = \alpha_1 = \frac{\frac{1}{n}\sum_{i=1}^{n-1}(n-i)x_{(i)}}{n-1} \tag{2.119}$$

$$M_{1,0,2} = bc^{-\left(1+\frac{1}{a}\right)}B\left(1+\frac{1}{a}, \frac{3}{c}-\frac{1}{a}\right) = \alpha_2 = \frac{\frac{1}{n}\sum_{i=1}^{n-2}(n-i)(n-i-1)x_{(i)}}{(n-1)(n-2)}. \tag{2.120}$$

2.5.6 Method of L-Moments

Hosking (1990) developed the method of L-moments, which is simpler than the method of PWMs. Following Hosking (1990), the L-moment is defined as a linear combination of PWMs. The L-moment is formulated as

$$\lambda_r = r^{-1}\sum_{k=0}^{r-1}(-1)^k\binom{r-1}{k}E(X_{r-k:r}) \tag{2.121}$$

where $X_{m:n}$ denotes the m-th-order statistic (i.e., the m-th smallest value) of the independent and identically distributed (IID) random variable X of size n.

Similar to the ordinary moments, the first, second, third, and fourth L-moments respectively represent the location, scale, skewness, and kurtosis of random variable X, which are given as

$$\lambda_1 = E(X) \tag{2.122}$$

$$\lambda_2 = \frac{E(X_{2:2}) - E(X_{1:2})}{2} \tag{2.123}$$

$$\lambda_3 = \frac{E(X_{3:3}) - 2E(X_{2:3}) + E(X_{1:3})}{3} \tag{2.124}$$

$$\lambda_4 = \frac{E(X_{4:4}) - 3E(X_{3:4}) + 3E(X_{2:4}) - E(X_{1:4})}{4}. \tag{2.125}$$

The corresponding sample L-moment for the IID random variable X of size n is given as

$$l_1 = \frac{1}{n} \sum_{i=1}^{n} x_{(i)} = \bar{x} = \alpha_0 \tag{2.126}$$

$$l_2 = \frac{1}{2} \binom{n}{2}^{-1} \sum_{i=1}^{n} \left(\binom{i-1}{1} - \binom{n-i}{1} \right) x_{(i)} = \alpha_0 - 2\alpha_1 \tag{2.127}$$

$$l_3 = \frac{1}{3} \binom{n}{3}^{-1} \sum_{i=1}^{n} \left(\binom{i-1}{2} - 2\binom{i-1}{1}\binom{n-i}{1} + \binom{n-i}{2} \right) x_{(i)} = \alpha_0 - 6\alpha_1 + 6\alpha_2 \tag{2.128}$$

$$l_4 = \frac{1}{4} \binom{n}{4}^{-1} \sum_{i=1}^{n} \left(\binom{i-1}{3} - 3\binom{i-1}{2}\binom{n-i}{1} + 3\binom{i-1}{1}\binom{n-i}{2} - \binom{n-i}{3} \right) x_{(i)}$$
$$= \alpha_0 - 12\alpha_1 = 30\alpha_2 - 20\alpha_3. \tag{2.129}$$

From Equations (2.122) to (2.124), the first three L-moments of the BSM distribution are given as

$$\lambda_1 = bc^{-\left(1+\frac{1}{a}\right)} B\left(1 + \frac{1}{a}, \frac{1}{c} - \frac{1}{a}\right) \tag{2.130}$$

$$\lambda_2 = bc^{-\left(1+\frac{1}{a}\right)} \left(B\left(1 + \frac{1}{a}, \frac{1}{c} - \frac{1}{a}\right) - 2B\left(1 + \frac{1}{a}, \frac{2}{c} - \frac{1}{a}\right) \right) \tag{2.131}$$

$$\lambda_3 = bc^{-\left(1+\frac{1}{a}\right)} \left(B\left(1 + \frac{1}{a}, \frac{1}{c} - \frac{1}{a}\right) - 6B\left(1 + \frac{1}{a}, \frac{2}{c} - \frac{1}{a}\right) + 6B\left(1 + \frac{1}{a}, \frac{3}{c} - \frac{1}{a}\right) \right). \tag{2.132}$$

Setting the population L-moments to its sample moments, the parameters can be estimated by solving the system of equations.

2.5.7 Method of Cumulative Moments

According to Burr (1942), the j-th cumulative moment about a given point a is defined as

$$
\begin{aligned}
M_j(a) &= \int_a^\infty (x-a)^j \big(1 - F(x)\big)\, dx - \int_{-\infty}^a (x-a)^j F(x)\, dx \\
&= \sum_{i=0}^j \binom{j}{i} (-a)^i M_{j-i} + \frac{(-a)^{j+1}}{j+1} \\
&= \frac{1}{j+1} \int_{-\infty}^\infty (x-a)^{j+1} f(x)\, dx \\
&= \frac{1}{j+1} \sum_{i=0}^{j+1} \binom{j+1}{i} (-a)^i \mu'_{j+1-i} = \frac{1}{j+1} \sum_{i=0}^{j+1} \binom{j+1}{i} (\mu'_1 - a)^i \mu_{j+1-i}
\end{aligned}
$$

$$(2.133)$$

where μ'_{j+1-i} denotes the $(j + 1 - i)$-th moment about the origin and μ_{j+1-i} denotes the $(j + 1 - i)$-th moment about the mean. If $a = 0$, Equation (2.133) becomes

$$
\begin{aligned}
M_j(0) = M_j &= \int_0^\infty x^j \big(1 - F(x)\big)\, dx - \int_{-\infty}^0 x^j F(x)\, dx = \frac{1}{j+1} \mu'_{j+1} \\
&= \frac{1}{j+1} \sum_{i=0}^{j+1} \binom{j+1}{i} \mu'^i_1 \mu_{j+1-i}.
\end{aligned}
$$

$$(2.134)$$

Furthermore, the j-th ordinary moment about point a can be expressed using the cumulative moment as

$$
\mu'_j = j \sum_{i=0}^{j-1} \binom{j-1}{i} (a - \mu'_1)^i M_{j-1-i}(a) + (a - \mu'_1)^j, \quad j > 1 \tag{2.135}
$$

$$
\begin{aligned}
\mu_j &= j \sum_{i=0}^{j-1} \binom{j-1}{i} (a - \mu'_1)^i m_{j-1-i}(a) + (a - \mu'_1)^j \\
&= j \sum_{i=0}^{j-1} \binom{j-1}{i} \big(-M_0(a)\big)^i M_{j-1-i}(a) + \big(-M_0(a)\big)^j, \quad j > 1.
\end{aligned}
$$

$$(2.136)$$

If $a = 0$, the above equations become

$$
\mu'_j = j M_{j-1}, \quad j > 0 \tag{2.137}
$$

$$
\mu_j = j \sum_{i=0}^{j-1} \binom{j-1}{i} (-M_0)^i M_{j-1-i} + (-M_0)^j, \quad j > 1. \tag{2.138}
$$

Now, we will discuss how to estimate the parameters of the BSM distribution using the method of cumulative moments. Substituting Equation (2.6) into Equation (2.134), we have

$$M_j = \int_0^\infty x^j \big(1 - F(x)\big)dx = \int_0^\infty x^j \left(1 + c\Big(\frac{x}{b}\Big)^a\right)^{-\frac{1}{c}} dx$$
$$= \frac{b^{j+1}}{a}c^{-\left(\frac{j+1}{a}\right)}B\left(\frac{j+1}{a}, \frac{1}{c} - \frac{j+1}{a}\right). \tag{2.139}$$

Substituting Equation (2.139) into Equation (2.137), we have

$$\mu'_1 = M_0 = \frac{b}{a}c^{-\frac{1}{a}}B\left(\frac{1}{a}, \frac{1}{c} - \frac{1}{a}\right) \tag{2.140}$$

$$\mu'_2 = 2M_1 = \frac{2b^2}{a}c^{-\frac{2}{a}}B\left(\frac{2}{a}, \frac{1}{c} - \frac{2}{a}\right) \tag{2.141}$$

$$\mu'_3 = 3M_2 = \frac{3b^3}{a}c^{-\frac{3}{a}}B\left(\frac{3}{a}, \frac{1}{c} - \frac{3}{a}\right). \tag{2.142}$$

Equating μ'_1, μ'_2, and μ'_3 to their sample moments about the origin, the parameters can be estimated by solving the system of equations. The moment diagram for the BSM distribution is graphed in Figure 2.3.

We have previously shown that the observed information matrix can be applied to construct the confidence interval for the parameters estimated using the MLE method, since the parameters so estimated are asymptotically normal. Parametric bootstrap method may be applied to construct the confidence

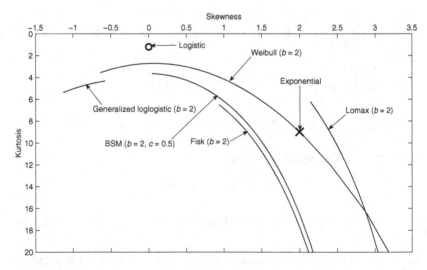

Figure 2.3 Moment diagram for BSM distribution and the limiting cases with different parameters.

interval for the parameters estimated with the use of methods of moments, L-moments, and cumulative moments and the entropy method as well as the MLE method. The detailed general procedure is provided in the Appendix.

2.6 Application

2.6.1 Synthetic Data

Here, synthetic data generated from the BSM distribution are applied first to illustrate the estimation method. The data are listed in Table 2.1. The random variable data so generated will guarantee the existence of the first three moments. All estimation methods, including MLE, moment, PWM, L-moment, cumulative moment, and entropy, are applied for comparison. Table 2.2 lists the parameters estimated with the use of each method, with the confidence interval constructed using the observed information for MLE

Table 2.1. *Synthetic data generated from BSM distribution* ($a = 8$, $b = 0.8$, $c = 1.8$).

1	0.568	26	0.782	51	0.734	76	0.628
2	0.946	27	0.847	52	0.961	77	0.883
3	0.543	28	0.791	53	0.934	78	0.695
4	0.583	29	0.927	54	0.854	79	0.775
5	0.844	30	0.907	55	0.958	80	0.879
6	0.608	31	0.730	56	0.934	81	0.709
7	1.084	32	0.798	57	0.668	82	0.729
8	1.083	33	0.476	58	0.634	83	0.900
9	0.979	34	1.886	59	3.584	84	0.716
10	0.649	35	0.661	60	0.663	85	1.093
11	0.929	36	0.617	61	0.525	86	1.851
12	0.843	37	0.769	62	0.866	87	0.986
13	1.675	38	0.680	63	1.199	88	0.755
14	0.922	39	0.827	64	0.936	89	0.880
15	1.061	40	0.753	65	0.675	90	0.618
16	0.809	41	1.469	66	0.767	91	1.264
17	0.798	42	1.312	67	0.813	92	1.194
18	1.095	43	0.559	68	1.827	93	1.084
19	0.596	44	0.993	69	0.654	94	0.714
20	0.638	45	0.718	70	1.144	95	0.886
21	0.665	46	0.794	71	0.919	96	0.500
22	0.778	47	0.859	72	0.771	97	0.795
23	1.104	48	1.414	73	0.675	98	0.740
24	1.064	49	0.791	74	0.796	99	0.657
25	0.570	50	1.860	75	0.823	100	0.668

Table 2.2. *Parameters estimated using different estimation methods for synthetic data and the corresponding confidence intervals.*

	Estimation method					
	MLE	MOM	PWM	L-moment[a]	Cumulative moment	Entropy[a]
a	10.796 [3.066, 12.991] [6.327, 14.06][b]	5.17 [5.071, 7.583]	19.812 [12.875, 28.301]	11.181 [0, 15.332]	5.17 [5.085, 7.296]	10.81 (0, 17.912)
b	0.751 [0.682, 0.810] [0.692, 0.81][b]	0.82 [0.678, 0.842]	0.694 [0.656, 0.711]	0.738 [0.658, 0.863]	0.82 [0.687, 0.845]	0.733 [0.68, 0.83]
c	2.778 [0, 3.966] [1.144, 4.411][b]	1.197 [1.297, 2.039]	6.077 [5.361, 7.870]	3.189 [0, 4.896]	1.197 [1.347, 2.068]	2.795 (0, 4.431)
λ_1						−9.036 [−18.236, 1][a]
λ_2						1.337 [0, 2.304][a]

Notes: [a] For the negative lower bound, it is forced to be 0; for the parameter with upper bound, the upper bound is forced if the upper confidence interval simulated is out of the range based on Equation (2.96).
[b] confidence interval constructed from Fisher information.

and parametric bootstrap for all the estimation methods. As shown in Table 2.2, there are five parameters for the entropy method in which λ_1 and λ_2 are Lagrange multipliers and a, b, and c are also embedded in the constraints to derive the entropy. Substituting the parameters estimated using the regular entropy method into the entropy-based distribution (i.e., Equation (2.68)), Equation (2.75) can be rewritten as

$$f(x) = 13.65 \left(\frac{x}{0.733}\right)^{9.936} \left(1 + 2.795 \left(\frac{x}{0.733}\right)^{10.81}\right)^{-1.337}. \tag{2.143}$$

Figure 2.4 compares the empirical frequency (histogram) and CDF with those obtained from the BSM distribution with the parameters estimated using different parameter estimation techniques. Table 2.3 lists the goodness-of-fit results evaluated using the Kolmogorov–Smirnov (KS) test with the parametric bootstrap approach. The test results show that while all the estimation

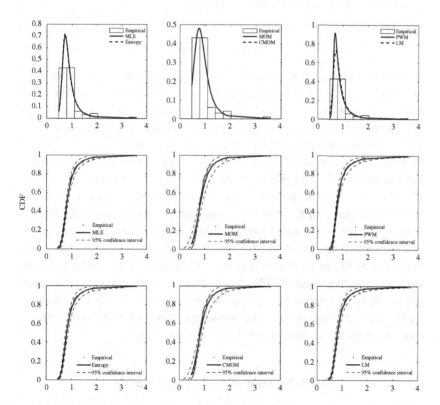

Figure 2.4 Comparison of the empirical frequency distribution with the frequency distribution obtained from the BSM distribution (synthetic data).

Table 2.3. *Goodness-of-fit test results for different parameter estimation methods.*

Parameter	MLE	MOM	PWM	L-moment	Cumulative moment	Entropy
D_n	0.046	0.094	0.098	0.059	0.094	0.057
P-value	0.783	0.19	0.095	0.509	0.18	0.592

methods may be applied, the PWM method yields the worst performance considering the computed KS tests. As shown in Figure 2.4, there exist minimal differences for the frequency distribution obtained using the parameters estimated with MLE and entropy methods. The L-moment method performs better than the PWM method, while the method of moments (MOM) performs exactly the same as the method of cumulative moments. Furthermore, compared with the empirical CDF, there are minimal differences in the case of how the parameters are estimated.

2.6.2 Peak Flow

In this section, the peak streamflow at USGS09239500 (Yampa River at Steamboat Springs, Colorado) is applied to evaluate the performance of different estimation methods. The data is retrieved from: nwis.waterdata.usgs .gov. Applying all six estimation methods discussed above, Table 2.4 lists the parameters estimated, Table 2.5 lists the goodness-of-fit results, and Figure 2.5 plots the frequency and CDF with the parameters estimated using different estimation methods. Compared with the analysis using synthetic data, analysis of the real-world peak flow data shows similar characteristics: (1) there is minimal difference in the case of the performance for the BSM distribution with the parameters estimated with MLE and entropy methods; (2) the moment and cumulative moment methods yield exactly the same estimates and same performance; (3) the performance of the PWM method is inferior to that of the L-moment method; and (4) PWM estimation yields the worst performance among all the estimation techniques. Hence, peak flow data may be modeled by the BSM distribution with the parameters estimated with MLE, entropy, moment, cumulative moment, and L-moment methods.

2.6.3 Maximum Daily Precipitation

In this section, the maximum daily rainfall is collected for Brenham, Texas (GHCND: USC00411048), from www.ncdc.noaa.gov/cdo-web. According to

Table 2.4. *Parameters estimated using different estimation methods for peak flow and the corresponding confidence intervals.*

	Estimation method					
	MLE	MOM	PWM	L-moment	Cumulative moment	Entropy
a	4.133 [2.917, 4.888] [3.054, 5.213][a]	4.132 [3.265, 4.788]	35.752 [0, 54.84]	4.245 [3.00, 5.116]	4.132 [3.197, 4.747]	3.911 [3.089, 6.64]
b	1.09E+02 [100.750, 117.356] [99.624, 118.885][a]	1.09E+02 [101.923, 116.173]	7.76E+01 [71.11, 86.33]	1.08E+02 [99.07, 116.571]	1.09E+02 [101.448, 116.287]	1.08E+02 [91.254, 121.429]
c	0.234 [0,0.467][b] [0, 0.658][a,b]	0.224 [0, 0.441][b]	10.596 [0, 16.41]	0.303 [0, 0.588][b]	0.224 [0, 0.441][b]	0.22 [0.038, 0.436][b]
λ_1						−3.255 [−4.091, 1][b]
λ_2						5.95 [0, 6.672][b]

Notes: [a] Confidence interval estimated based on Fisher information.
[b] Due to the limits of the population parameters, 0 (or 1) is forced as the lower (or upper) bound for parameter c and Lagrange multiplier $\lambda_2(\lambda_1)$.

Table 2.5. *Goodness-of-fit test results for the probability distributions fitted to peak flow with different parameter estimation methods.*

Parameter	MLE	MOM	PWM	L-moment	Cumulative moment	Entropy
D_n	0.06	0.06	0.197	0.056	0.062	0.059
P-value	0.282	0.39	<0.05	0.331	0.39	0.368

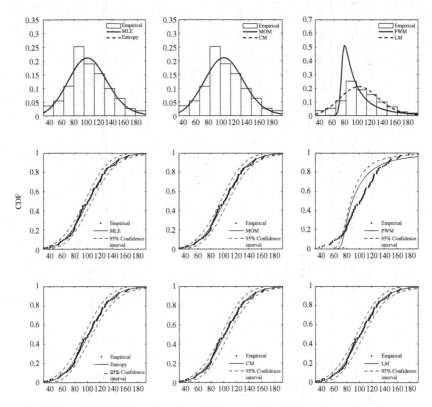

Figure 2.5 Comparison of the empirical frequency distribution with the frequency distribution obtained from the BSM distribution for the peak flow data at USGS09239500.

the analysis with synthetic data and real-world peak flow data, MLE and entropy methods are applied to analyze the maximum daily rainfall data. Table 2.6 lists the parameters estimated using MLE and entropy methods. Table 2.7 lists the goodness-of-fit results for the fitted distribution. Table 2.7 indicates that both estimation methods can be applied, based on the goodness-

Table 2.6. *Parameters estimated for maximum daily rainfall with different estimation methods.*

Parameter estimated	a	b	c	λ_1	λ_2
MLE	3.885 [2.363, 4.72]	90.883 [79.853, 100.241]	0.844 [0.097, 1.734]		
Entropy	0.774 [0.512, 1.205]	90.452 (0, 158.01]	1.109 [0.899, 2.138]	-14.199 [-19.028, 4.339]	37.591 (0, 4.339)[a]

Note: [a] due to the limits of the population parameters, 0 (or 1) is forced as the lower (or upper) bound for parameter c and Lagrange multiplier $\lambda_2(\lambda_1)$.

.

Table 2.7. *Goodness-of-fit test results for the distribution fitted to maximum daily rainfall.*

Estimation method	KS goodness-of-fit test	
	D_n	P-value
MLE	0.0579	0.254
Entropy	0.0457	0.781

of-fit results. Figure 2.6a and b visually indicate that both methods may be adequate to model the maximum daily precipitation dataset.

2.6.4 Drought (Total Flow Deficit)

The drought variable (total flow deficit: S) at Tilden, Texas, is applied as a case study. The annual maximum total flow deficit is computed as the maximum of the summation of the continuous flow deficit below the average monthly streamflow with the scheme shown in Figure 2.7. To evaluate the applicability of the BSM distribution to the total flow deficit, the MLE and entropy methods are applied to estimate the parameters.

Table 2.8 lists the parameters and the corresponding 95% confidence bound estimated for the total flow deficit. The goodness-of-fit results listed in Table 2.9 indicate that both MLE and entropy methods may be applied to model the total flow deficit. The KS test statistic values show that the test statistic is larger for the BSM distribution with parameters

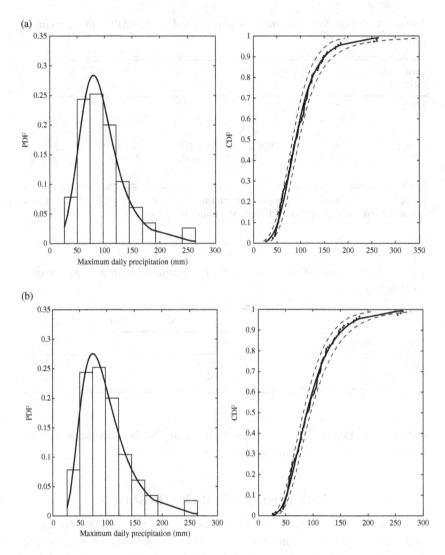

Figure 2.6 Comparison of empirical and fitted BSM distributions for maximum daily precipitation: (a) MLE; (b) entropy.

estimated using MLE than the maximum entropy-based BSM distribution. Figure 2.8a and b compare the empirical frequency (distribution) with the fitted BSM distribution. Comparisons agree with the goodness-of-fit analysis, and the BSM distribution may be applied to model the total flow deficit (drought).

Table 2.8. *Parameters and 95% confidence bound estimated for total flow deficit.*

Parameter estimated	a	b	c	λ_1	λ_2
MLE	0.859	4.675E+04	0.1680		
	[0.617, 1.007]	[3.007E+04, 6.053E+04]	$(0, 0.333)^a$		
Entropy	0.358	5.68E+04	0.158	−0.64	38.912
	[0.296, 0.467]	[5.679E+04, 5.695E+04]	[0.066, 0.305]	[−0.778, 0.095]	(0, 48.699)

[a] From the parameter constraint, $c > 0$.

Table 2.9. *Goodness-of-fit results for the distribution fitted to total flow deficit.*

	KS goodness-of-fit test	
Estimation method	D_n	P-value
MLE	0.072	0.08
Entropy	0.053	0.583

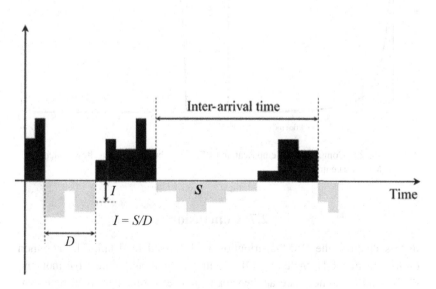

Figure 2.7 Description of drought variable.

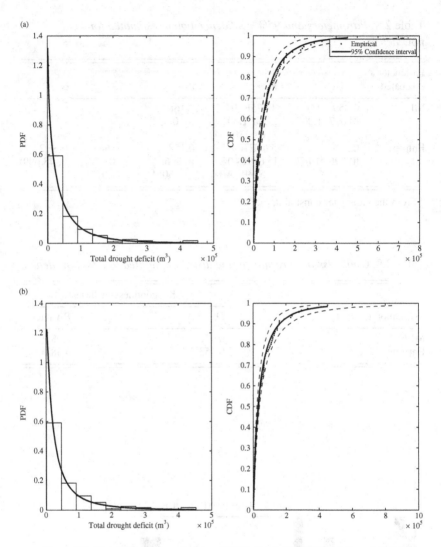

Figure 2.8 Comparison of empirical and BSM distribution for total flow deficit:
(a) MLE; (b) entropy.

2.7 Conclusion

In this chapter, the BSM distribution is discussed in detail. The common
estimation methods, including MLE, entropy, moment, cumulative moment,
PWM, and L-moment, are applied for parameter estimation. With both synthetic data (generated from the BSM distribution) and the real-world peak flow

data from USGS09239500, the PWM method yields the worst performance while all other methods yield similar performances. It can be concluded that one may choose any method (i.e., MLE, entropy, MOM, cumulative MOM, or L-moment) to estimate the parameters of the BSM distribution. Additionally, we also studied maximum daily precipitation data of Brenham, Texas, and the total flow deficit at Tilden, Texas. The MLE and entropy methods are applied for parameter estimation. The goodness-of-fit study and visual comparison with empirical distribution indicate that the BSM distribution may be properly applied to model the maximum daily precipitation and total flow deficit.

References

Brouers, F. (2014a). Statistical foundations of empirical isotherms. *Open Journal of Statistics,* Vol. 4, pp. 687–701.

Brouers, F. (2014b). The fractal (BSF) kinetics equation and its applications. *Journal of Modern Physics,* Vol. 5, pp. 1954–1998.

Brouers, F. (2015). The Burr XII distribution family and the maximum entropy principle: Power-law phenomena are not necessarily "nonextensive." *Open Journal of Statistics,* Vol. 5, pp. 730–741.

Brouers, F. and Al-Musawi, T.J. (2015). On the optimum use of isotherms model for the characterization of biosorption of lead onto algae. *Journal of Molecular Liquids,* Vol. 212, pp. 46–51.

Brouers, F. and Sotolongo-Costa, O. (2005). Relaxation in heterogeneous systems: A rare events dominated phenomenon. *Physica A: Statistical Mechanics and Its Applications,* Vol. 356, pp. 359–374.

Burr, I.W. (1942). Cumulative frequency functions. *The Annals of Mathematical Statistics,* Vol. 13, No. 2, pp. 215–232.

Dubey, M.J. and Gove, J.H. (2015). Size-based distribution in the generalized beta distribution family, with applications to forestry. *Forestry,* Vol. 88, pp. 141–151.

Fisk, P.R. (1961). The graduation of income distributions. *Econometrica,* Vol. 29, No. 2, pp. 171–185.

Greenwood, J.A., Landwehr, J.M., and Matalas, N.C. (1979). Probability weighted moments: definition and relation to parameters of several distributions expressable in inverse form. *Water Resources Research,* Vol. 15, No. 5, pp. 1049–1054.

Hosking, J.R.M. (1990). L-moments: Analysis and estimation of distribution using linear combinations of order statistics. *Journal of the Royal Statistical Society: Series B (Methodological),* Vol. 52, No. 1, 105–124.

Papalexiou, S.M. and Koutsoyiannis, D. (2012). Entropy-based derivation of probability distributions: A case study to daily rainfall. *Advances in Water Resources,* Vol. 45, pp. 51–57.

Singh, S.K. and Maddala, G.S. (1976). A function for the size distribution of incomes. *Econometrica,* Vol. 44, pp. 963–970.

Singh, V.P. and Rajagopal, A.K. (1986). A new method of parameter estimation for hydrologic frequency analysis. *Hydrological Science and Technology*, Vol. 2, No. 3, pp. 33–40.

Sornette, D. (2003). *Critical Phenomena in Natural Sciences*. 2nd edition, chapter 14, Springer, Heidelberg.

Tsallis, C. (1988). Possible generalization of Boltzmann–Gibbs statistics. *Journal of Statistical Physics* Vol. 52, Nos. 1–2, pp. 479–487.

Weron, K. and Kotulski, M. (1997). On the equivalence of the parallel channel and the correlated cluster relaxation models. *Journal of Statistical Physics*, Vol. 88, Nos. 5/6, pp. 1241–1256.

3

Halphen Type A Distribution

3.1 Introduction

The Halphen type A (Hal-A) distribution is a member of the Halphen family proposed by Halphen in the 1940s (Halphen 1941, 1955) and published by Morlat in 1956. This family comprises type A distribution, type B distribution, and type inverse B distribution. Sheshadri (1993, 1997) sketched a brief historical account of the Halphen family and Perreault et al. (1999) provided a short history and the rationale that led Halphen to derive his family of distributions.

Dvorak et al. (1988) discussed the mathematical and statistical properties of the Halphen distributions and related systems of frequency functions. Perreault et al. (1999) presented the Hal-A distribution for frequency analysis of hydro-meteorological extremes. They concluded that this distribution provided sufficient flexibility to fit a large variety of datasets. The Hal-A distribution fits observations that are independent and identically distributed (Fateh et al. 2010). This distribution represents a potential alternative to the generalized extreme value distribution to model extreme hydrometeorological and hydrological events (Fateh et al. 2010). This chapter derives the Hal-A distribution using the entropy theory and estimates its parameters using methods of entropy, moments, and maximum likelihood estimation (MLE). It draws heavily from Perreault et al. (1999).

3.2 Hal-A Distribution and Its Characteristics

The probability density function (PDF) of the Hal-A distribution is given by

$$f(x) = \frac{1}{2m^v K_v(2a)} x^{v-1} \exp\left(-a\left(\frac{x}{m} + \frac{m}{x}\right)\right); \quad x > 0, \tag{3.1}$$

where $K_v(\cdot)$ is the modified Bessel function of the second kind of order v, $v \in R$; $m > 0$ is a scale parameter, and v and $a > 0$ are the shape parameters. The Hal-A distribution is unimodal and exhibits exponentially decaying right and left tails. The modified Bessel function of the second kind can be expressed as (Watson, 1966; Abramowitz and Stegun, 1972)

$$K_v(s) = \frac{1}{2} \int_0^\infty x^{v-1} \exp\left(-\frac{s}{2}\left(x + \frac{1}{x}\right)\right) dx; \quad s > 0. \tag{3.2}$$

For small values of s, Equation (3.2) yields

$$K_v(s) \approx \frac{2^{v-1}}{s^v} \Gamma(v), \quad v > 0. \tag{3.3}$$

For large values of s,

$$K_v(s) = \sqrt{\frac{\pi}{2}} s^{-\frac{1}{2}} \exp(-s) \left(1 + \frac{u-1}{8s} + \frac{(u-1)(u-9)}{2!(8s)^2} + \frac{(u-1)(u-9)(u-25)}{3!(8s)^3} + \cdots\right), \tag{3.4}$$

where $u = 4v^2$.

For large values of v,

$$K_v(s) = \sqrt{\frac{\pi}{2}} 2^v v^{v-\frac{1}{2}} \exp(-v) s^{-v}. \tag{3.5}$$

It is interesting to note that if X follows HAL-A distribution $[f(x; m, a, v)]$, then its inverse $Y = 1/X$ also follows HAL-A distribution $[f(y; m^{-1}, a, -v)]$ as

$$f(y) = \frac{1}{2m^v K_v(2a)} y^{-v-1} \exp\left(-a\left(\frac{y}{m^{-1}} + \frac{m^{-1}}{y}\right)\right); \quad y > 0. \tag{3.6}$$

Equation (3.6) occurs because of the symmetry of the modified Bessel function of the second kind with respect to order v, that is, $K_v(2a) = K_{-v}(2a)$.

The Hal-A distribution leads to gamma and inverse gamma distributions as limiting distributions. To show this property, let $\theta = am$ and $w = a/m$; then the Hal-A PDF becomes

$$f(x) = \frac{\left(\frac{w}{\theta}\right)^{\frac{v}{2}}}{2K_v(2\sqrt{\theta w})} x^{v-1} \exp\left(-\left(wx + \frac{\theta}{x}\right)\right); \quad x > 0. \tag{3.7}$$

Using the symmetric properties of the Bessel function, it can be shown that

$$\lim_{\theta \to 0} \frac{\left(\frac{w}{\theta}\right)^{\frac{v}{2}}}{2K_v(\sqrt{\theta w})} = \lim_{\theta \to 0} \frac{\left(\frac{w}{\theta}\right)^{\frac{v}{2}}}{\Gamma(v)(\theta w)^{-\frac{v}{2}}} = \frac{w^v}{\Gamma(v)}. \tag{3.8}$$

Applying the property defined in Equation (3.8), Equation (3.7) can be expressed as

$$\lim_{\theta \to 0} f(x) = \lim_{\theta \to 0} \frac{\left(\frac{w}{\theta}\right)^{\frac{v}{2}}}{2K_v(2\sqrt{\theta w})} x^{v-1} \exp\left(-\left(wx + \frac{\theta}{x}\right)\right) = \frac{w^v}{\Gamma(v)} x^{v-1} \exp(-wx); \quad x > 0. \tag{3.9}$$

Equation (3.9) is a gamma distribution with $w > 0$ as a scale parameter and $v > 0$ as a shape parameter.

If $v < 0$, we have $K_{-v}(s) = K_v(s)$ based on the symmetry of the modified Bessel function of the second kind. Furthermore let $w \to 0$; Equation (3.7) becomes

$$\lim_{w \to 0} f(x) = \lim_{w \to 0} \frac{\left(\frac{w}{\theta}\right)^{-\frac{|v|}{2}}}{2K_{|v|}(2\sqrt{\theta w})} \left(\frac{1}{x}\right)^{|v|+1} \exp\left(-\frac{\theta}{x}\right) = \frac{\theta^{|v|}}{\Gamma(|v|)} \left(\frac{1}{x}\right)^{|v|+1} \exp\left(-\frac{\theta}{x}\right). \tag{3.10}$$

Equation (3.10) represents the inverse gamma distribution with scale parameter θ and shape parameter $|v|$. The inverse gamma distribution may also be obtained with the given conditions $v < 0$, $\alpha \to 0$, $m \to \infty$ with $\alpha m \to \theta$ in which θ is a constant.

The Hal-A distribution also yields the normal distribution as a limiting distribution for fixed v if α tends to infinity. Consider a linear transformation of X as

$$Z = \sqrt{2\alpha}\left(\frac{X}{m} - 1\right). \tag{3.11}$$

Then, Z tends to $N(0,1)$. The PDF of Z becomes

$$f(z) = \frac{\exp(-2\alpha)}{2\sqrt{2\alpha}K_v(2\alpha)} \left(\frac{z}{\sqrt{2\alpha}} + 1\right)^{v-1} \exp\left(-\left(\frac{1}{2}\left(\frac{z^2}{1 + \frac{z}{\sqrt{2\alpha}}}\right)\right)\right); \quad z \in (-\sqrt{2\alpha}, \infty). \tag{3.12}$$

With $\alpha \to \infty$, we have $\lim_{\alpha \to 0} \frac{\exp(-2\alpha)}{\sqrt{2\alpha}K_\nu(2\alpha)} \to \frac{1}{\sqrt{\pi/2}}$, and Equation (3.12) may be expressed as

$$\lim_{\alpha \to \infty} f(z) = \frac{1}{\sqrt{2\pi}} \exp\left(\frac{z^2}{2}\right), \quad z \in (-\infty, +\infty). \tag{3.13}$$

Furthermore, in the case of Hal-A distribution, there are three sufficient statistics:

$$T_1 = \sum_{i=1}^{n} \ln x_i = n \ln \bar{G}, \quad T_2 = \sum_{i=1}^{n} x_i = n\bar{X}, \quad T_3 = \sum_{i=1}^{n} \frac{1}{x_i} = n\bar{H}^{-1}. \tag{3.14}$$

In Equation (3.14), n is the sample size and $\bar{G}, \bar{X}$, and $\bar{H}$ are the sample geometric, arithmetic, and harmonic means, respectively.

3.3 Differential Equation for Derivation of Hal-A Distribution

Following Dvorak et al. (1988) and taking the logarithm of Equation (3.1), we obtain

$$\ln f(x) = \ln\left(\frac{1}{2m^\nu K_\nu(2\alpha)}\right) + (\nu - 1)\ln x - \alpha\frac{x}{m} - \alpha\frac{m}{x}. \tag{3.15}$$

Differentiating Equation (3.15) with respect to x, we have

$$\frac{1}{f(x)}\frac{df}{dx} = \frac{\nu - 1}{x} - \frac{\alpha}{m} + \alpha\frac{m}{x^2} = \frac{-\left(\frac{\alpha}{m}\right)x^2 + (\nu - 1)x + \alpha m}{x^2}. \tag{3.16}$$

Comparing with the differential equations for the Pearson family, it is seen that Equation (3.16) is expressed in the same manner as the Pearson family, that is,

$$\frac{1}{f(x)}\frac{df}{dx} = \frac{a_0 + a_1 x + a_2 x^2}{x^2}, \tag{3.17a}$$

where

$$a_0 = \alpha m, \quad a_1 = \nu - 1, \quad \text{and} \quad a_2 = -\frac{\alpha}{m}. \tag{3.17b}$$

Solution of Equation (3.17a) or (3.17b) yields the Hal-A distribution.

Equating Equation (3.17a) to zero, we can obtain the modes and antimodes by solving

$$-\frac{\alpha}{m}x^2 + (v-1)x + \alpha m = 0. \tag{3.18}$$

The solutions of Equation (3.18) are given as

$$x_m = \frac{m(v-1)}{2\alpha} \pm \sqrt{\left(\frac{m(v-1)}{2\alpha}\right)^2 + m^2} = m\left[\frac{v-1}{2\alpha} \pm \sqrt{\frac{(v-1)^2}{4\alpha^2} + 1}\right]. \tag{3.19}$$

According to the mode existence summary provided in Perreault et al. (1999), $\frac{a_0}{a_2} = \frac{\alpha m}{-(\alpha/m)} = -m^2 < 0$. Thus, the Hal-A distribution is a unimodal distribution. Figure 3.1 plots some examples of Hal-A distribution with different parameters.

Recalling the definition of recurrence interval T $\left(\text{i.e., } T = 1/(1 - F(x))\right)$, the asymptotic relationship for $T(x)$ given by Gumbel (1958) can be expressed for large x as

$$T(x) \propto \frac{df(x)/dx}{\left(f(x)\right)^2}. \tag{3.20}$$

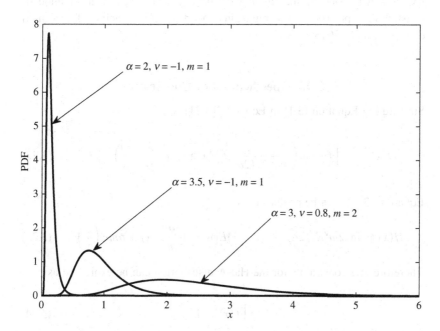

Figure 3.1 Hal-A PDFs.

Similar to gamma and Gumbel distributions, Ouarda et al. (1994) showed that the Hal-A distribution exhibits the right tail behavior as $x \propto \ln T$.

3.4 Derivation of Hal-A Distribution by Entropy Theory

Four steps are involved in deriving the distribution using the Shannon entropy (Singh, 1998): (1) a set of constraints is defined; (2) the entropy is maximized using the method of Lagrange multipliers; (3) an entropy-based distribution is obtained, which is least biased in accordance with the principle of maximum entropy; and (4) the Lagrange multipliers are related to the constraints, which are then related to the distribution parameters.

For a continuous random variable X, such as streamflow, the Shannon entropy of X, $H(X)$, can be expressed as (Shannon, 1948)

$$H(x) = -\int_0^\infty f(x)\ln f(x)dx, \qquad (3.21)$$

where $f(x)$ is the PDF of X, which ranges from 0 to ∞. The objective is to derive $f(x)$, which is done by maximizing entropy subject to given information or constraints, expressed as constraints, in concert with the principle of maximum entropy (Singh, 2013).

3.4.1 Specification of Constraints

Substituting Equation (3.1) in Equation (3.21), we get

$$H(x) = -\int_0^\infty f(x) \ln\left(\frac{1}{2m^\nu K_\nu(2\alpha)}x^{\nu-1}\exp\left(-\alpha\left(\frac{x}{m}+\frac{m}{x}\right)\right)\right)dx. \qquad (3.22)$$

Equation (3.22) can be evaluated as

$$H(x) = \ln\left(2m^\nu K_\nu(2\alpha)\right) - (\nu-1)E[\ln x] + \frac{\alpha}{m}E(x) + \alpha m E\left(\frac{1}{x}\right). \qquad (3.23)$$

Therefore, the constraints for the Hal-A distribution can be expressed as

$$\int_0^\infty f(x) = 1 \qquad (3.24)$$

$$\int_0^\infty f(x)\ln x\,dx = E[\ln x] \tag{3.25}$$

$$\int_0^\infty \left(\frac{x}{m}+\frac{m}{x}\right)f(x)dx = E\left[\frac{x}{m}+\frac{m}{x}\right]. \tag{3.26}$$

3.4.2 Entropy Maximizing

Using the method of Lagrange multipliers, the Lagrangian function L can be expressed as

$$L = -\int_0^\infty f(x)\ln f(x)dx - (\lambda_0 - 1)\left(\int_0^\infty f(x)dx - 1\right) - \lambda_1\left(\int_0^\infty f(x)\ln x\,dx - E(\ln x)\right)$$
$$- \lambda_2\left(\int_0^\infty \left(\frac{x}{m}+\frac{m}{x}\right)f(x)dx - E\left(\frac{x}{m}+\frac{m}{x}\right)\right). \tag{3.27}$$

Differentiating Equation (3.27) with respect to $f(x)$ and equating the derivative to zero, we obtain

$$\frac{\partial L}{\partial f(x)} = -\ln f(x) - \lambda_0 - \lambda_1 \ln x - \lambda_2\left(\frac{x}{m}+\frac{m}{x}\right) = 0. \tag{3.28}$$

Equation (3.28) yields the entropy-based least-biased PDF of the Hal-A distribution:

$$f(x) = \exp\left(-\lambda_0 - \lambda_1\ln x - \lambda_2\left(\frac{x}{m}+\frac{m}{x}\right)\right). \tag{3.29}$$

3.4.3 Relation between Lagrange Multipliers and Distribution Parameters

Substituting Equation (3.29) in Equation (3.24), we get

$$\int_0^\infty f(x)dx = \int_0^\infty \exp\left(-\lambda_0 - \lambda_1\ln x - \lambda_2\left(\frac{x}{m}+\frac{m}{x}\right)\right)dx = 1. \tag{3.30}$$

From Equation (3.30), we can express λ_0 as

$$\exp(\lambda_0) = \int_0^\infty \exp\left(-\lambda_1 \ln x - \lambda_2\left(\frac{x}{m}+\frac{m}{x}\right)\right)dx = \int_0^\infty x^{-\lambda_1}\exp\left(-\lambda_2\left(\frac{x}{m}+\frac{m}{x}\right)\right)dx$$

$$= 2\left[\left(\frac{1}{2}\right)\int_0^\infty x^{-\lambda_1}\exp\left(-\lambda_2\left(\frac{x}{m}+\frac{m}{x}\right)\right)dx\right].$$

$$(3.31)$$

Let $y = x/m \Rightarrow dx = m\,dy$. Equation (3.31) can be re-evaluated as

$$\exp(\lambda_0) = 2\left[\left(\frac{1}{2}\right)\int_0^\infty (my)^{-\lambda_1}\exp\left(-\lambda_2\left(y+\frac{1}{y}\right)\right)m\,dy\right]$$

$$= 2\left[\left(\frac{1}{2}\right)m^{-\lambda_1+1}\int_0^\infty y^{-\lambda_1}\exp\left(-\lambda_2\left(y+\frac{1}{y}\right)\right)dy\right] = 2m^{1-\lambda_1}K_{1-\lambda_1}(2\lambda_2).$$

$$(3.32)$$

Now, Equation (3.32) can be expressed as a modified Bessel function of the second kind.

Substituting Equation (3.32) in Equation (3.29), we obtain

$$f(x) = \frac{1}{2m^{1-\lambda_1}K_{1-\lambda_1}(2\lambda_2)}x^{-\lambda_1}\exp\left(-\lambda_2\left(\frac{x}{m}+\frac{m}{x}\right)\right). \qquad (3.33)$$

Comparing Equation (3.33) with Equation (3.1), it is seen that

$$\lambda_1 = 1 - \nu, \quad \lambda_2 = \alpha. \qquad (3.34)$$

Substituting Equation (3.34) in Equation (3.33), Equation (3.33) may be expressed exactly in the same form as Equation (3.1).

3.5 Parameter Estimation

The Hal-A distribution parameters are estimated using the regular entropy method, parameter space expansion method, and method of moments (MOM) and MLE method.

3.5.1 Regular Entropy Method

This method involves deriving (1) relations between Lagrange multipliers and constraints, (2) relations between Lagrange multipliers and parameters, and (3) relations between parameters and constraints.

3.5.1.1 Relations between Lagrange Multipliers and Constraints

The Lagrange multiplier λ_0 from Equation (3.32) can be expressed as

$$\lambda_0 = \lambda_0(\lambda_1, \lambda_2) = \ln 2 + (1 - \lambda_1) \ln m + \ln K_{1-\lambda_1}(2\lambda_2). \tag{3.35}$$

From Equation (3.31), λ_0 can also be defined as

$$\lambda_0 = \lambda_0(\lambda_1, \lambda_2) = \ln \int_0^\infty \exp\left(-\lambda_1 \ln x - \lambda_2\left(\frac{x}{m} + \frac{m}{x}\right)\right) dx. \tag{3.36}$$

Differentiating Equation (3.35) with respect to λ_1 and λ_2, we obtain

$$\ln 2 + (1 - \lambda_1) \ln m + \ln K_{1-\lambda_1}(2\lambda_2)$$

$$\frac{\partial \lambda_0}{\partial \lambda_1} = -\ln m + \frac{1}{K_{1-\lambda_1}(2\lambda_2)} \frac{\partial K_{1-\lambda_1}(2\lambda_2)}{\partial \lambda_1} \tag{3.37}$$

$$\frac{\partial \lambda_0}{\partial \lambda_2} = \frac{2}{K_{1-\lambda_1}(2\lambda_2)} \left(\frac{1-\lambda_1}{2\lambda_2} K_{1-\lambda_1}(2\lambda_2) - K_{2-\lambda_1}(2\lambda_2)\right) = \frac{1-\lambda_1}{\lambda_2} - \frac{2K_{2-\lambda_1}(2\lambda_2)}{K_{1-\lambda_1}(2\lambda_2)}. \tag{3.38}$$

From Equation (3.36) we know there are three parameters; thus, one more equation is needed as

$$\frac{\partial^2 \lambda_0}{\partial \lambda_2^2} = -\frac{1-\lambda_1}{\lambda_2^2} - 2 \frac{\dfrac{\partial K_{2-\lambda_1}(2\lambda_2)}{\partial \lambda_2} K_{1-\lambda_1}(2\lambda_2) - K_{2-\lambda_1}(2\lambda_2) \dfrac{\partial K_{1-\lambda_1}(2\lambda_2)}{\partial \lambda_2}}{\left(K_{1-\lambda_1}(2\lambda_2)\right)^2}. \tag{3.39}$$

Substituting $\frac{\partial K_\nu(Z)}{\partial Z} = -\frac{1}{2}(K_{\nu-1}(Z) + K_{\nu+1}(Z))$ in Equation (3.39), Equation (3.39) can be rewritten as

$$\frac{\partial^2 \lambda_0}{\partial \lambda_2^2} = -\frac{1-\lambda_1}{\lambda_2^2} - 2\left(-\frac{K_{1-\lambda_1}(2\lambda_2) + K_{3-\lambda_1}(2\lambda_2)}{K_{1-\lambda_1}(2\lambda_2)} + \frac{K_{2-\lambda_1}(2\lambda_2)\left(K_{-\lambda_1}(2\lambda_2) + K_{2-\lambda_1}(2\lambda_2)\right)}{K_{1-\lambda_1}^2(2\lambda_2)}\right). \tag{3.40}$$

Now differentiating Equation (3.36) with respect to λ_1 and λ_2, the results are

$$\frac{\partial \lambda_0}{\partial \lambda_1} = -\frac{\int_0^\infty (\ln x) \exp\left(-\lambda_1 \ln x - \lambda_2 \left(\frac{x}{m} + \frac{m}{x}\right)\right) dx}{\int_0^\infty \exp\left(-\lambda_1 \ln x - \lambda_2 \left(\frac{x}{m} + \frac{m}{x}\right)\right) dx} = -E(\ln x) \qquad (3.41)$$

$$\frac{\partial \lambda_0}{\partial \lambda_2} = -\frac{\int_0^\infty \left(\frac{x}{m} + \frac{m}{x}\right) \exp\left(-\lambda_1 \ln x - \lambda_2 \left(\frac{x}{m} + \frac{m}{x}\right)\right) dx}{\int_0^\infty \exp\left(-\lambda_1 \ln x - \lambda_2 \left(\frac{x}{m} + \frac{m}{x}\right)\right) dx} = -E\left(\frac{x}{m} + \frac{m}{x}\right) \qquad (3.42)$$

$$\frac{\partial^2 \lambda_0}{\partial \lambda_2^2} = \frac{\int_0^\infty \left(\frac{x}{m} + \frac{m}{x}\right)^2 \exp\left(-\lambda_1 \ln x - \lambda_2 \left(\frac{x}{m} + \frac{m}{x}\right)\right) dx}{\int_0^\infty \exp\left(-\lambda_1 \ln x - \lambda_2 \left(\frac{x}{m} + \frac{m}{x}\right)\right) dx} = \mathrm{var}\left[\left(\frac{x}{m} + \frac{m}{x}\right)\right]. \qquad (3.43)$$

Equating Equations (3.37)–(3.39) to Equations (3.41)–(3.43), we have

$$-\ln m + \frac{1}{K_{1-\lambda_1}(2\lambda_2)} \frac{\partial K_{1-\lambda_1}(2\lambda_2)}{\partial \lambda_1} = -E(\ln(x)) \qquad (3.44)$$

$$\frac{1-\lambda_1}{\lambda_2} - \frac{2K_{2-\lambda_1}(2\lambda_2)}{K_{1-\lambda_1}(2\lambda_2)} = -E\left(\frac{x}{m} + \frac{m}{x}\right) \qquad (3.45)$$

$$-\frac{1-\lambda_1}{\lambda_2^2} - 2\left(-1 - \frac{K_{3-\lambda_1}(2\lambda_2)}{K_{1-\lambda_1}(2\lambda_2)} + \frac{K_{2-\lambda_1}(2\lambda_2)(K_{-\lambda_1}(2\lambda_1) + K_{2-\lambda_1}(2\lambda_2))}{K_{1-\lambda_1}^2(2\lambda_2)}\right)$$
$$= E\left[\left(\frac{x}{m} + \frac{m}{x}\right)^2\right].$$

$$(3.46)$$

The Lagrange multipliers can then be estimated by solving the system of equations with the Newton–Raphson method.

3.5.2 Parameter Space Expansion Method

From Equation (3.21), the constraints can be defined as

$$\int_0^\infty f(x)dx = 1 \qquad (3.47)$$

$$\int_0^\infty \alpha\left(\frac{x}{m}+\frac{m}{x}\right)f(x)dx = E\left(\alpha\left(\frac{x}{m}+\frac{m}{x}\right)\right) \tag{3.48}$$

$$\int_0^\infty \ln\left(x^{\nu-1}\right)f(x)dx = E\left(\ln(x^{\nu-1})\right). \tag{3.49}$$

Following the same procedure as before, the entropy-based PDF of Hal-A distribution can be written as

$$f(x) = \exp\left(-\lambda_0 - \lambda_1\alpha\left(\frac{x}{m}+\frac{m}{x}\right) - \lambda_2(\nu-1)\ln x\right). \tag{3.50}$$

Substituting Equation (3.50) in Equation (3.47), we get

$$\exp(\lambda_0) = \int_0^\infty \exp\left(-\lambda_1\alpha\left(\frac{x}{m}+\frac{m}{x}\right) - \lambda_2(\nu-1)\ln x\right)dx. \tag{3.51}$$

Equation (3.51) can be solved as

$$\exp(\lambda_0) = \int_0^\infty x^{-\lambda_2(\nu-1)} \exp\left(-\lambda_1\alpha\left(\frac{x}{m}+\frac{m}{x}\right)\right)dx$$
$$= 2m^{-\lambda_2(\nu-1)+1} K_{-\lambda_2(\nu-1)+1}(2\lambda_1\alpha). \tag{3.52}$$

Therefore,

$$\lambda_0 = \ln 2 + \left(-\lambda_2(\nu-1)+1\right)\ln m + \ln\left(K_{-\lambda_2(\nu-1)+1}(2\lambda_1\alpha)\right). \tag{3.53}$$

Equation (3.50) can be rewritten as

$$f(x) = \frac{1}{2m^{-\lambda_2(\nu-1)+1}K_{-\lambda_2(\nu-1)+1}(2\lambda_1\alpha)} \exp\left(-\lambda_1\alpha\left(\frac{x}{m}+\frac{m}{x}\right) - \lambda_2(\nu-1)\ln x\right). \tag{3.54}$$

Now the entropy of Equation (3.54) can be written as

$$H(f(x)) = \ln 2 + \left(-\lambda_2(\nu-1)+1\right)\ln m + \ln\left(K_{-\lambda_2(\nu-1)+1}(2\lambda_1\alpha)\right)$$
$$+ \lambda_1\alpha E\left(\frac{x}{m}+\frac{m}{x}\right) + E\left(\lambda_2(\nu-1)\ln x\right). \tag{3.55}$$

Differentiating Equation (3.55) with respect to each parameter separately and equating the derivative to zero,

$$\frac{\partial H}{\partial \lambda_1} = \frac{1}{K_{-\lambda_2(\nu-1)+1}(2\lambda_1\alpha)} \frac{\partial K_{-\lambda_2(\nu-1)+1}(2\lambda_1\alpha)}{\partial \lambda_1} + \alpha E\left(\frac{x}{m} + \frac{m}{x}\right) = 0$$

$$\Rightarrow -\alpha \frac{K_{-\lambda_2(\nu-1)}(2\lambda_1\alpha) + K_{-\lambda_2(\nu-1)+2}(2\lambda_1\alpha)}{K_{-\lambda_2(\nu-1)+1}(2\lambda_1\alpha)} + \alpha\left(E\left(\frac{x}{m} + \frac{m}{x}\right)\right) = 0$$

$$\Rightarrow E\left(\frac{x}{m} + \frac{m}{x}\right) = \frac{K_{-\lambda_2(\nu-1)}(2\lambda_1\alpha) + K_{-\lambda_2(\nu-1)+2}(2\lambda_1\alpha)}{K_{-\lambda_2(\nu-1)+1}(2\lambda_1\alpha)}$$

$$(3.56)$$

$$\frac{\partial H}{\partial \lambda_2} = (1-\nu)\ln m + \frac{1}{K_{-\lambda_2(\nu-1)+1}(2\lambda_1\alpha)} \frac{\partial K_{-\lambda_2(\nu-1)+1}(2\lambda_1\alpha)}{\partial \lambda_2} - (1-\nu)E(\ln x) = 0.$$

$$(3.57)$$

In Equation (3.57), let $\nu1 = -\lambda_2(\nu-1) + 1$. Equation (3.57) may be rewritten as

$$\frac{\partial H}{\partial \lambda_2} = (1-\nu)\left[\frac{1}{K_{\nu1}(2\lambda_1\alpha)} \frac{\partial K_{\nu1}(2\lambda_1\alpha)}{\partial \nu1} - E\left(\ln\left(\frac{x}{m}\right)\right)\right] = 0 \qquad (3.58)$$

$$\frac{\partial H}{\partial \nu} = -\lambda_2 \ln m + \frac{1}{K_{-\lambda_2(\nu-1)+1}(2\lambda_1\alpha)} \frac{\partial K_{-\lambda_2(\nu-1)+1}(2\lambda_1\alpha)}{\partial \nu} + E(\lambda_2 \ln x)$$

$$\Rightarrow \frac{1}{K_{-\lambda_2(\nu-1)+1}(2\lambda_1\alpha)} \frac{\partial K_{-\lambda_2(\nu-1)+1}(2\lambda_1\alpha)}{\partial \nu} + \lambda_2 E\left(\ln\left(\frac{x}{m}\right)\right) = 0.$$

$$(3.59)$$

Again, let $\nu1 = -\lambda_2(\nu-1) + 1$. Equation (3.59) may be rewritten as

$$\frac{\partial H}{\partial \nu} = \lambda_2\left[\frac{1}{K_{\nu1}(2\lambda_1\alpha)} \frac{\partial K_{\nu1}(2\lambda_1\alpha)}{\partial \nu1} - E\left(\ln\left(\frac{x}{m}\right)\right)\right] = 0 \qquad (3.60)$$

$$\frac{\partial H}{\partial m} = \frac{-\lambda_2(\nu-1) + 1}{m} + \lambda_1\alpha E\left(-\frac{x}{m^2} + \frac{1}{x}\right) = 0$$

$$\Rightarrow \frac{-\lambda_2(\nu-1) + 1}{m} + \frac{\lambda_1\alpha}{m} E\left(\frac{m}{x} - \frac{x}{m}\right) = 0$$

$$(3.61)$$

$$\frac{\partial H}{\partial \alpha} = \frac{1}{K_{-\lambda_2(\nu-1)+1}(2\lambda_1\alpha)} \frac{\partial K_{-\lambda_2(\nu-1)+1}(2\lambda_1\alpha)}{\partial \alpha} + \lambda_1 E\left(\frac{x}{m} + \frac{m}{x}\right) = 0$$

$$\Rightarrow \lambda_1\left(\frac{K_{-\lambda_2(\nu-1)}(2\lambda_1\alpha) + K_{-\lambda_2(\nu-1)+2}(2\lambda_1\alpha)}{K_{-\lambda_2(\nu-1)+1}(2\lambda_1\alpha)}\right) + \lambda_1\left(E\left(\frac{x}{m} + \frac{m}{x}\right)\right) = 0$$

$$\Rightarrow \lambda_1\left[\frac{K_{-\lambda_2(\nu-1)}(2\lambda_1\alpha) + K_{-\lambda_2(\nu-1)+2}(2\lambda_1\alpha)}{K_{-\lambda_2(\nu-1)+1}(2\lambda_1\alpha)} - E\left(\frac{x}{m} + \frac{m}{x}\right)\right] = 0.$$

$$(3.62)$$

Comparing Equation (3.57) with Equation (3.62), we have $\lambda_1 = a \Rightarrow d\lambda_1 = da$. Comparing Equation (3.58) with Equation (3.60), we have $\lambda_2 = 1 - v \Rightarrow d\lambda_2 = -dv$.

Now Equation (3.60) can be rewritten as

$$\frac{\partial H}{\partial v} = \frac{1}{K_{1+(1-v)^2}(2\alpha^2)} \frac{\partial K_{1+(1-v)^2}(2\alpha^2)}{\partial v} - E\left[\ln\left(\frac{x}{m}\right)\right] = 0$$

$$\Rightarrow E\left[\ln\left(\frac{x}{m}\right)\right] = \frac{1}{K_{1+(1-v)^2}(2\alpha^2)} \frac{\partial K_{1+(1-v)^2}(2\alpha^2)}{\partial v}.$$

(3.63)

Equation (3.62) can be rewritten as

$$\frac{\partial H}{\partial \alpha} = \frac{K_{-\lambda_2(v-1)}(2\lambda_1\alpha) + K_{-\lambda_2(v-1)+2}(2\lambda_1\alpha)}{K_{-\lambda_2(v-1)+1}(2\lambda_1\alpha)} - E\left(\frac{x}{m} + \frac{m}{x}\right) = 0$$

$$\Rightarrow E\left(\frac{x}{m} + \frac{m}{x}\right) = \frac{K_{-\lambda_2(v-1)}(2\lambda_1\alpha) + K_{-\lambda_2(v-1)+2}(2\lambda_1\alpha)}{K_{-\lambda_2(v-1)+1}(2\lambda_1\alpha)}.$$

(3.64)

Equation (3.61) may be rewritten as

$$\frac{\partial H}{\partial m} = \frac{-\lambda_2(v-1)+1}{m} + \frac{\lambda_1\alpha}{m}E\left(\frac{m}{x} - \frac{x}{m}\right) = 0 \Rightarrow E\left(\frac{x}{m} - \frac{m}{x}\right) = \frac{1+(1-v)^2}{\alpha^2}.$$

(3.65)

To this end, the parameters may be estimated by solving Equations (3.63)–(3.65) simultaneously using the Newton–Raphson method.

3.5.3 MOM

The rth moment of the Hal-A distribution about the origin can be written as

$$M_r = \int_0^\infty x^r f(x)dx = \frac{1}{2m^v k_v(2\alpha)} \int_0^\infty x^{r+v-1} \exp\left(\alpha\left(\frac{x}{m} + \frac{m}{x}\right)\right)dx = \frac{m^r K_{v+r}(2\alpha)}{K_v(2\alpha)}.$$

(3.66)

Recalling the relation between moments about the origin and those about the centroid (Singh, 1998),

$$M_r^\mu = \sum_{j=0}^r \binom{r}{j} M_j(-\mu)^{r-j},$$

(3.67)

where μ is the first moment or centroid equal to

$$\mu = M_1 = \frac{m K_{v+1}(2\alpha)}{K_v(2\alpha)}. \tag{3.68}$$

Thus,

$$M_2^{\mu} = \frac{m^2 K_{v+2}(2\alpha)}{K_v(2\alpha)} - \frac{m^2 \left(K_{v+1}(2\alpha)\right)^2}{\left(K_v(2\alpha)\right)^2} \tag{3.69}$$

$$M_3^{\mu} = \frac{m^3 K_{v+3}(2\alpha)}{K_v(2\alpha)} - \frac{3m^3 K_{v+1}(2\alpha) K_{v+2}(2\alpha)}{\left(K_v(2\alpha)\right)^2} + \frac{2m^3 \left(K_{v+1}(2\alpha)\right)^3}{\left(K_v(2\alpha)\right)^3} \tag{3.70}$$

$$M_4^{\mu} = \frac{m^4 K_{v+4}(2\alpha)}{K_v(2\alpha)} - \frac{4m^4 K_{v+1}(2\alpha) K_{v+3}(2\alpha)}{\left(K_v(2\alpha)\right)^2} + \frac{6m^4 \left(K_{v+1}(2\alpha)\right)^2 K_{v+2}(2\alpha)}{\left(K_v(2\alpha)\right)^3}$$
$$- \frac{3m^4 \left(K_{v+1}(2\alpha)\right)^4}{\left(K_v(2\alpha)\right)^4}$$

$$\tag{3.71}$$

$$C_v = \frac{(M_2^{\mu})^{\frac{1}{2}}}{M_1^{\mu}} = \frac{\sqrt{K_{v+2}(2\alpha) K_{v(2\alpha)} - \left(K_{v+1}(2\alpha)\right)^2}}{K_{v+1}(2\alpha)} \tag{3.72}$$

$$C_s = \frac{M_3^{\mu}}{(M_2^{\mu})^{\frac{3}{2}}}$$
$$= \frac{\left(K_v(2\alpha)\right)^2 K_{v+3}(2\alpha) - 3 K_v(2\alpha) K_{v+1}(2\alpha) K_{v+2}(2\alpha) + 2 \left(K_{v+1}(2\alpha)\right)^3}{\left(K_v(2\alpha) K_{v+2}(2\alpha) - \left(K_{v+1}(2\alpha)\right)^2\right)^{\frac{3}{2}}}$$

$$\tag{3.73}$$

$$C_g = \frac{M_4^{\mu}}{(M_2^{\mu})^2}$$
$$= \frac{\left(K_v(2\alpha)\right)^3 K_{v+4}(2\alpha) - 4(K_v(2\alpha)^2 K_{v+1}(2\alpha) K_{v+3}(2\alpha) + 6 K_v(2\alpha) \left(K_{v+1}(2\alpha)\right)^2 K_{v+2}(2\alpha) - 3 \left(K_{v+1}(2\alpha)\right)^4}{K_v(2\alpha) K_{v+2}(2\alpha) - \left(K_{v+1}(2\alpha)\right)^2}. \tag{3.74}$$

The moment ratio diagram for the Hal-A distribution is shown in Figure 3.2 It is seen from the figure that the moment ratio of the Hal-A distribution is bounded by the limiting gamma and inverse gamma distributions.

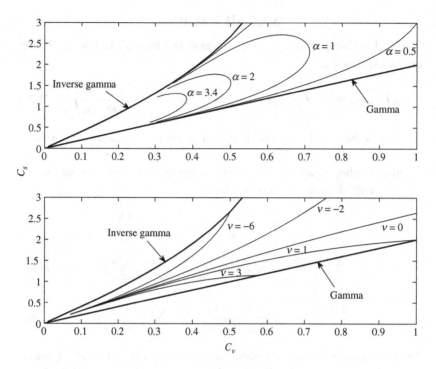

Figure 3.2 Moment ratio diagram of Hal-A distribution.

According to Fitzgerald (2000) and El Adlouni and Bobée (2017), the parameters may be expressed explicitly for the MOM as

$$m^2 = \frac{E(X^{-1})\text{var}(X) - E(X)\left(E(X)E(X^{-1}) - 1\right)}{E(X)\text{var}(X^{-1}) - E(X^{-1})\left(E(X)E(X^{-1}) - 1\right)} \tag{3.75}$$

$$\alpha = \frac{m^{-1}E(X) - mE(X^{-1})}{m^{-2}\text{var}(X) - m^2\text{var}(X^{-1})} \tag{3.76}$$

$$\nu = \frac{\left(m^{-1}E(X) - mE(X^{-1})\right)^2}{m^{-2}\text{var}(X) - m^2\text{var}(X^{-1})}. \tag{3.77}$$

For the parameters estimated with entropy and the MOM, the corresponding confidence interval can be estimated with the parametric bootstrap method. The detailed procedure is given in the Appendix.

3.5.4 MLE Method

The log-likelihood function for the sample drawn from the Hal-A distribution can be expressed as

$$\ln L = -n \ln\left(2m^{\nu} K_{\nu}(2\alpha)\right) + (\nu - 1) \sum_{i=1}^{n} \ln x_i - \alpha \sum_{i=1}^{n} \left(\frac{x_i}{m} + \frac{m}{x_i}\right), \quad (3.78)$$

where n is the sample size and ν, α, and m are the parameters that need to be estimated.

Differentiating Equation (3.78) with respect to parameters ν, α, and m and equating each derivative to zero, we have

$$\frac{\partial \ln L}{\partial m} = 0 = -\frac{n\nu}{m} - \alpha \sum_{i=1}^{n} \left(-\frac{x_i}{m^2} + \frac{1}{x_i}\right) \quad (3.79)$$

$$\frac{\partial \ln L}{\partial \nu} = 0 = -n\ln m - \frac{n}{K_{\nu}(2\alpha)} \frac{\partial K_{\nu}(2\alpha)}{\partial \nu} + \sum_{i=1}^{n} \ln x_i \quad (3.80)$$

$$\frac{\partial \ln L}{\partial \alpha} = 0 = -\frac{n}{K_{\nu}(2\alpha)} \frac{\partial K_{\nu}(2\alpha)}{\partial \alpha} - \sum_{i=1}^{n} \left(\frac{x_i}{m} + \frac{m}{x_i}\right). \quad (3.81)$$

Then, the parameters can be estimated by solving the system of Equations (3.79)–(3.81) numerically. Besides the parametric bootstrap method, the observed information matrix can also be applied to construct the confidence interval for the parameters estimated with MLE. As discussed in Chapter 2, the observed information is evaluated as the inverse of the negative observed Hessian matrix. For Hal-A distribution, the elements of the matrix may be expressed as

$$\frac{\partial^2 \ln L}{\partial m^2} = \frac{n\nu}{m^2} - \alpha \sum_{i=1}^{n} \frac{2x_i}{m^3} \quad (3.82)$$

$$\frac{\partial^2 \ln L}{\partial m \partial \nu} = -\frac{n}{m} \quad (3.83)$$

$$\frac{\partial^2 \ln L}{\partial m \partial \alpha} = -\sum_{i=1}^{n} \left(-\frac{x_i}{m^2} + \frac{1}{x_i}\right) \quad (3.84)$$

$$\frac{\partial^2 \ln L}{\partial \nu^2} = \frac{n}{[K_{\nu}(2\alpha)]^2} \left(\frac{\partial K_{\nu}(2\alpha)}{\partial \nu}\right)^2 - \frac{n}{K_{\nu}(2\alpha)} \frac{\partial^2 K_{\nu}(2\alpha)}{\partial \nu^2} \quad (3.85)$$

$$\frac{\partial^2 \ln L}{\partial \nu \partial \alpha} = \frac{n}{[K_{\nu}(2\alpha)]^2} \frac{\partial K_{\nu}(2\alpha)}{\partial \alpha} \frac{\partial K_{\nu}(2\alpha)}{\partial \nu} - \frac{n}{K_{\nu}(2\alpha)} \frac{\partial^2 K_{\nu}(2\alpha)}{\partial \nu \partial \alpha} \quad (3.86)$$

$$\frac{\partial^2 \ln L}{\partial \alpha^2} = \frac{n}{[K_v(2\alpha)]^2} \left[\frac{\partial K_v(2\alpha)}{\partial \alpha}\right]^2 - \frac{n}{K_v(2\alpha)} \frac{\partial^2 K_v(2\alpha)}{\partial \alpha^2}. \qquad (3.87)$$

For the parameters as asymptotically normal, the critical value of the confidence interval can then be estimated from the standard normal distribution.

3.6 Application

3.6.1 Simulating Hal-A Distributed Random Variable with Fixed Parameters

Following El Adlouni and Bobée (2017), we first briefly discuss how to generate the Hal-A distributed random variable using the acceptance-rejection method. An instrumental distribution is needed to apply the acceptance-rejection method. Given the gamma distribution as the limiting distribution of the Hal-A distribution, the gamma distribution is applied as the instrumental distribution. Then the acceptance-rejection method is described as follows:

Step 1: Generate random variable Y from the instrumental distribution $g(y)$, that is, gamma distribution.

Step 2: Generate uniform distributed random variable U.

Step 3: For the properly selected scale factor c, compute $f(y)/cg(y)$, where $f(y)$ is the density of the Hal-A distribution.

Step 4: If $U \leq \frac{f(y)}{cg(y)} \in (0,1]$, then we set $X = Y$, which is the random variable simulated for the Hal-A distribution. Otherwise, we need to go back to Step 1.

Furthermore, let the gamma distribution be given as

$$g(y; \alpha, \beta) = \frac{1}{\beta^\alpha \Gamma(\alpha)} y^{\alpha-1} \exp\left(-\frac{1}{\beta}y\right). \qquad (3.88)$$

The relation of the first two moments for random variable Y (i.e., following the gamma distribution) and random variable X (i.e., following the Hal-A distribution) may be expressed as

$$E(Y) = E(X) \qquad (3.89)$$

$$\text{var}(Y) = 2\text{var}(X). \qquad (3.90)$$

From Equations (3.89) and (3.90), $E(X)$ and $\text{var}(X)$ may be computed using Equations (3.75) and (3.76). The parameters in Equation (3.88) can then be computed using

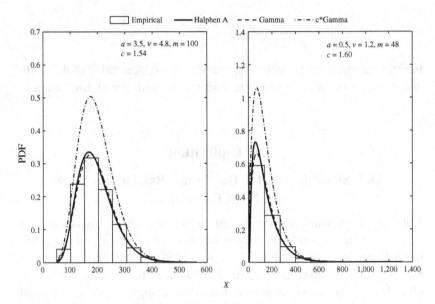

Figure 3.3 Simulated Hal-A random variables with fixed parameters.

$$\alpha = \frac{(E(X))^2}{2\mathrm{var}(X)}; \qquad \beta = \frac{E(X)}{\alpha}. \tag{3.91}$$

The scale factor c can be numerically estimated from the ratio of $f(x)/g(x)$ as

$$c = \max\left(\frac{f(x)}{g(x)}\right) + 0.1. \tag{3.92}$$

Figure 3.3 illustrates the simulated Hal-A distributed random variables with fixed parameters using the acceptance-rejection method. It is shown that the variables are properly simulated for the Hal-A distribution using the acceptance-rejection method with the gamma distribution as the instrumental distribution.

3.6.2 Parameter Estimation Using Simulated Random Variables

In this section, we will apply the MLE method, MOM, and entropy method to estimate the parameters for Hal-A distributed random variables. Additionally, before we apply the real-world data, we will first use the simulated random variables (sample size = 200) from Section 3.6.1. Table 3.1 lists the parameters estimated using the above three methods. Table 3.2 lists the confidence

Table 3.1. *Parameters estimated for synthetic data.*

Method		Parameter		
Synthetic dataset 1 ($\alpha = 3.5$, $v = 4.8$, $m = 100$)				
	α	v	m	
MLE	4.4	3.99	115.87	
MOM	4.46	3.56	121.59	
Entropy	λ_0	λ_1	λ_2	M
	8.536	−2.545	4.481	121.962
Equation (3.23): $\alpha = 4.48$, $v = 3.545$, $m = 121.962$				
Synthetic dataset 2 ($\alpha = 0.5$, $v = 1.2$, $m = 48$)				
MLE	0.44	1.1	46.86	
MOM	0.45	1.21	44.99	
Entropy	λ_0	λ_1	λ_2	m
	6.46	−0.48	0.30	27.37
Equation (3.23): $\alpha = 0.30$, $v = 1.48$, $m = 27.37$				

Table 3.2. *GoF study for synthetic data and corresponding 95% confidence intervals for synthetic datasets.*

		Dataset 1		
		α	v	m
MLE	(1)	[2.04, 6.76]	[−7.32, 15.3]	$(0, 268.55)^a$
	(2)	[0.07, 5.52]	[−7.67, 11.47]	[1.30, 443.11]
GoF		$D_n = 0.0453$, $P = 0.332$		
MOM	(2)	[1.15, 5.54]	[−7.6, 10.68]	[22.23, 426.54]
GoF		$D_n = 0.0451$, $P = 0.319$		
Entropy		λ_1	λ_2	m
(regular method)		[−8.04, 4.53]	[3.22, 7.86]	[31.96, 225.14]
		$D_n = 0.045$, $P = 0.42$		
		Dataset 2		
		α	v	m
MLE		[0.22, 0.65]	[0.32, 1.88]	[11.28, 82.44]
		[0.21, 0.66]	[0.42, 1.98]	[2.82, 75.44]
		$D_n = 0.029$, $P = 0.93$		
MOM		[0.192, 0.70]	[0.26, 2.06]	[14.65, 92.27]
		$D_n = 0.032$, $P = 0.90$		
Entropy		λ_1	λ_2	λ_3
(regular method)		[−0.94, 0.81]	[0.01, 0.61]	[1.13, 98.82]
		$D_n = 0.032$, $P = 0.88$		

intervals approximated for the parameters as well as the goodness-of-fit (GoF) results. The results indicate that (1) MOM and the MLE method yield similar performance for both datasets; (2) the entropy method performs better for dataset 1 than for dataset 2; (3) all methods pass the GoF study. Figures 3.4

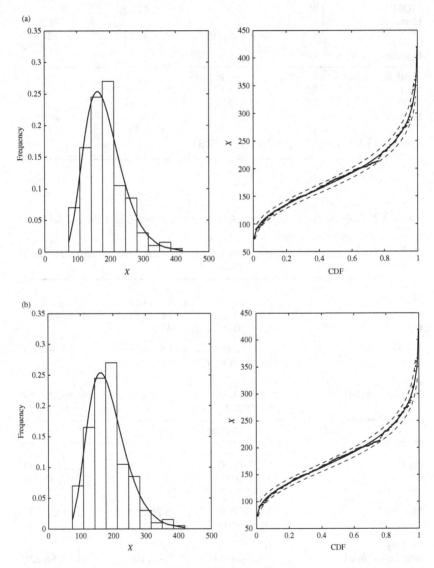

Figure 3.4 Comparison of fitted distribution to the nonparametric distribution for dataset 1: (a) MLE; (b) MOM; and (c) entropy.

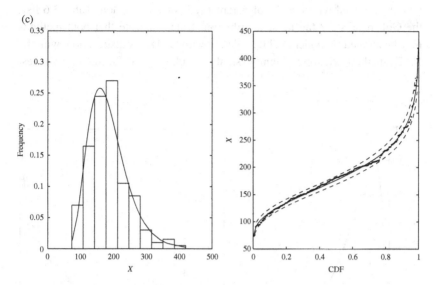

Figure 3.4 (*cont.*)

and 3.5 compare the fitted distribution with nonparametric estimates, which further confirms the GoF findings listed in Table 3.2

3.6.3 Peak Flow

Previously, we have shown that the acceptance-rejection method may be applied to simulate Hal-A distributed random variables. All three methods (i.e., MLE, MOM, and entropy) may be successfully applied for parameter estimation. In this section, the peak flow at USGS09239500 is applied as an example to illustrate its application. Applying MLE, moment, and entropy methods, Table 3.3 lists the estimated parameters. Table 3.4 lists the confidence intervals approximated for the fitted parameters and the GoF study results. Figure 3.6 compares the parametric distribution with nonparametric distribution. The comparison shows that the performance of the entropy method is inferior to the MLE and moment methods. The finding is consistent with the GoF study results listed in Table 3.4, though all three methods pass the GoF study.

3.6.4 Total Flow Deficit

The MLE and entropy methods are applied to model the total flow deficit. Table 3.5 lists the parameters estimated and the corresponding 95% confidence bound simulated with the use of parametric bootstrap method. Table 3.6 lists the GoF results for the fitted distribution, which indicate that both methods may be applied to model total flow deficit with the Hal-A distribution, with the MLE method yielding slightly smaller Kolmogorov–Smirnov (KS) test

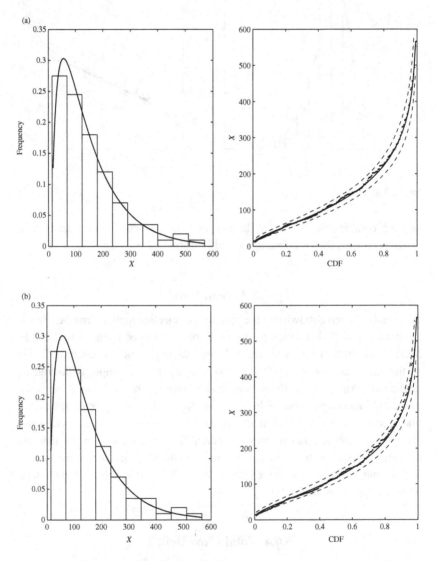

Figure 3.5 Comparison of fitted distribution to the nonparametric distribution for dataset 2: (a) MLE; (b) MOM; and (c) entropy.

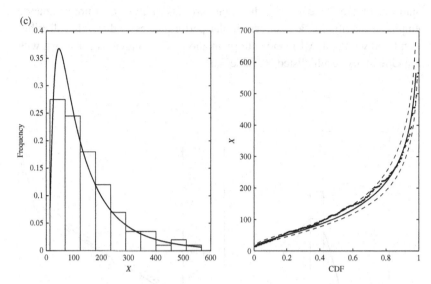

Figure 3.5 (*cont.*)

Table 3.3. *Parameters estimated: peak flow.*

Method	Parameter			
	α	ν	M	
MLE	0.35	10.07	3.58	
MOM	4.2	7.04	47.09	
Entropy	λ_0	λ_1	λ_2	M
	39.14	−9.85	0.37	3.54
Equation (3.23): $\alpha = 0.37$, $\nu = 10.85$, $m = 3.54$				

Table 3.4. *95% confidence intervals approximated and GoF study results: peak flow.*

Method		Parameter		
		α	ν	m
MLE	(1)	$(0, 5.51]^a$	[7.31, 12.83]	$(0, 56.48]^a$
	(2)	[0.02, 5.68]	[−5.31, 12.54]	[0.18, 168.37]
GoF		$D_n = 0.059$, $P = 0.38$		
MOM	(2)	[0.95, 6.51]	[−10.68, 12.31]	[9.03, 328.42]
GoF		$D_n = 0.058$, $P = 0.41$		
Entropy		λ_1	λ_2	λ_3
	(2)	[−18.74, −6.21]	$(0, 0.73]^a$	$(0, 6.94]^a$
GoF		$D_n = 0.048$, $P = 0.77$		

^{a} The lower limit is forced to be greater than 0, according to the parameter constraint.

statistic. Figure 3.7 compares the parametric distribution with nonparametric empirical distribution and shows similar performances of the Hal-A distribution fitted with both MLE and entropy methods. This finding is consistent with the GoF study results listed in Table 3.6.

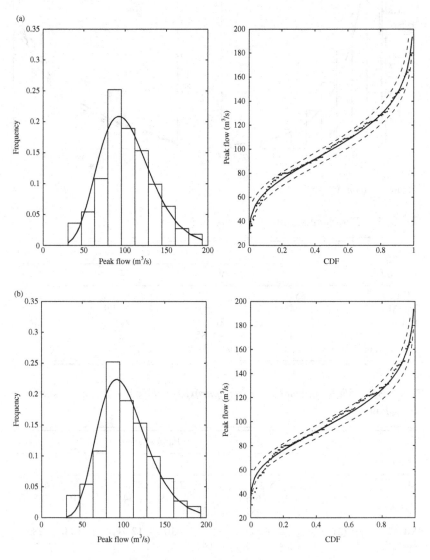

Figure 3.6. Comparison of parametric distribution versus nonparametric distribution for peak flow data: (a) MLE; (b) MOM; and (c) entropy.

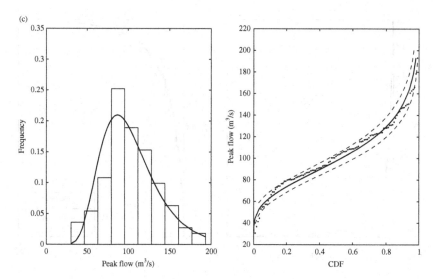

Figure 3.6. (*cont.*)

Table 3.5. *Parameters estimated and 95% confidence bound: total flow deficit.*

Method	α	v	m
MLE	0.108	0.226	1.674E+04
	[0.057, 0.143]	[−0.089, 0.566]	[884.272, 2.495E+04]
Entropy	λ_1	λ_2	λ_3
	0.580	0.086	1.01E+04
	[0.222, 0.852]	[0.027, 0.158]	(0, 1.893E+04)

Table 3.6. *GoF test results: total flow deficit.*

	D_n	P-value
MLE	0.0504	0.725
Entropy	0.0518	0.692

3.6.5 Maximum Daily Precipitation

The MLE and entropy methods are applied to model the maximum daily precipitation at Brenham, Texas. Table 3.7 lists the parameters estimated and the corresponding 95% confidence bound simulated with the use of the

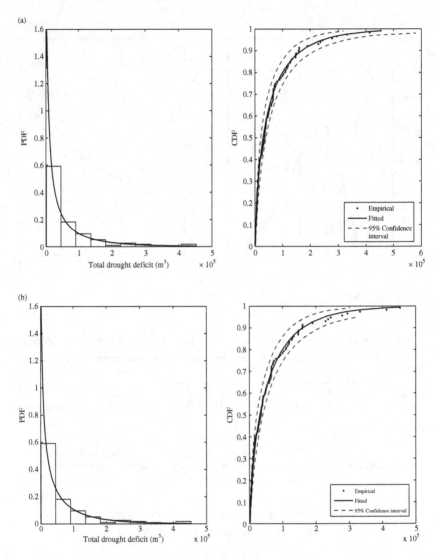

Figure 3.7 Comparison of parametric distribution versus nonparametric distribution for total flow deficit: (a) MLE and (b) entropy.

parametric bootstrap method. Table 3.8 lists the GoF results for the fitted distribution, which indicate that both methods may be applied to model maximum daily precipitation, with the Hal-A distribution with the entropy method yielding slightly smaller KS test statistic. Figure 3.8 compares the parametric distribution with nonparametric empirical distribution and shows similar performances of the Hal-A distribution fitted with both MLE and entropy methods. This finding is consistent with the GoF study results listed in Table 3.8.

Table 3.7. *Parameters estimated and 95% confidence bound: maximum daily precipitation.*

Method	α	v	m
MLE	2.526	−0.260	93.175
	[1.808, 4.269]	[−6.024, 5.304]	(0, 164.643]
Entropy	λ_1	λ_2	λ_3
	1.1	2.466	90.56
	[−0.596, 3.096]	[1.624, 3.079]	[58.146, 117.975]

Table 3.8. *GoF test results: maximum daily precipitation.*

	D_n	P-value
MLE	0.0415	0.964
Entropy	0.0389	0.845

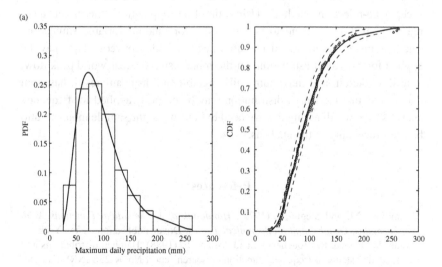

Figure 3.8 Comparison of parametric distribution versus nonparametric distribution for maximum daily precipitation: (a) MLE and (b) entropy.

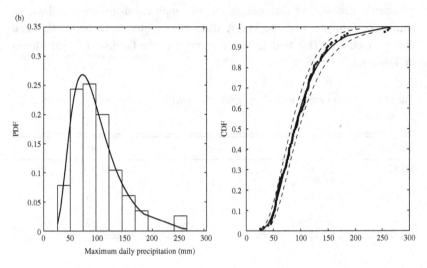

Figure 3.8 (*cont.*)

3.7 Conclusion

In this chapter, we discussed the Hal-A distribution, its parameter estimation with the use of entropy, MOM, MLE, and random variable simulation with the acceptance-rejection method. Using the true population, the acceptance-rejection method is found to be successful for random variable simulation. The parameters estimated using all three methods converge and may be applied for parameter estimation. Furthermore, using the real-world peak flow, total flow deficit, and maximum daily precipitation as examples, it has been determined that the Hal-A distribution may be properly applied for frequency analysis. It is worth noting that we need to investigate the sample moment ratio before proceeding to the analysis.

References

Abramovitz, M. and Stegun, I. (1972). *Handbook of Mathematical Functions. With Formulas, Graphs, and Mathematical Tables*. Dover Publications, New York.

Dvorak, V., Bobée, B., Boucher, S., and Ashkar, F. (1988). Halphen distributions and related systems of frequency functions research report No. R-236 INRS-Eau.Ste-Foy. Quebec, Canada, 81 pp.

El Adlouni, S. and Bobée, B. (2017). *Halphen Distribution Family: With Application in Hydrological Frequency Analysis.* Water Resources Publications, LLC, Highland Ranch, CO.

Fateh, C., Salaheddine E.A., and Bernard, B. (2010). Mixed estimation methods for Halphen distributions with applications in extreme hydrologic events. *Stochastic Environmental Research and Risk Assessment,* Vol. 24, No. 3, pp. 359–376.

Fitzgerald, D.L. (2000). Statistical aspects of Tricomiis function and modified Bessel functions of the second kind. *Stochastic Environmental Research and Risk Assessment,* Vol. 14, pp. 139–158.

Gumbel, E.J. (1958). *Statistics of Extremes.* Columbia University Press, New York. doi: 10.7312/gumb92958.

Halphen, E. (1941). Sur un nouveau type de couBRe de fréquence. *Comptes Rendus de l'Académie des Sciences,* Vol. 213, pp. 633–635. Published under the name of "Dugué" due to war constraints.

Halphen, E. (1955). Les fonctions factorielles. *Publications de l'Institut de Statistique de l'Université de Paris,* Vol. IV, Fascicule I, pp. 21–39.

Ouarda, T.B.M.J., Ashkar, F., Ben Said, E.M., and Hourani, L. (1994). Distribution statistiques utilisees on hydrologie: Transformation et proprieties asymptotiques. Rapport de recherché STAT-13. University of Moncton, N.B., Canada.

Perreault, L., Bobée, B., and Rasmussen, P. (1999). Halphen distribution system. I: Mathematical and statistical properties. *Journal of Hydrologic Engineering,* 4, No. 3, pp. 189–199.

Shannon, C.E. (1948). A mathematical theory of communication. *Bell System Technical Journal,* Vol. 27, No. 3, pp. 379–423.

Sheshadri, V. (1993). *The Inverse Gaussian Distribution.* Clarendon Press, Oxford.

Sheshadri, V. (1997). Halphen's laws. In S. Kotz and N.L. Johnson (eds.), *Encyclopedia of Statistical Sciences,* Update, Vol. 1. John Wiley & Sons, New York, pp. 302–306.

Singh, V.P. (1998). *Entropy Based Parameter Estimation in Hydrology.* Kluwer Academic Publishers, Dordrecht.

Singh, V.P. (2013). *Entropy Theory and Its Application in Environmental and Water Engineering.* Wiley-Blackwell, Hoboken, NJ.

Watson, G.N.A. (1966). *Treatise on the Theory of Bessel Functions,* 2nd ed. Cambridge University Press, Cambridge, England, 804 pp.

4

Halphen Type B Distribution

4.1 Introduction

Perreault et al. (1999) discussed the Halphen type B (Hal-B) distribution for frequency analysis of hydrometeorological extremes. They concluded that this distribution provided sufficient flexibility to fit a large variety of datasets. The Hal-B distribution exhibits greater flexibility in modeling smaller values and fits observations that are independent and identically distributed (Fateh et al. 2010). It has recently been shown that this distribution represents a potential alternative to the generalized extreme value distribution to model extreme hydrometeorological and hydrological events (Fateh et al. 2010). In this chapter the Hal-B distribution is derived using the entropy theory.

4.2 Hal-B Distribution and Its Characteristics

The probability density function (PDF) of the Hal-B distribution is given by

$$f(x) = \frac{2}{m^{2v} ef_v(\alpha)} x^{2v-1} \exp\left(-\left(\frac{x}{m}\right)^2 + \alpha\left(\frac{x}{m}\right)\right); \quad m > 0, \ \alpha \in \mathbb{R}, \ v > 0,$$

$$(4.1)$$

where $ef_v()$ is the exponential factorial function. For $v > 0$, $ef_v(\alpha)$ is defined as

$$ef_v(\alpha) = 2 \int_0^\infty x^{2v-1} \exp(-x^2 + \alpha x) dx. \qquad (4.2)$$

In general, the $ef_\nu(\alpha)$ function can be represented by an infinite series with the use of gamma functions as

$$ef_\nu(\alpha) = \Gamma(\nu) + \Gamma\left(\nu + \frac{1}{2}\right)\alpha + \Gamma(\nu + 1)\frac{\alpha}{2!} + \cdots + \Gamma\left(\nu + \frac{r}{2}\right)\frac{\alpha}{r!} + \cdots.$$

(4.3)

The $ef_\nu(\alpha)$ function is related to the parabolic cylinder function $U(\gamma, \beta)$ and the confluent hypergeometric function $M(a, b, z)$ (Abramowitz and Stegun, 1972) as

$$ef_\nu(\alpha) = \frac{\Gamma(2\nu)\exp\left(\dfrac{\alpha^2}{8}\right)}{2^{\nu-1}}U\left(2\nu - \frac{1}{2}, -\frac{\alpha}{\sqrt{2}}\right)$$

(4.4)

$$ef_\nu(\alpha) = \Gamma(\nu)M\left(\nu, \frac{1}{2}, \frac{\alpha^2}{4}\right) + \alpha\Gamma\left(\nu + \frac{1}{2}\right)M\left(\nu + \frac{1}{2}, \frac{3}{2}, \frac{\alpha^2}{4}\right),$$

(4.5)

where

$$U(\gamma, \beta) = \exp\frac{\left(-\dfrac{\beta^2}{4}\right)}{\Gamma\left(\gamma + \dfrac{1}{2}\right)}\int_0^\infty t^{\gamma - \frac{1}{2}}\exp\left(-\beta t - \frac{t^2}{2}\right)dt$$

(4.6)

$$M(a, b, z) = \frac{\Gamma(b)}{\Gamma(b - a)\Gamma(a)}\int_0^1 \exp(zt)t^{a-1}(1 - t)^{b-a-1}dt.$$

(4.7)

The $ef_\nu(\alpha)$ function follows the recurrence relations as

$$ef_{\nu+1}(\alpha) = \frac{\alpha}{2}ef_{\nu+\frac{1}{2}}(\alpha) + ef_\nu(\alpha)$$

(4.8)

$$\frac{\partial^n ef_\nu(\alpha)}{\partial \alpha^n} = ef_{\nu+\frac{n}{2}}(\alpha).$$

(4.9)

The $ef_\nu(\alpha)$ function can be asymptotically approximated (Halphen, 1955) for small values of α as

$$ef_\nu(\alpha) = 2\Gamma(2\nu)\frac{1}{|\alpha|^{2\nu}}\left(1 - \frac{2\nu(2\nu + 1)}{1}\frac{1}{\alpha^2} + \cdots\right)$$

(4.10)

and for large values of α as

$$ef_\nu(\alpha) = 2\sqrt{\pi}\left(\frac{\alpha}{2}\right)^{2\nu-1}\exp\left(\frac{\alpha^2}{4}\right)\left(1 + \frac{(2\nu - 1)(2\nu - 2)}{\alpha^2} + \cdots\right)$$

(4.11)

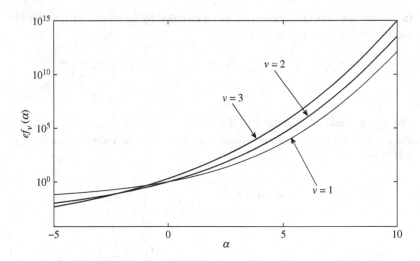

Figure 4.1 Exponential factorial function.

Figure 4.1 illustrates the exponential factorial function with different v and α. It is seen that the ef_v function shows exponential growth.

4.3 Differential Equation for Hal-B Distribution and Distribution Characteristics

The differential equation for deriving the Hal-B distribution can be written as

$$\frac{1}{f(x)}\frac{df(x)}{dx} = \frac{(2v-1) + \left(\frac{\alpha}{m}\right)x - \left(\frac{2}{m^2}\right)x^2}{x}. \tag{4.12}$$

Equating the derivative term $(df(x)/dx)$ to zero and solving the quadratic equation, we obtain

$$x = \frac{m\left(\alpha \pm \sqrt{\alpha^2 + 8(2v-1)}\right)}{4}. \tag{4.13}$$

Equation (4.13) prescribes the mode (with the plus sign in the numerator) and antimode (with the minus sign) of the Hal-B distribution, that is,

$$x = m\left(\frac{\alpha}{4} + \sqrt{\left(\frac{\alpha}{4}\right)^2 + v - \frac{1}{2}}\right) \tag{4.14}$$

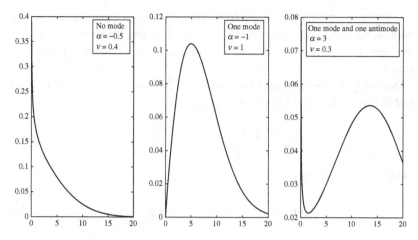

Figure 4.2 Hal-B distributions.

and

$$x = m\left(\frac{\alpha}{4} - \sqrt{\left(\frac{\alpha}{4}\right)^2 + v - \frac{1}{2}}\right). \tag{4.15}$$

Equation (4.14) shows that if $v = 0.5$, then the mode occurs at $m\alpha/2$. As shown in Figure 4.2, the Hal-B distribution exhibits three types of shapes as follows: (1) the distribution has no mode if $\alpha < 0$ and $v < 0.5$; (2) the distribution is unimodal if $v > 0.5$ or $v = 0.5$ and $\alpha > 0$; and (3) the distribution has one mode and one antimode if $\alpha > 0$, $v < 0.5$, and $(-\alpha/4)^2 < 0.5 - v$.

Perreault et al. (1999) have shown that gamma and normal distributions can be obtained as the limiting forms of the Hal-B distribution in what follows. Let $a = -\alpha m$ and $b = -\alpha/m$, the PDF of the Hal-B distribution, that is, Equation (4.1), can be rewritten as

$$f(x) = \frac{2}{\left(\frac{a}{b}\right)^v ef_v(-\sqrt{ab})} x^{2v-1} \exp\left(-\frac{b}{a}x^2 - bx\right). \tag{4.16}$$

As $a \to \infty$, applying Equation (4.10) we have

$$\lim_{a \to \infty} \frac{2}{\left(\frac{a}{b}\right)^v ef_v(-\sqrt{ab})} = \frac{b^{2v}}{\Gamma(2v)} \tag{4.17}$$

$$\lim_{a \to \infty} f(x) = \frac{b^{2v}}{\Gamma(2v)} x^{2v-1} \exp(-bx). \tag{4.18}$$

As shown in Equation (4.18), the limiting distribution is a gamma distribution with scale parameter $b = -\alpha/m$, $\alpha < 0$ and shape parameter $2v$, $v > 0$. In other words, the limiting gamma distribution is obtained if $\alpha m \to -\infty$ (i.e., $\alpha \to -\infty$).

Applying the linear transformation $y = \sqrt{2}((x/m) - (\alpha/2))$, the PDF of Hal-B distribution may be rewritten as

$$f(y) = \frac{\sqrt{2}}{ef_v(\alpha)} \left(\frac{y}{\sqrt{2}} + \frac{\alpha}{2} \right)^{2v-1} \exp\left(-\frac{y^2}{2} + \frac{\alpha^2}{4} \right). \tag{4.19}$$

Now if $\alpha \to \infty$, we have

$$\lim_{\alpha \to \infty} ef_v(\alpha) = 2\sqrt{\pi} \left(\frac{\alpha}{2} \right)^{2v-1} \exp\left(\frac{\alpha^2}{4} \right). \tag{4.20}$$

Substituting Equation (4.20) in Equation (4.19), we obtain the limiting distribution as

$$\lim_{\alpha \to \infty} f(y) = \lim_{\alpha \to \infty} \frac{1}{\sqrt{2\pi}} \left(\frac{\frac{y}{\sqrt{2}} + \frac{\alpha}{2}}{\frac{\alpha}{2}} \right)^{2v-1} \exp\left(-\frac{\alpha^2}{4} \right) \exp\left(-\frac{y^2}{2} + \frac{\alpha^2}{4} \right)$$

$$= \lim_{\alpha \to \infty} \frac{1}{\sqrt{2\pi}} \left(\frac{\sqrt{2}y}{\alpha} \right)^{2v-1} \exp\left(-\frac{y^2}{2} \right) \tag{4.21}$$

$$\Rightarrow \lim_{\alpha \to \infty} f(y) = \frac{1}{\sqrt{2\pi}} \exp\left(-\frac{y^2}{2} \right).$$

It is shown in Equation (4.21) that the Hal-B distribution converges to the standard normal distribution if $\alpha \to \infty$.

4.4 Derivation of Hal-B Distribution by Entropy Theory

Four steps are involved in deriving the distribution using the Shannon entropy (Singh, 1998): (1) a set of constraints is defined; (2) the entropy is maximized using the method of Lagrange multipliers; (3) an entropy-based distribution is obtained, which is least biased in accordance with the principle of maximum entropy; and (4) the Lagrange multipliers are related to the distribution parameters.

For a continuous positive random variable X, such as streamflow, the Shannon entropy of X, $H(X)$, can be expressed as (Shannon, 1948)

$$H(x) = -\int_0^\infty f(x)\ln f(x)dx. \tag{4.22}$$

where $f(x)$ is the PDF of X, which ranges from 0 to ∞. The objective is to derive $f(x)$ by maximizing entropy subject to given information, expressed as constraints, in concert with the principle of maximum entropy (Singh, 2013).

4.4.1 Specification of Constraints

Substituting Equation (4.1) in Equation (4.22), we get

$$H(x) = -\int_0^\infty f(x)\ln\left(\frac{2}{m^{2v}ef_v(\alpha)}x^{2v-1}\exp\left(-\left(\frac{x}{m}\right)^2 + \alpha\left(\frac{x}{m}\right)\right)\right)dx. \quad (4.23)$$

Equation (4.23) may be rewritten as

$$H(x) = -\ln\left(\frac{2}{m^{2v}ef_v(\alpha)}\right) - (2v-1)E(\ln x) + \frac{1}{m^2}E(x^2) - \frac{\alpha}{m}E(x). \quad (4.24)$$

Therefore, the constraints for the Hal-B distribution can be expressed as

$$\int_0^\infty f(x)dx = 1 \quad (4.25)$$

$$\int_0^\infty xf(x)dx = E(x) \quad (4.26)$$

$$\int_0^\infty (\ln x)f(x)dx = E(\ln x) \quad (4.27)$$

$$\int_0^\infty x^2 f(x) = E(x^2) = \text{var}(x) + (E(x))^2. \quad (4.28)$$

4.4.2 Entropy Maximizing

Using the method of Lagrange multipliers, the Lagrangian function L can be expressed as

$$L = -\int_0^\infty f(x)\ln f(x)dx - (\lambda_0 - 1)\left(\int_0^\infty f(x)dx - 1\right) - \lambda_1\left(\int_0^\infty xf(x)dx - E(x)\right)$$
$$- \lambda_2\left(\int_0^\infty (\ln x)f(x)dx - E(\ln x)\right) - \lambda_3\left(\int_0^\infty x^2 f(x)dx - E(x^2)\right).$$

$$(4.29)$$

Differentiating Equation (4.29) with respect to $f(x)$ and equating the derivative to zero, we obtain

$$\frac{\partial L}{\partial f(x)} = 0 = -\ln(f(x)) - \lambda_0 - \lambda_1 x - \lambda_2(\ln x) - \lambda_3 x^2. \tag{4.30}$$

Equation (4.30) yields the entropy-based least-biased PDF of the Hal-B distribution:

$$f(x) = \exp(-\lambda_0 - \lambda_1 x - \lambda_2 \ln x - \lambda_3 x^2). \tag{4.31}$$

4.4.3 Relation between Lagrange Multipliers and Distribution Parameters

Substituting Equation (4.31) in Equation (4.25), we get

$$\int_0^\infty f(x)dx = \int_0^\infty \exp(-\lambda_0 - \lambda_1 x - \lambda_2 \ln x - \lambda_3 x^2)dx = 1. \tag{4.32}$$

From Equation (4.32), we have

$$\exp(\lambda_0) = \int_0^\infty \exp(-\lambda_1 x - \lambda_2 \ln x - \lambda_3 x^2)dx = \int_0^\infty x^{-\lambda_2} \exp(-\lambda_1 x - \lambda_3 x^2)dx.$$

$$\tag{4.33}$$

Let $t = \sqrt{\lambda_3} x \Rightarrow x = t/\sqrt{\lambda_3} \Rightarrow dx = (1/\sqrt{\lambda_3})dt$. With the use of these quantities, Equation (4.33) can be expressed as

$$\exp(\lambda_0) = (\lambda_3)^{\frac{\lambda_2-1}{2}} \int_0^\infty t^{-\lambda_2} \exp\left(-t^2 - \frac{\lambda_1}{\sqrt{\lambda_3}}t\right)dt. \tag{4.34}$$

Equation (4.34) may be expressed using the exponential factorial function as

$$\exp(\lambda_0) = \frac{1}{2}(\lambda_3)^{\frac{\lambda_2-1}{2}} ef_{\frac{1-\lambda_2}{2}}\left(-\frac{\lambda_1}{\sqrt{\lambda_3}}\right). \tag{4.35}$$

In Equation (4.35), if we set $v = (1 - \lambda_2)/2$ and $\alpha = -\lambda_1/\sqrt{\lambda_3}$, Equation (4.26) is rewritten as

$$\exp(\lambda_0) = \frac{1}{2}(\lambda_3)^{-v} ef_v(\alpha). \tag{4.36}$$

Substitution of Equation (4.35) in Equation (4.31) yields

$$f(x) = \frac{2\lambda_3^{\frac{1-\lambda_2}{2}} x^{-\lambda_2}}{ef_{\frac{1-\lambda_2}{2}}\left(-\frac{\lambda_1}{\sqrt{\lambda_3}}\right)} \exp(-\lambda_1 x - \lambda_3 x^2). \tag{4.37}$$

Let $\lambda_1 = -\alpha/m$, $\lambda_3 = 1/m^2$; the entropy-based Hal-B distribution, that is, Equation (4.37), can be expressed in exactly the same way as Equation (4.1). Thus, the relations between Lagrange multipliers and distribution parameters can be summarized as

$$\lambda_1 = -\frac{\alpha}{m} \tag{4.38}$$

$$\lambda_2 = 1 - 2v \tag{4.39}$$

$$\lambda_3 = \frac{1}{m^2}. \tag{4.40}$$

4.5 Parameter Estimation

The Hal-B distribution parameters are estimated using the regular entropy method, parameter space expansion method, method of moments (MOM), and method of maximum likelihood estimation (MLE).

4.5.1 Regular Entropy Method

This method involves deriving (1) relations between Lagrange multipliers and constraints, (2) relations between Lagrange multipliers and parameters, and (3) relations between parameters and constraints.

4.5.1.1 Relations between Lagrange Multipliers and Constraints
The Lagrange multiplier λ_0 from Equation (4.35) can be expressed as

$$\lambda_0 = -\ln 2 + \frac{\lambda_2 - 1}{2} \ln \lambda_3 + \ln\left(ef_{\frac{1-\lambda_2}{2}}\left(-\frac{\lambda_1}{\sqrt{\lambda_3}}\right)\right). \tag{4.41}$$

From Equation (4.31), λ_0 can also be defined as

$$\lambda_0 = \ln\left(\int_0^\infty \exp(-\lambda_1 x - \lambda_2 \ln x - \lambda_3 x^2) dx\right). \tag{4.42}$$

Differentiating Equation (4.41) with respect to λ_1, λ_2, and λ_3, we obtain

$$\frac{\partial \lambda_0}{\partial \lambda_1} = \frac{1}{ef_{\frac{1-\lambda_2}{2}}\left(-\frac{\lambda_1}{\sqrt{\lambda_3}}\right)} \frac{\partial ef_{\frac{1-\lambda_2}{2}}\left(-\frac{\lambda_1}{\sqrt{\lambda_3}}\right)}{\partial \lambda_1} = -\frac{1}{\sqrt{\lambda_3}}\left(\frac{ef_{\left(\frac{2-\lambda_2}{2}\right)}\left(-\frac{\lambda_1}{\sqrt{\lambda_3}}\right)}{ef_{\frac{1-\lambda_2}{2}}\left(-\frac{\lambda_1}{\sqrt{\lambda_3}}\right)}\right)$$

(4.43)

$$\frac{\partial \lambda_0}{\partial \lambda_2} = \frac{1}{2}\ln \lambda_3 + \frac{1}{ef_{\frac{1-\lambda_2}{2}}\left(-\frac{\lambda_1}{\sqrt{\lambda_3}}\right)} \frac{\partial ef_{\frac{1-\lambda_2}{2}}\left(-\frac{\lambda_1}{\sqrt{\lambda_3}}\right)}{\partial \lambda_2}$$

(4.44)

$$\frac{\partial \lambda_0}{\partial \lambda_3} = \frac{\lambda_2-1}{2\lambda_3} + \frac{1}{ef_{\frac{1-\lambda_2}{2}}\left(-\frac{\lambda_1}{\sqrt{\lambda_3}}\right)} \frac{\partial ef_{\frac{1-\lambda_2}{2}}\left(-\frac{\lambda_1}{\sqrt{\lambda_3}}\right)}{\partial \lambda_3} = \frac{\lambda_2-1}{2\lambda_3} + \frac{\lambda_1 \lambda_3^{-\frac{3}{2}}}{2}\left(\frac{ef_{\frac{2-\lambda_2}{2}}\left(-\frac{\lambda_1}{\sqrt{\lambda_3}}\right)}{ef_{\frac{1-\lambda_2}{2}}\left(-\frac{\lambda_1}{\sqrt{\lambda_3}}\right)}\right).$$

(4.45)

In Equation (4.44), $\left(\partial ef_{\frac{1-\lambda_2}{2}}(-\lambda_1/\sqrt{\lambda_3})\right)/\partial \lambda_2$ can be expressed using Equation (4.2) or (4.3) as

$$\frac{\partial ef_{\frac{1-\lambda_2}{2}}\left(-\frac{\lambda_1}{\sqrt{\lambda_3}}\right)}{\partial \lambda_2} = -2\int_0^\infty x^{-\lambda_2}\ln x \exp\left(-x^2 - \frac{\lambda_1}{\sqrt{\lambda_3}}x\right)dx.$$

(4.46)

Now differentiating Equation (4.33) with respect to λ_1, λ_2, and λ_3, the result is

$$\frac{\partial \lambda_0}{\partial \lambda_1} = -\frac{\int_0^\infty x\exp(-\lambda_1 x - \lambda_2\ln x - \lambda_3 x^2)dx}{\int_0^\infty \exp(-\lambda_1 x - \lambda_2\ln x - \lambda_3 x^2)dx} = -E(x)$$

(4.47)

$$\frac{\partial \lambda_0}{\partial \lambda_2} = -\frac{\int_0^\infty (\ln x)\exp(-\lambda_1 x - \lambda_2\ln x - \lambda_3 x^2)dx}{\int_0^\infty \exp(-\lambda_1 x - \lambda_2\ln x - \lambda_3 x^2)dx} = -E(\ln x)$$

(4.48)

$$\frac{\partial \lambda_0}{\partial \lambda_2} = -\frac{\int_0^\infty x^2 \exp(-\lambda_1 x - \lambda_2\ln x - \lambda_3 x^2)dx}{\int_0^\infty \exp(-\lambda_1 x - \lambda_2\ln x - \lambda_3 x^2)dx} = -E(x^2).$$

(4.49)

Equating Equations (4.43)–(4.45) to Equations (4.47)–(4.49), the relations between Lagrange multipliers and constraints are obtained, respectively, as

$$-\frac{1}{\sqrt{\lambda_3}}\left(\frac{ef_{\left(\frac{2-\lambda_2}{2}\right)}\left(-\frac{\lambda_1}{\sqrt{\lambda_3}}\right)}{ef_{\frac{1-\lambda_2}{2}}\left(-\frac{\lambda_1}{\sqrt{\lambda_3}}\right)}\right) = -E(x) \tag{4.50}$$

$$\frac{1}{2}\ln\lambda_3 + \frac{1}{ef_{\frac{1-\lambda_2}{2}}\left(-\frac{\lambda_1}{\sqrt{\lambda_3}}\right)}\frac{\partial ef_{\frac{1-\lambda_2}{2}}\left(-\frac{\lambda_1}{\sqrt{\lambda_3}}\right)}{\partial\lambda_2} = -E(\ln x) \tag{4.51}$$

$$\frac{\lambda_2 - 1}{2\lambda_3} + \frac{\lambda_1\lambda_3^{-\frac{3}{2}}}{2}\left(\frac{ef_{\frac{2-\lambda_2}{2}}\left(-\frac{\lambda_1}{\sqrt{\lambda_3}}\right)}{ef_{\frac{1-\lambda_2}{2}}\left(-\frac{\lambda_1}{\sqrt{\lambda_3}}\right)}\right) = -E(x^2). \tag{4.52}$$

The Lagrange multipliers may then be optimized by solving the system of equations. Additionally, Equations (4.38)–(4.40) denote the relation between the Lagrange multipliers and the distribution parameters.

4.5.2 Parameter Space Expansion Method

In the case of the parameter space expansion method, the constraints are given as

$$\int_0^\infty f(x)dx = 1 \tag{4.53}$$

$$\int_0^\infty \left(\frac{a}{m}x\right) f(x)dx = E\left(\frac{a}{m}x\right) \tag{4.54}$$

$$\int_0^\infty ((2v-1)\ln x)f(x)dx = E((2v-1)\ln x) \tag{4.55}$$

$$\int_0^\infty \left(\frac{x}{m}\right)^2 f(x)dx = E\left(\left(\frac{x}{m}\right)^2\right). \tag{4.56}$$

Following the same procedure as in the regular entropy method, the entropy-based PDF of the Hal-B distribution can be written as

$$f(x) = \exp\left(-\lambda_0 - \lambda_1\left(\frac{\alpha}{m}x\right) - \lambda_2(2v-1)\ln x - \lambda_3\left(\frac{x}{m}\right)^2\right). \qquad (4.57)$$

Substituting Equation (4.57) in Equation (4.53), we have

$$\exp(\lambda_0) = \int_0^\infty x^{-\lambda_2(2v-1)} \exp\left(-\lambda_3\left(\frac{x}{m}\right)^2 - \lambda_1\alpha\left(\frac{x}{m}\right)\right) dx. \qquad (4.58)$$

In Equation (4.58), given $\lambda_3 > 0$ and let $y = \sqrt{\lambda_3}(x/m) \Rightarrow x = (m/\sqrt{\lambda_3})y \Rightarrow dx = (m/\sqrt{\lambda_3})dy$, Equation (4.58) can then be solved as

$$\exp(\lambda_0) = \frac{1}{2}\left(\frac{m}{\sqrt{\lambda_3}}\right)^{\left(1-\lambda_2(2v-1)\right)} ef_{\frac{-\lambda_2(2v-1)+1}{2}}\left(-\frac{\lambda_1\alpha}{\sqrt{\lambda_3}}\right). \qquad (4.59)$$

Substituting Equation (4.57) in Equation (4.22), we get

$$H(x) = -\ln 2 + \left(1 - \lambda_2(2v-1)\right)\ln\left(\frac{m}{\sqrt{\lambda_3}}\right) + \ln ef_{\frac{-\lambda_2(2v-1)+1}{2}}\left(-\frac{\lambda_1\alpha}{\sqrt{\lambda_3}}\right) + \lambda_1\alpha E\left(\frac{x}{m}\right)$$
$$+ \lambda_2(2v-1)E(\ln x) + \lambda_3 E\left(\left(\frac{x}{m}\right)^2\right). \qquad (4.60)$$

Differentiating Equation (4.60) with respect to the Lagrange multipliers and the distribution parameters, we have

$$\frac{\partial H}{\partial \lambda_1} = 0 = \frac{1}{ef_{\frac{-\lambda_2(2v-1)+1}{2}}\left(-\frac{\lambda_1\alpha}{\sqrt{\lambda_3}}\right)} \frac{\partial ef_{\frac{-\lambda_2(2v-1)+1}{2}}\left(-\frac{\lambda_1\alpha}{\sqrt{\lambda_3}}\right)}{\partial \lambda_1} + E\left(\frac{\alpha x}{m}\right)$$

$$\Rightarrow -\frac{\alpha}{\sqrt{\lambda_3}}\left(\frac{ef_{\frac{-\lambda_2(2v-1)+2}{2}}\left(-\frac{\lambda_1\alpha}{\sqrt{\lambda_3}}\right)}{ef_{\frac{-\lambda_2(2v-1)+1}{2}}\left(-\frac{\lambda_1\alpha}{\sqrt{\lambda_3}}\right)}\right) + \alpha E\left(\frac{x}{m}\right) = 0 \qquad (4.61)$$

$$\frac{\partial H}{\partial \lambda_2} = 0 = -(2v-1)\ln\left(\frac{m}{\sqrt{\lambda_3}}\right) + \frac{\partial ef_{\frac{-\lambda_2(2v-1)+1}{2}}\left(-\frac{\lambda_1\alpha}{\sqrt{\lambda_3}}\right)}{\partial \lambda_2} + (2v-1)E(\ln x).$$
$$\frac{ef_{\frac{-\lambda_2(2v-1)+1}{2}}\left(-\frac{\lambda_1\alpha}{\sqrt{\lambda_3}}\right)} \qquad (4.62)$$

Let $V1 = (-\lambda_2(2v-1)+1)/2$. Equation (4.62) may be rewritten as

$$\frac{\partial H}{\partial \lambda_2} = 0 = -(2v-1)\ln\left(\frac{m}{\sqrt{\lambda_3}}\right) - \frac{2v-1}{2} \frac{\frac{\partial ef_{V1}\left(-\frac{\lambda_1 \alpha}{\sqrt{\lambda_3}}\right)}{\partial V1}}{ef_{\frac{-\lambda_2(2v-1)+1}{2}}\left(-\frac{\lambda_1 \alpha}{\sqrt{\lambda_3}}\right)} + (2v-1)E(\ln x)$$

$$\Rightarrow (2v-1)E(\ln x) - (2v-1)\ln\left(\frac{m}{\sqrt{\lambda_3}}\right) - \frac{2v-1}{2} \frac{\frac{\partial ef_{V1}\left(-\frac{\lambda_1 \alpha}{\sqrt{\lambda_3}}\right)}{\partial V1}}{ef_{\frac{-\lambda_2(2v-1)+1}{2}}\left(-\frac{\lambda_1 \alpha}{\sqrt{\lambda_3}}\right)} = 0$$

(4.63)

$$\frac{\partial H}{\partial \lambda_3} = 0 = -\frac{1-\lambda_2(2v-1)}{2\lambda_3} + \frac{1}{ef_{\frac{-\lambda_2(2v-1)+1}{2}}\left(-\frac{\lambda_1 \alpha}{\sqrt{\lambda_3}}\right)} \frac{\partial ef_{\frac{-\lambda_2(2v-1)+1}{2}}\left(-\frac{\lambda_1 \alpha}{\sqrt{\lambda_3}}\right)}{\partial \lambda_3} + E\left(\left(\frac{x}{m}\right)^2\right)$$

$$\Rightarrow -\frac{1-\lambda_2(2v-1)}{2\lambda_3} + \frac{\lambda_1}{2\lambda_3\sqrt{\lambda_3}}\left(\frac{\partial ef_{\frac{-\lambda_2(2v-1)+2}{2}}\left(-\frac{\lambda_1 \alpha}{\sqrt{\lambda_3}}\right)}{\partial ef_{\frac{-\lambda_2(2v-1)+1}{2}}\left(-\frac{\lambda_1 \alpha}{\sqrt{\lambda_3}}\right)}\right) + E\left(\left(\frac{x}{m}\right)^2\right) = 0$$

(4.64)

$$\frac{\partial H}{\partial \alpha} = 0 = \frac{1}{ef_{\frac{-\lambda_2(2v-1)+1}{2}}\left(-\frac{\lambda_1 \alpha}{\sqrt{\lambda_3}}\right)} \frac{\partial ef_{\frac{-\lambda_2(2v-1)+1}{2}}\left(-\frac{\lambda_1 \alpha}{\sqrt{\lambda_3}}\right)}{\partial \alpha} + \lambda_1 E\left(\frac{x}{m}\right)$$

$$\Rightarrow -\frac{\lambda_1}{\sqrt{\lambda_3}}\left(\frac{ef_{\frac{-\lambda_2(2v-1)+2}{2}}\left(-\frac{\lambda_1 \alpha}{\sqrt{\lambda_3}}\right)}{ef_{\frac{-\lambda_2(2v-1)+1}{2}}\left(-\frac{\lambda_1 \alpha}{\sqrt{\lambda_3}}\right)}\right) + \lambda_1 E\left(\frac{x}{m}\right) = 0$$

(4.65)

$$\frac{\partial H}{\partial m} = 0 = -\frac{1-\lambda_2(2v-1)}{m} - \frac{\lambda_1 \alpha}{m}E\left(\frac{x}{m}\right) - \frac{2\lambda_3}{m}E\left(\left(\frac{x}{m}\right)^2\right)$$

(4.66)

$$\frac{\partial H}{\partial v} = -2\lambda_2\ln\left(\frac{m}{\sqrt{\lambda_3}}\right) + \frac{1}{ef_{\frac{-\lambda_2(2v-1)+1}{2}}\left(-\frac{\lambda_1 \alpha}{\sqrt{\lambda_3}}\right)} \frac{\partial ef_{\frac{-\lambda_2(2v-1)+1}{2}}\left(-\frac{\lambda_1 \alpha}{\sqrt{\lambda_3}}\right)}{\partial v} + 2\lambda_2 E(\ln x).$$

(4.67)

Let $V1 = (-\lambda_2(2v-1)+1)/2$. Equation (4.67) may be rewritten as

$$\frac{\partial H}{\partial v} = -2\lambda_2 \ln\left(\frac{m}{\sqrt{\lambda_3}}\right) - \lambda_2 \frac{\partial ef_{V1}\left(-\frac{\lambda_1\alpha}{\sqrt{\lambda_3}}\right)}{ef_{-\frac{\lambda_2(2v-1)}{2}+1}\left(-\frac{\lambda_1\alpha}{\sqrt{\lambda_3}}\right)} + 2\lambda_2 E(\ln x). \qquad (4.68)$$

Substituting Equation (4.65) in Equation (4.66), we have

$$\frac{\partial H}{\partial m} = \frac{1 - \lambda_2(2v-1)}{m} - \frac{\alpha}{m\sqrt{\lambda_3}}\left(\frac{\partial ef_{-\frac{\lambda_2(2v-1)}{2}+2}\left(-\frac{\lambda_1\alpha}{\sqrt{\lambda_3}}\right)}{\partial ef_{-\frac{\lambda_2(2v-1)}{2}+1}\left(-\frac{\lambda_1\alpha}{\sqrt{\lambda_3}}\right)}\right) - \frac{2\lambda_3}{m}E\left(\left(\frac{x}{m}\right)^2\right) = 0.$$

$$(4.69)$$

Comparing Equation (4.64) with Equation (4.69), we have $\lambda_3 = 2m$.
Comparing Equation (4.61) with Equation (4.65), we have $\lambda_1 = \alpha$.
Comparing Equation (4.63) with Equation (4.68), we have $\lambda_2 = 2v - 1$.

To this end, six equations may be reduced to three equations as

$$E\left(\frac{x}{m}\right) = \frac{\alpha}{\sqrt{2m}}\left(\frac{ef_{-\frac{(2v-1)^2}{2}+2}\left(-\frac{\alpha^2}{\sqrt{2m}}\right)}{ef_{-\frac{(2v-1)^2}{2}+1}\left(-\frac{\alpha^2}{\sqrt{2m}}\right)}\right) \qquad (4.70)$$

$$E(\ln(x)) = -\frac{1}{2}\ln 2 + \frac{1}{2}\frac{\partial ef_{V1}\left(-\frac{\alpha^2}{\sqrt{2m}}\right)}{ef_{-\frac{(2v-1)^2}{2}+1}\left(-\frac{\alpha^2}{\sqrt{2m}}\right)}; \quad V1 = \frac{-(2v-1)^2 + 1}{2}$$

$$(4.71)$$

$$E\left(\left(\frac{x}{m}\right)^2\right) + \frac{\alpha}{4m}E\left(\frac{x}{m}\right) = \frac{1 - (2v-1)^2}{4m}. \qquad (4.72)$$

Solving the system of Equations (4.70)–(4.72) numerically, the Lagrange multipliers and distribution parameters can be estimated using the Newton–Raphson method.

4.5.3 MOM

The rth moment of x/m for the Hal-B distribution, Equation (4.1), about the origin may be given as

$$E\left(\left(\frac{x}{m}\right)^r\right) = \int_0^\infty \left(\frac{x}{m}\right)^r f(x)dx = \frac{2}{m^{2v}ef_v(\alpha)} \int_0^\infty \left(\frac{2}{m}\right)^r x^{2v-1} \exp\left(-\left(\frac{x}{m}\right)^2 + \alpha\left(\frac{x}{m}\right)\right)dx$$

$$= \frac{2}{ef_v(\alpha)} \int_0^\infty \left(\frac{x}{m}\right)^{2v+r-1} \exp\left(-\left(\frac{x}{m}\right)^2 + \alpha\left(\frac{x}{m}\right)\right)d\left(\frac{x}{m}\right) = \frac{ef_{v+\frac{r}{2}}(\alpha)}{ef_v(\alpha)}.$$

$$(4.73)$$

Equivalently, Equation (4.73) may be rewritten as

$$E(x^r) = \frac{m^r ef_{v+\frac{r}{2}}(\alpha)}{ef_v(\alpha)}. \qquad (4.74)$$

Applying the general recursive relation for the exponential factorial function (El Adlouni and Bobée, 2017),

$$ef_{v+r}(\alpha) = \frac{\alpha}{2}ef_{v+r-\frac{1}{2}}(\alpha) + (v+r-1)ef_{v+r-1}(\alpha). \qquad (4.75)$$

Substituting Equation (4.75) in Equation (4.74), we have

$$E\left(\left(\frac{x}{m}\right)^r\right) = \frac{\frac{\alpha}{2}ef_{v+\frac{r}{2}-\frac{1}{2}}(\alpha) + \left(v+\frac{r}{2}-1\right)ef_{v+\frac{r}{2}-1}(\alpha)}{ef_v(\alpha)}$$

$$= \frac{\alpha}{2}E\left(\left(\frac{x}{m}\right)^{r-1}\right) + \left(v+\frac{r}{2}-1\right)E\left(\left(\frac{x}{m}\right)^{r-2}\right). \qquad (4.76)$$

Now from Equation (4.76), we have

$$E\left(\left(\frac{x}{m}\right)\right) = \frac{\alpha}{2} + \left(v-\frac{1}{2}\right)E\left(\left(\frac{x}{m}\right)^{-1}\right) \qquad (4.77)$$

$$E\left(\left(\frac{x}{m}\right)^2\right) = \frac{\alpha}{2}E\left(\left(\frac{x}{m}\right)\right) + v \qquad (4.78)$$

$$E\left(\left(\frac{x}{m}\right)^3\right) = \frac{\alpha}{2}E\left(\left(\frac{x}{m}\right)^2\right) + \left(v+\frac{1}{2}\right)E\left(\left(\frac{x}{m}\right)\right). \qquad (4.79)$$

Solving Equations (4.77)–(4.79), we have the parameters

$$m^2 = \frac{2\text{var}(x)}{2v\left(1 - E(x)E(x^{-1})\right) + E(x)E(x^{-1})} \qquad (4.80)$$

$$v = \frac{E(x)E(x^{-1})\left(E(x^3)E(x) - \left(E(x^2)\right)^2\right) - \text{var}(x)\left(E(x)\right)^2}{2\left(1 - E(x)E(x^{-1})\right)\left(\left(E(x^2)\right)^2 - E(x^3)E(x)\right) - \left(\text{Var}(x)\right)^2} \qquad (4.81)$$

$$\alpha = \frac{m\left(2v\left(E(x) - E(x^2)E(x^{-1})\right) + E(x^2)E(x^{-1})\right)}{\mathrm{var}(x)}. \qquad (4.82)$$

The gamma distribution again is the limiting distribution for the Hal-B distribution, as it is for the Halphen type A (Hal-A) distribution.

4.5.4 MLE Method

The log-likelihood function of the Hal-B distribution is expressed as

$$\ln L = n\left(\ln 2 - 2v\ln m - \ln ef_v(\alpha)\right) + (2v-1)\sum_{i=1}^{n}\ln x_i - \sum_{i=1}^{n}\left(\frac{x_i}{m}\right)^2 + \alpha\sum_{i=1}^{n}\left(\frac{x_i}{m}\right). \qquad (4.83)$$

Taking the derivative with respect to the distribution parameters and setting them equal to zero, we have

$$\frac{\partial \ln L}{\partial \alpha} = 0 = -\frac{n}{ef_v(\alpha)}\frac{\partial ef_v(\alpha)}{\partial \alpha} + \sum_{i=1}^{n}\left(\frac{x_i}{m}\right) \Rightarrow \frac{ef_{v+\frac{1}{2}}(\alpha)}{ef_v(\alpha)} = \frac{\bar{x}}{m} \qquad (4.84)$$

$$\frac{\partial \ln L}{\partial v} = 0 = -2n\ln m - \frac{n}{ef_v(\alpha)}\frac{\partial ef_v(\alpha)}{\partial v} + 2\sum_{i=1}^{n}\ln x_i \Rightarrow \ln m + \frac{2}{ef_v(\alpha)}\frac{\partial ef_v(\alpha)}{\partial v} = \overline{\ln x} \qquad (4.85)$$

$$\frac{\partial \ln L}{\partial m} = 0 = -\frac{2vn}{m} + \frac{2}{m^3}\sum_{i=1}^{n}x_i^2 - \frac{\alpha}{m^2}\sum_{i=1}^{n}x_i \Rightarrow -2v + \frac{2}{m^2}\overline{x^2} - \frac{\alpha}{m}\bar{x}. \qquad (4.86)$$

Substituting Equation (4.78) in Equation (4.86), Equation (4.86) may be rewritten as

$$\frac{\partial \ln L}{\partial m} = 0 \Rightarrow \overline{x^2} = m^2\left(v + \frac{\alpha}{2m}\bar{x}\right) = m^2\frac{ef_{v+1}(\alpha)}{ef_v(\alpha)}. \qquad (4.87)$$

Solving the system of equations (4.84)–(4.87), we may obtain the parameters.

4.6 Application

4.6.1 Simulating Hal-B Distributed Random Variables with Fixed Parameters

Similar to the discussion for the Hal-A distribution, the acceptance-rejection method is applied for random number simulation. From Equation (4.74), we obtain the coefficients of variation, skewness, and kurtosis as

$$C_\nu = \frac{\sqrt{ef_\nu(\alpha)ef_{\nu+1}(\alpha) - ef^2_{\nu+\frac{1}{2}}}}{ef_{\nu+\frac{1}{2}}} \tag{4.88}$$

$$C_s = \frac{ef^2_\nu(\alpha)ef_{\nu+\frac{3}{2}}(\alpha) - 3ef_\nu(\alpha)ef_{\nu+\frac{1}{2}}(\alpha)ef_{\nu+1}(\alpha) + 2ef^3_{\nu+\frac{1}{2}}}{\left(ef_\nu(\alpha)ef_{\nu+1}(\alpha) - ef^2_{\nu+\frac{1}{2}}\right)^{\frac{3}{2}}} \tag{4.89}$$

$$C_g = \frac{ef^3_\nu(\alpha)ef_{\nu+2}(\alpha) - 4ef^2_\nu(\alpha)ef_{\nu+\frac{1}{2}}(\alpha)ef_{\nu+\frac{3}{2}}(\alpha) + 6ef_\nu(\alpha)ef^2_{\nu+\frac{1}{2}}(\alpha)ef_{\nu+1}(\alpha) - 3ef^4_{\nu+\frac{1}{2}}(\alpha)}{\left(ef_\nu(\alpha)ef_{\nu+1}(\alpha) - ef^2_{\nu+\frac{1}{2}}\right)^2}.$$
$$\tag{4.90}$$

Using Equations (4.88) and (4.89), Figure 4.3 plots the moment ratio diagram of the Hal-B distribution. It is seen from the figure that the gamma distribution is again the limiting distribution for the Hal-B distribution.

According to El Adlouni and Bobée (2017), the mean and variance of the limiting distribution, that is, Equation (3.88) in Chapter 3, are determined based on the parameter ν as follows:

$$\begin{cases} E(Y) = E(X) \\ \text{var}(Y) = 5\text{var}(X) \end{cases}, \quad \text{if } \nu \leq 1 \tag{4.91}$$

$$\begin{cases} E(Y) = E(X) \\ \text{var}(Y) = 2\text{var}(X) \end{cases}, \quad \text{if } \nu > 1. \tag{4.92}$$

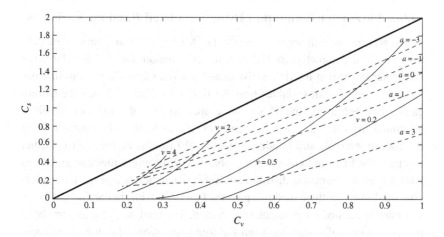

Figure 4.3 Moment ratio diagram of Hal-B distribution.

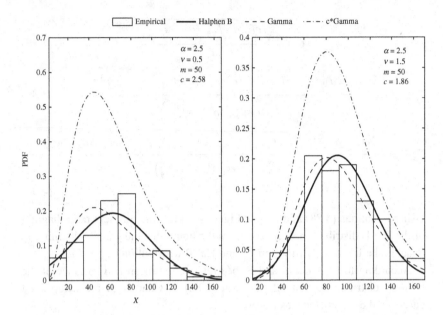

Figure 4.4 Random variables simulated from Hal-B distribution.

The simulation procedure is exactly the same as that discussed for the Hal-A distribution. To this end, with a different parameter v, Figure 4.4 illustrates the random variable simulated from the Hal-B distribution using the acceptance-rejection method.

4.6.2 Parameter Estimation Using Simulated Random Variables

In this section, we will apply the method of MLE, MOM, and entropy method to estimate the parameters for Hal-B distributed random variable. Similar to the parameter estimation for Hal-A distributed random variables, we will first use the simulated random variables from Section 4.6.1. Table 4.1 lists the parameters estimated using the above three methods for the random variables generated from $\alpha = 2.5$, $v = 0.5$, and $m = 50$, as well as the corresponding confidence intervals and goodness-of-fit (GoF) results. Figure 4.5a–c further compare the PDF and cumulative probability distribution function evaluated with the fitted parametric distribution. Table 4.1 indicates that the parameters estimated using all three methods pass the GoF tests with similar GoF statistics and may be applied for parameter estimation. The random variables simulated may be successfully sampled from the true population using the acceptance-rejection method. Figure 4.5a–c further confirm the results obtained.

Table 4.1. *Parameters estimated, confidence interval, and GoF results: synthetic dataset.*

		α	ν	m	
MLE		2.86	0.48	49.64	
Confidence interval	(1)	[2.57, 3.13]	[0.478, 0.484]	[49.55, 49.73]	
Confidence interval	(2)	[2.06, 4.54]	[−0.03, 0.69]	[34.76, 56.51]	
	GoF	$D_n = 0.0424, P = 0.414$			
MOM		3.08	0.50	46.91	
Confidence interval	(2)	[2.10, 3.54]	[0.30, 0.70]	[43.46, 54.97]	
	GoF	$D_n = 0.047, P = 0.75$			
		$\lambda_0{}^a$	λ_1	λ_2	λ_3
Entropy		−1.18	−0.06	0.0040	4.55E−04
Equations (4.38)–(4.40):		$\alpha = 2.86$, $\nu = 0.48$, and $m = 49.64$			
	(2)		[−0.08, −0.031]	[−0.39, 0.51]	[0.0003, 0.0005]
	GoF	$D_n = 0.042, P = 0.45$			

Notes: (1) Confidence interval constructed using Fisher's information.
(2) Confidence interval constructed using parametric bootstrap method.
[a] λ_0 is a function of $\{\lambda_1, \lambda_2, \lambda_3\}$ so its confidence interval is not evaluated.

4.6.3 Peak Flow

Previously, we have shown that the acceptance-rejection method may be applied to simulate the Hal-B distributed random variables. All three estimation methods (i.e., MLE, MOM, and entropy) may be successfully applied for parameter estimation. In this section, the peak flow data from USGS09239500 is used as an example to illustrate its application. Applying moment, MLE, and entropy methods, Table 4.2 lists the parameters estimated, the corresponding confidence intervals, and the GoF study results. The GoF study shows all three methods may be successfully used to model peak flow. Figure 4.6a–c compare the fitted parametric distribution with the empirical distribution. They indicate similar performances among the three estimation methods.

4.6.4 Maximum Daily Precipitation

In this section, the maximum daily precipitation at Brenham, Texas, is applied for analysis. The MLE and entropy methods are applied for parameter estimation. Table 4.3 lists the parameters estimated, the corresponding confidence

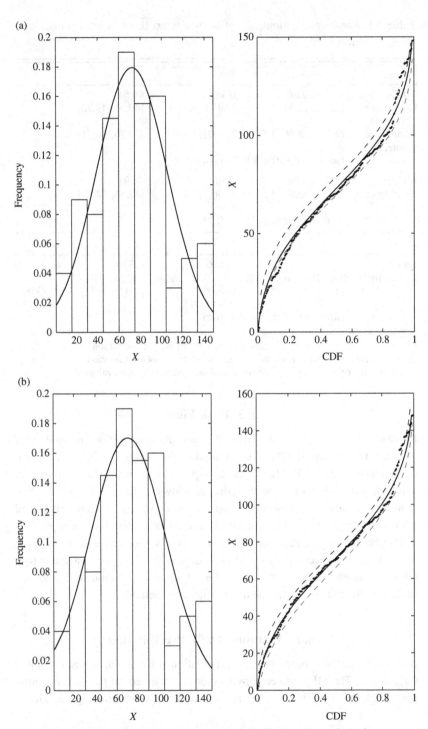

Figure 4.5 Comparison of fitted and empirical distributions for synthetic dataset:
(a) MLE; (b) MOM; and (c) entropy.

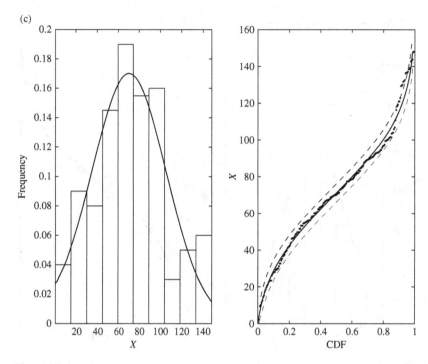

Figure 4.5 (*cont.*)

Table 4.2. *Parameters estimated, confidence intervals, and GoF results: peak flow.*

		α	ν	m	
MLE		3.768	1.049	47.460	
	(1)	[3.485, 4.051]	[1.048, 1.0494]	[47.39, 47.52]	
	(2)	[1.102, 5.056]	[0.24, 2.223]	[40.725, 63.021]	
	GoF	$D_n = 0.065, P = 0.31$			
MOM		3.740	1.050	47.730	
	(2)	[2.251, 3.868]	[1.049, 2.579]	[39.13, 56.29]	
	GoF	$D_n = 0.063, P = 0.332$			
		λ_0[a]	λ_1	λ_2	λ_3
Entropy		−3.275	−0.079	−1.099	4.44E−04
Equations (4.38)–(4.40): $\alpha = 3.161$, $\quad \nu = 1.396$, and $m = 49.815$					
	(2)		[−0.11, −0.042]	[−2.62, 1.24]	[0.0003, 0.0006]
	GoF	$D_n = 0.064, P = 0.35$			

[a] λ_0 is a function of $\{\lambda_1, \lambda_2, \lambda_3\}$ so its confidence interval is not evaluated.

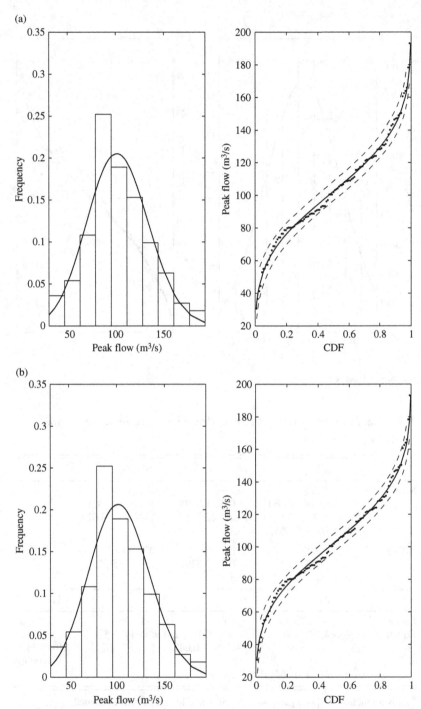

Figure 4.6 Comparison of fitted and empirical distribution for peak flow: (a) MLE; (b) MOM; and (c) entropy.

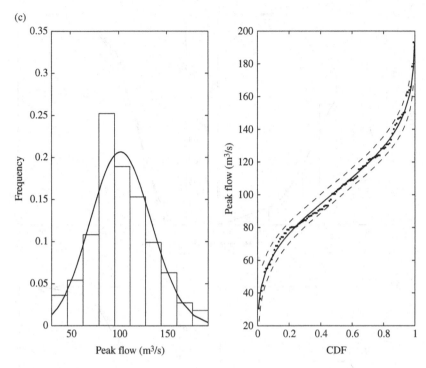

Figure 4.6 (*cont.*)

Table 4.3. *Parameters estimated, confidence intervals, and GoF results: maximum daily precipitation.*

		α	ν	m
MLE		−2.35	1.765	118.936
		[−5.429, −0.263]	[1.121, 2.420]	[76.9255, 147.2116]
	GoF	$D_n = 0.0743, P = 0.108$		
		λ_1	λ_2	λ_3
Entropy		0.0198	−2.53	7.07E−05
		[−4.94E−05, 0.035]	[−3.393, −1.005]	[1.80E−06, 1.28E−04]
	GoF	$D_n = 0.0745, P = 0.104$		

intervals, and the GoF study results. The GoF study shows that both methods yield similar Kolmogorov–Smirnov (KS) statistic values. Figure 4.7a and b compare the fitted parametric distribution with the empirical distribution. The comparison visually confirms that the Hal-B distribution may be applied to model maximum daily precipitation.

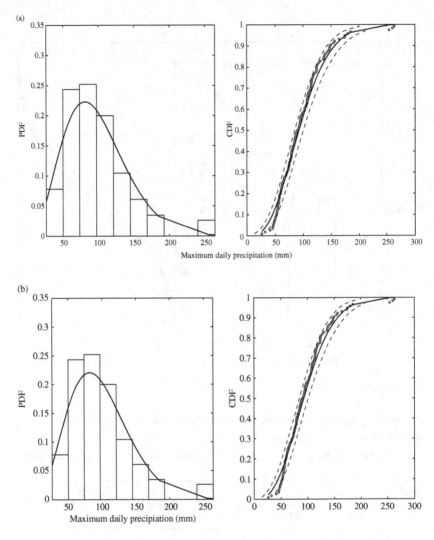

Figure 4.7 Comparison of fitted and empirical distributions for maximum daily precipitation: (a) MLE and (b) entropy.

4.6.5 Total Flow Deficit

Here, the total flow deficit at Tilden, Texas, is applied as a case study. Similar to maximum daily precipitation, MLE and entropy methods are applied for parameter estimation. Table 4.4 lists the parameters estimated, the corresponding confidence intervals, and the GoF study results. The GoF study shows that (1) the Hal-B distribution fitted using MLE outperforms the distribution fitted using the entropy estimation method and (2) the maximum entropy-based Hal-B distribution barely passes the KS GoF test. Figure 4.8a and b compare the fitted parametric distribution with the empirical distribution and indicate similar performances.

Table 4.4. *Parameters estimated, confidence intervals, and GoF results: total flow deficit.*

		α	ν	m
MLE		−6.701	0.348	655.66
		[−7.033, −4.749]	[0.282, 0.350]	[630.724, 684.604]
	GoF	$D_n = 0.094, P = 0.116$		
		λ_1	λ_2	λ_3
Entropy		0.098	0.29	1.93E−06
		[0.0082, 0.019]	[0.131, 0.401]	(0, 2.281E−06]
	GoF	$D_n = 0.111, P = 0.061$		

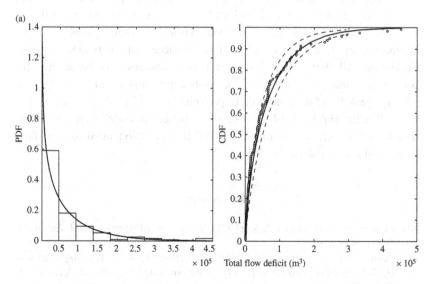

Figure 4.8 Comparison of fitted and empirical distributions for total flow deficit: (a) MLE and (b) entropy.

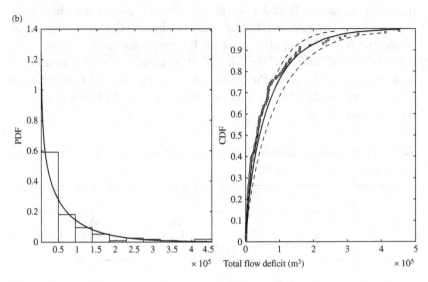

Figure 4.8 (*cont.*)

4.7 Conclusion

In this chapter, we discussed the Hal-B distribution, its parameter estimation with the use of entropy, MOM, and MLE, and the random variable simulation with the acceptance-rejection method. Using the true population, the acceptance-rejection method is found to be successful for random variable simulation. All three parameter estimation techniques may be applied for parameter estimation. We also evaluated its application with real-world data, namely, peak flow, maximum daily precipitation, and total flow deficit. We found that the Hal-B distribution may be applied to model these real-world datasets. However, the entropy-based Hal-B distribution fitted to total flow deficit just barely passes the GoF test.

References

Abramowitz, M. and Stegun, I.A. (1972). *Handbook of Mathematical Functions*. Dover, New York.

El Adlouni, S. and Bobée, B. (2017). *Halphen Distribution Family: With Application in Hydrological Frequency Analysis*. Water Resources Publications, LLC, Highlands Ranch, CO.

Fateh, C., Salaheddine, E.A., and Bernard, B. (2010). Mixed estimation methods for Halphen distributions with applications in extreme hydrologic events. *Stochastic Environmental Research and Risk Assessment*, Vol. 24, No. 3, pp. 359–376.

Halphen, E. (1955). Les fonctions factorielles. *Publications de l'Institut de Statistique de l'Université de Paris*, Vol. IV, Fascicule I, pp. 21–39.

Perreault, L., Bobée, B., and Rasmussen, P. (1999). Halphen distribution system. I: Mathematical and statistical properties. *Journal of Hydrologic Engineering*, Vol. 4, No. 3, pp. 189–199.

Shannon, C.E. (1948). A mathematical theory of communication. *Bell System Technical Journal*, Vol. 27, No. 3, pp. 379–423.

Singh, V.P. (1998). *Entropy Based Parameter Estimation in Hydrology*. Kluwer Academic Publishers, Dordrecht, the Netherlands.

Singh, V.P. (2013). *Entropy Theory and Its Application in Environmental and Water Engineering*. Wiley-Blackwell, Hoboken, NJ.

5

Halphen Inverse B Distribution

5.1 Introduction

The probability density function (PDF) of the Halphen type inverse B (Hal-IB) distribution is given by

$$f(x) = \frac{2}{m^{-2v}ef_v(\alpha)} x^{-2v-1} \exp\left(-\left(\frac{m}{x}\right)^2 + \alpha\left(\frac{m}{x}\right)\right), \quad x > 0, \qquad (5.1)$$

where $ef_v(\cdot)$ is the exponential factorial function defined in the previous chapter, $m > 0$ is the scale parameter, $\alpha \in \mathbb{R}$ is a shape parameter, and $v > 0$ is also a shape parameter. The Hal-IB distribution can be obtained from the Halphen type B (Hal-B) distribution using a simple transformation. If X follows the Hal-B distribution then $1/X$ follows the Hal-IB distribution. Specifically, if $f(x; \alpha, v, m)$ is the PDF of the Hal-B distribution of random variable X then $f(x; \alpha, v, m^{-1})$ will be the PDF of the Hal-IB distribution of random variable $Y = 1/X$.

5.2 Differential Equation for Hal-IB Distribution and Distribution Characteristics

The differential equation for generating the Hal-IB distribution can be written as

$$\frac{1}{f(x)}\frac{df(x)}{dx} = \frac{2m^2 - \alpha mx - (2v+1)x^2}{x^2}. \qquad (5.2)$$

140

Equating $df(x)/dx$ to zero, we can obtain the modal characteristics of the Hal-IB distribution as

$$x = \frac{m\left(-\alpha \pm \sqrt{\alpha^2 + 8(2v+1)}\right)}{2(2v+1)}. \tag{5.3}$$

From Equation (5.1), it is seen that the position of mode occurs at

$$x = m\left(-\frac{\alpha}{2(2v+1)} + \frac{\sqrt{\alpha^2 + 8(2v+1)}}{2(2v+1)}\right). \tag{5.4}$$

Equation (5.4) shows that the Hal-IB distribution is unimodal.

Perreault et al. (1999) showed that the inverse gamma distribution is the limiting distribution for the Hal-IB distribution. Let $a = -\alpha m$ and $b = -\alpha/m$. Equation (5.1) can be rewritten as

$$f(x) = \frac{2}{\left(\frac{a}{b}\right)^{-2v} ef_v(-\sqrt{ab})} x^{-2v-1} \exp\left(-\frac{a}{b}\frac{1}{x^2} - \frac{a}{x}\right), \quad x > 0. \tag{5.5}$$

Let $b \to \infty$. Using the asymptotic expression of the exponential factorial function, Equation (5.5) reduces to

$$\lim_{b \to \infty} f(x) = \frac{a^{2v}}{\Gamma(2v)} \left(\frac{1}{x}\right)^{2v+1} \exp\left(-\frac{a}{x}\right). \tag{5.6}$$

Equation (5.6) is the PDF of the inverse gamma distribution with a as the scale parameter and $2v$ as the shape parameter. The inverse gamma distribution is obtained from Equation (5.1) if $\alpha \to -\infty$ and $m \to \infty$ as well as $\alpha/m \to -\infty$ and $\alpha m \to 0^-$. In this case, $-\alpha m$ is the scale parameter and $2v$ is the shape parameter.

5.3 Derivation Using Entropy Theory

For a continuous random variable X, the Shannon entropy of X, $H(X)$, can be expressed as (Shannon, 1948)

$$H(X) = -\int_0^\infty f(x) \ln f(x) dx, \tag{5.7}$$

where $f(x)$ is the PDF of X. The objective is to derive $f(x)$, which is done by maximizing entropy subject to given information, expressed as constraints, in

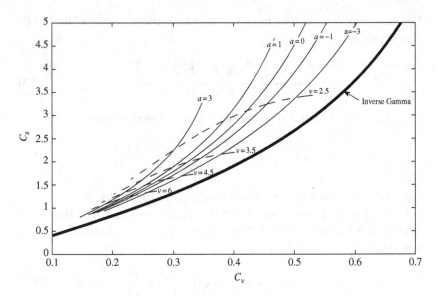

Figure 5.1 Moment diagram for Hal-IB distribution.

concert with the principle of maximum entropy (Singh 2011, 2013). Using the moment diagram, Figure 5.1 graphically shows that the inverse gamma distribution is the limiting distribution of the Hal-IB distribution.

5.3.1 Specification of Constraints

Substituting Equation (5.1) in Equation (5.7), we have

$$H(x) = -\int_0^\infty f(x)\left(\ln\left(\frac{2}{m^{-2v}ef_v(\alpha)}\right) - (2v+1)\ln x - \left(\frac{m}{x}\right)^2 + \alpha\left(\frac{m}{x}\right)\right)dx.$$

(5.8)

Equation (5.8) yields the following constraints:

$$\int_0^\infty f(x)dx = C_0 = 1$$

(5.9)

$$\int_0^\infty f(x)\ln f(x)dx = C_1 = E(\ln x)$$

(5.10)

$$\int_0^\infty \frac{1}{x} f(x) dx = C_2 = E\left(\frac{1}{x}\right) \tag{5.11}$$

$$\int_0^\infty \frac{1}{x^2} f(x) dx = C_3 = E\left(\frac{1}{x^2}\right). \tag{5.12}$$

5.3.2 Entropy Maximizing

The Lagrangian function L can be constructed as

$$L = -\int_0^\infty f(x) \ln f(x) dx - (\lambda_0 - 1)\left(\int_0^\infty f(x) dx - 1\right) - \lambda_1\left(\int_0^\infty (\ln x) f(x) dx - E(\ln x)\right)$$
$$- \lambda_2\left(\int_0^\infty \frac{1}{x} f(x) dx - E\left(\frac{1}{x}\right)\right) - \lambda_3\left(\int_0^\infty \frac{1}{x^2} f(x) dx - E\left(\frac{1}{x^2}\right)\right). \tag{5.13}$$

Differentiating Equation (5.13) with respect to $f(x)$ and equating the derivative to zero, we obtain

$$\frac{\partial L}{\partial f(x)} = -\ln f(x) - \lambda_0 - \lambda_1 \ln x - \lambda_2\left(\frac{1}{x}\right) - \lambda_3\left(\frac{1}{x^2}\right). \tag{5.14}$$

The entropy-based PDF can be defined from Equation (5.14) as

$$f(x) = x^{-\lambda_1} \exp\left(-\lambda_0 - \frac{\lambda_2}{x} - \frac{\lambda_3}{x^2}\right). \tag{5.15}$$

Substituting Equation (5.15) in Equation (5.9), we have

$$\exp(\lambda_0) = \int_0^\infty x^{-\lambda_1} \exp\left(-\frac{\lambda_2}{x} - \frac{\lambda_3}{x^2}\right) dx. \tag{5.16}$$

For $\lambda_3 > 0$ and let $x = \sqrt{\lambda_3}/t$, then $dx = -(\sqrt{\lambda_3}/t^2)$. Thus, Equation (5.16) may be rewritten as

$$\exp(\lambda_0) = \lambda_3^{\frac{-\lambda_1+1}{2}} \int_0^\infty t^{\lambda_1-2} \exp\left(-t^2 - \frac{\lambda_2}{\sqrt{\lambda_3}} t\right) dt. \tag{5.17}$$

Applying the exponential factorial function with $\lambda_1 > 1$, Equation (5.17) can be expressed as

$$\exp(\lambda_0) = \frac{\lambda_3^{\frac{-\lambda_1+1}{2}}}{2} \, ef_{\frac{\lambda_1-1}{2}}\left(-\frac{\lambda_2}{\sqrt{\lambda_3}}\right).$$
(5.18)

Substitution of Equation (5.18) in Equation (5.15) yields

$$f(x) = \frac{2}{\lambda_3^{\frac{-\lambda_1+1}{2}} \, ef_{\frac{\lambda_1-1}{2}}\left(-\frac{\lambda_2}{\sqrt{\lambda_3}}\right)} \, x^{-\lambda_1} \exp\left(-\frac{\lambda_2}{x} - \frac{\lambda_3}{x^2}\right).$$
(5.19)

Comparison of Equation (5.19) with Equation (5.1) shows that

$$\lambda_1 = 2v + 1$$
(5.20)

$$\lambda_2 = -am$$
(5.21)

$$\lambda_3 = m^2.$$
(5.22)

Furthermore, comparing with the population parameters, it is shown that $\lambda_1 > 1$ and $\lambda_3 > 0$.

5.4 Parameter Estimation

5.4.1 Regular Entropy Method

From Equation (5.18), λ_0 can be expressed as

$$\lambda_0 = -\ln 2 + \frac{-\lambda_1 + 1}{2} \ln \lambda_3 + \ln ef_{\frac{\lambda_1-1}{2}}\left(-\frac{\lambda_2}{\sqrt{\lambda_3}}\right).$$
(5.23)

This equation shows that λ_0 is a function of λ_1, λ_2, and λ_3. Differentiating Equation (5.23) with respect to λ_1, λ_2, and λ_3 individually, we have

$$\frac{\partial \lambda_0}{\partial \lambda_1} = -\frac{1}{2}\ln \lambda_3 + \frac{1}{ef_{\frac{\lambda_1-1}{2}}\left(-\frac{\lambda_2}{\sqrt{\lambda_3}}\right)} \frac{\partial ef_{\frac{\lambda_1-1}{2}}\left(-\frac{\lambda_2}{\sqrt{\lambda_3}}\right)}{\partial \lambda_1}$$
(5.24)

$$\frac{\partial \lambda_0}{\partial \lambda_2} = \frac{1}{ef_{\frac{\lambda_1-1}{2}}\left(-\frac{\lambda_2}{\sqrt{\lambda_3}}\right)} \frac{\partial ef_{\frac{\lambda_1-1}{2}}\left(-\frac{\lambda_2}{\sqrt{\lambda_3}}\right)}{\partial \lambda_2} = -\frac{ef_{\frac{\lambda_1}{2}}\left(-\frac{\lambda_2}{\sqrt{\lambda_3}}\right)}{\sqrt{\lambda_3}\, ef_{\frac{\lambda_1-1}{2}}\left(-\frac{\lambda_2}{\sqrt{\lambda_3}}\right)}$$
(5.25)

$$\frac{\partial \lambda_0}{\partial \lambda_3} = \frac{-\lambda_1 + 1}{2\lambda_3} + \frac{1}{ef_{\frac{\lambda_1 - 1}{2}}\left(-\frac{\lambda_2}{\sqrt{\lambda_3}}\right)} \frac{\partial ef_{\frac{\lambda_1 - 1}{2}}\left(-\frac{\lambda_2}{\sqrt{\lambda_3}}\right)}{\partial \lambda_3}$$

$$= \frac{-\lambda_1 + 1}{2\lambda_3} + \frac{\lambda_2 ef_{\frac{\lambda_1}{2}}\left(-\frac{\lambda_2}{\sqrt{\lambda_3}}\right)}{2\lambda_3^{\frac{3}{2}} ef_{\frac{\lambda_1 - 1}{2}}\left(-\frac{\lambda_2}{\sqrt{\lambda_3}}\right)}. \tag{5.26}$$

Furthermore, differentiating Equation (5.16) with respect to λ_1, λ_2, and λ_3, we have

$$\frac{\partial \lambda_0}{\partial \lambda_1} = -\frac{\int_0^\infty (\ln x) x^{-\lambda_1} \exp\left(-\frac{\lambda_2}{x} - \frac{\lambda_3}{x^2}\right) dx}{\int_0^\infty x^{-\lambda_1} \exp\left(-\frac{\lambda_2}{x} - \frac{\lambda_3}{x^2}\right) dx} = -E(\ln X) \tag{5.27}$$

$$\frac{\partial \lambda_0}{\partial \lambda_2} = -\frac{\int_0^\infty \left(\frac{1}{x}\right) x^{-\lambda_1} \exp\left(-\frac{\lambda_2}{x} - \frac{\lambda_3}{x^2}\right) dx}{\int_0^\infty x^{-\lambda_1} \exp\left(-\frac{\lambda_2}{x} - \frac{\lambda_3}{x^2}\right) dx} = -E\left(\frac{1}{X}\right) \tag{5.28}$$

$$\frac{\partial \lambda_0}{\partial \lambda_3} = -\frac{\int_0^\infty \left(\frac{1}{x^2}\right) x^{-\lambda_1} \exp\left(-\frac{\lambda_2}{x} - \frac{\lambda_3}{x^2}\right) dx}{\int_0^\infty x^{-\lambda_1} \exp\left(-\frac{\lambda_2}{x} - \frac{\lambda_3}{x^2}\right) dx} = -E\left(\frac{1}{X^2}\right). \tag{5.29}$$

Equating Equations (5.24)–(5.26) to Equations (5.27)–(5.29) accordingly, we have

$$-\frac{1}{2}\ln \lambda_3 + \frac{1}{ef_{\frac{\lambda_1 - 1}{2}}\left(-\frac{\lambda_2}{\sqrt{\lambda_3}}\right)} \frac{\partial ef_{\frac{\lambda_1 - 1}{2}}\left(-\frac{\lambda_2}{\sqrt{\lambda_3}}\right)}{\partial \lambda_1} = -E(\ln X) \approx -\overline{\ln X} \tag{5.30}$$

$$-\frac{ef_{\frac{\lambda_1}{2}}\left(-\frac{\lambda_2}{\sqrt{\lambda_3}}\right)}{\sqrt{\lambda_3}\, ef_{\frac{\lambda_1 - 1}{2}}\left(-\frac{\lambda_2}{\sqrt{\lambda_3}}\right)} = -E\left(\frac{1}{X}\right) \approx -\frac{\overline{1}}{x} \tag{5.31}$$

$$\frac{-\lambda_1 + 1}{2\lambda_3} + \frac{\lambda_2 ef_{\frac{\lambda_1}{2}}\left(-\frac{\lambda_2}{\sqrt{\lambda_3}}\right)}{2\lambda_3^{\frac{3}{2}} ef_{\frac{\lambda_1 - 1}{2}}\left(-\frac{\lambda_2}{\sqrt{\lambda_3}}\right)} = -E\left(\frac{1}{X^2}\right) \approx -\frac{\overline{1}}{x^2}. \tag{5.32}$$

The Lagrange multipliers can be obtained by solving the system of Equations (5.30)–(5.32).

5.4.2 Parameter Space Expansion Method

From Equation (5.8), the following constraints, in addition to Equation (5.9), can be specified as

$$\int_0^\infty ((2v-1)\ln x)f(x)dx = E\big((2v-1)\ln x\big) = C_1 \tag{5.33}$$

$$\int_0^\infty \frac{am}{x}f(x)dx = E\left(\frac{am}{x}\right) = C_2 \tag{5.34}$$

$$\int_0^\infty \frac{m^2}{x^2}f(x)dx = E\left(\frac{m^2}{x^2}\right) = C_3. \tag{5.35}$$

Following the same procedure as for the regular entropy method, the entropy-based PDF can be constructed as

$$f(x) = x^{-\lambda_1(2v-1)}\exp\left(-\lambda_0 - \frac{am\lambda_2}{x} - \frac{m^2\lambda_3}{x^2}\right). \tag{5.36}$$

Substitution of Equation (5.36) in Equation (5.9) yields

$$\exp(\lambda_0) = \int_0^\infty x^{-\lambda_1(2v-1)}\exp\left(-\frac{am\lambda_2}{x} - \frac{m^2\lambda_3}{x^2}\right)dx. \tag{5.37}$$

Let $t = (m\sqrt{\lambda_3}/x) \Rightarrow x = (m\sqrt{\lambda_3}/t) \Rightarrow dx = -(m\sqrt{\lambda_3}/t^2)dt$. Equation (5.37) can be rewritten as

$$\exp(\lambda_0) = (m\sqrt{\lambda_3})^{-\lambda_1(2v-1)+1}\int_0^\infty t^{\lambda_1(2v-1)-2}\exp\left(-t^2 - \frac{a\lambda_2}{\sqrt{\lambda_3}}t\right)dt. \tag{5.38}$$

Again, applying the exponential factorial function with the condition of $\lambda_3 > 0$, Equation (5.38) can be written as

$$\exp(\lambda_0) = \frac{(m\sqrt{\lambda_3})^{-\lambda_1(2v-1)+1}}{2}ef_{\frac{\lambda_1(2v-1)-1}{2}}\left(-\frac{a\lambda_2}{\sqrt{\lambda_3}}\right). \tag{5.39}$$

Substitution of Equation (5.39) into Equation (5.36) yields

$$f(x) = \frac{2}{(m\sqrt{\lambda_3})^{\lambda_1(2\nu-1)-1} ef_{\frac{\lambda_1(2\nu-1)-1}{2}}\left(-\frac{\alpha\lambda_2}{\sqrt{\lambda_3}}\right)} x^{-\lambda_1(2\nu-1)} \exp\left(-\frac{\alpha m\lambda_2}{x} - \frac{m^2\lambda_3}{x^2}\right).$$

(5.40)

Now the entropy of Equation (5.40) can be written as

$$H(X) = -\ln 2 + \left(\lambda_1(2\nu-1)-1\right)\ln(m\sqrt{\lambda_3}) + \ln\left(ef_{\frac{\lambda_1(2\nu-1)-1}{2}}\left(-\frac{\alpha\lambda_2}{\sqrt{\lambda_3}}\right)\right)$$

$$+ \lambda_1(2\nu-1)E(\ln(x)) + \alpha m\lambda_2 E\left(\frac{1}{x}\right) + m^2\lambda_3 E\left(\frac{1}{x^2}\right).$$

(5.41)

Differentiating Equation (5.41) with respect to Lagrange multipliers λ_1, λ_2, and λ_3 as well as distribution parameters individually and equating each derivative to zero yields

$$\frac{\partial H}{\partial \lambda_1} = (2\nu-1)\ln(m\sqrt{\lambda_3}) + \frac{1}{ef_{\frac{\lambda_1(2\nu-1)-1}{2}}\left(-\frac{\alpha\lambda_2}{\sqrt{\lambda_3}}\right)} \frac{\partial ef_{\frac{\lambda_1(2\nu-1)-1}{2}}\left(-\frac{\alpha\lambda_2}{\sqrt{\lambda_3}}\right)}{\partial \lambda_1}$$

$$+ 2\nu E(\ln x) = 0$$

(5.42)

$$\Rightarrow (2\nu-1)\ln(m\sqrt{\lambda_3}) + \frac{2\nu-1}{2} \frac{\frac{\partial ef_{V1}\left(-\frac{\alpha\lambda_2}{\sqrt{\lambda_3}}\right)}{\partial V1}}{ef_{V1}\left(-\frac{\alpha\lambda_2}{\sqrt{\lambda_3}}\right)}$$

$$+ (2\nu-1)E(\ln x) = 0, \quad V1 = \frac{\lambda_1(2\nu-1)-1}{2}$$

$$\frac{\partial H}{\partial \lambda_2} = \frac{1}{ef_{\frac{\lambda_1(2\nu-1)-1}{2}}\left(-\frac{\alpha\lambda_2}{\sqrt{\lambda_3}}\right)} \frac{\partial ef_{\frac{\lambda_1(2\nu-1)-1}{2}}\left(-\frac{\alpha\lambda_2}{\sqrt{\lambda_3}}\right)}{\partial \lambda_2} + \alpha m E\left(\frac{1}{x}\right)$$

$$= -\frac{\alpha ef_{\frac{\lambda_1(2\nu-1)}{2}}\left(-\frac{\alpha\lambda_2}{\sqrt{\lambda_3}}\right)}{\sqrt{\lambda_3} ef_{\frac{\lambda_1(2\nu-1)-1}{2}}\left(-\frac{\alpha\lambda_2}{\sqrt{\lambda_3}}\right)} + \alpha m E\left(\frac{1}{x}\right) = 0$$

(5.43)

$$\Rightarrow \alpha m E\left(\frac{1}{x}\right) = \frac{\alpha ef_{\frac{\lambda_1(2\nu-1)}{2}}\left(-\frac{\alpha\lambda_2}{\sqrt{\lambda_3}}\right)}{\sqrt{\lambda_3} ef_{\frac{\lambda_1(2\nu-1)-1}{2}}\left(-\frac{\alpha\lambda_2}{\sqrt{\lambda_3}}\right)}$$

$$\frac{\partial H}{\partial \lambda_3} = \frac{\lambda_1(2v-1)-1}{2\lambda_3} + \frac{1}{ef_{\frac{\lambda_1(2v-1)-1}{2}}\left(-\frac{\alpha\lambda_2}{\sqrt{\lambda_3}}\right)} \frac{\partial ef_{\frac{\lambda_1(2v-1)-1}{2}}\left(-\frac{\alpha\lambda_2}{\sqrt{\lambda_3}}\right)}{\partial \lambda_3} + m^2 E\left(\frac{1}{x^2}\right)$$

$$= \frac{\lambda_1(2v-1)-1}{2\lambda_3} + \frac{\alpha\lambda_2 ef_{\frac{\lambda_1(2v-1)}{2}}\left(-\frac{\alpha\lambda_2}{\sqrt{\lambda_3}}\right)}{2\lambda_3^{\frac{3}{2}} ef_{\frac{\lambda_1(2v-1)-1}{2}}\left(-\frac{\alpha\lambda_2}{\sqrt{\lambda_3}}\right)} + m^2 E\left(\frac{1}{x^2}\right) = 0$$

$$(5.44)$$

$$\frac{\partial H}{\partial \alpha} = -\frac{\lambda_2 ef_{\frac{\lambda_1(2v-1)}{2}}\left(-\frac{\alpha\lambda_2}{\sqrt{\lambda_3}}\right)}{\sqrt{\lambda_3} ef_{\frac{\lambda_1(2v-1)-1}{2}}\left(-\frac{\alpha\lambda_2}{\sqrt{\lambda_3}}\right)} + \lambda_2 m E\left(\frac{1}{x}\right) = 0 \qquad (5.45)$$

$$\frac{\partial H}{\partial m} = \frac{\lambda_1(2v-1)-1}{m} + \lambda_2 \alpha E\left(\frac{1}{x}\right) + 2m\lambda_3 E\left(\frac{1}{x^2}\right) = 0 \qquad (5.46)$$

$$\frac{\partial H}{\partial v} = 2\lambda_1 \ln(m\sqrt{\lambda_3}) + \frac{1}{ef_{\frac{\lambda_1(2v-1)-1}{2}}\left(-\frac{\alpha\lambda_2}{\sqrt{\lambda_3}}\right)} \frac{\partial ef_{\frac{\lambda_1(2v-1)-1}{2}}\left(-\frac{\alpha\lambda_2}{\sqrt{\lambda_3}}\right)}{\partial v} + \lambda_1 E(2\ln x) = 0$$

$$\Rightarrow 2\lambda_1 \ln(m\sqrt{\lambda_3}) + \lambda_1 \frac{\frac{\partial ef_{V1}\left(-\frac{\alpha\lambda_2}{\sqrt{\lambda_3}}\right)}{\partial V1}}{e \, f_{V1}\left(-\frac{\alpha\lambda_2}{\sqrt{\lambda_3}}\right)}$$

$$+2\lambda_1 E\left(\ln(x)\right) = 0, \quad V1 = \frac{\lambda_1(2v-1)-1}{2}.$$

$$(5.47)$$

Comparing Equation (5.42) with Equation (5.47), we have $\lambda_1 = 2v - 1$.
Comparing Equation (5.43) with Equation (5.45), we have $\lambda_2 = \alpha$.
Substituting Equation (5.43) in Equation (5.46), we have

$$\frac{\lambda_1(2v-1)-1}{m} + \lambda_2 \frac{\alpha ef_{\frac{\lambda_1(2v-1)}{2}}\left(-\frac{\alpha\lambda_2}{\sqrt{\lambda_3}}\right)}{m\sqrt{\lambda_3} ef_{\frac{\lambda_1(2v-1)-1}{2}}\left(-\frac{\alpha\lambda_2}{\sqrt{\lambda_3}}\right)} + 2m\lambda_3 E\left(\frac{1}{x^2}\right) = 0. \quad (5.48)$$

Comparing Equation (5.44) with Equation (5.48), we have $m = 2\lambda_3$.

Now with the established relations, Equations (5.45)–(5.47) can be rewritten as

$$\frac{\partial H}{\partial \alpha} = -\frac{\alpha ef_{\frac{(2\nu-1)^2}{2}}\left(-\frac{\alpha^2}{\sqrt{\lambda_3}}\right)}{\sqrt{2m}ef_{\frac{(2\nu-1)^2}{2}-1}\left(-\frac{\alpha^2}{\sqrt{\lambda_3}}\right)} + \alpha mE\left(\frac{1}{x}\right) = 0 \qquad (5.49)$$

$$\frac{\partial H}{\partial \nu} = (2\nu-1)\ln\left((2\nu-1)\sqrt{2m}\right) + \frac{2\nu-1}{2}\frac{\partial ef_{V1}\left(-\frac{\alpha^2}{\sqrt{2m}}\right)}{ef_{V1}\left(-\frac{\alpha^2}{\sqrt{2m}}\right)} + (2\nu-1)E(\ln x) = 0,$$

$$V1 = \frac{(2\nu-1)^2-1}{2}$$

$$(5.50)$$

$$\frac{\partial H}{\partial m} = \frac{(2\nu-1)^2-1}{m} + \frac{\alpha^2 ef_{\frac{(2\nu-1)^2}{2}}\left(-\frac{\alpha^2}{\sqrt{\lambda_3}}\right)}{2\left(\frac{m}{2}\right)^{\frac{3}{2}}ef_{\frac{(2\nu-1)^2}{2}-1}\left(-\frac{\alpha^2}{\sqrt{m/2}}\right)} + m^2 E\left(\frac{1}{x^2}\right) = 0. \quad (5.51)$$

To this end, the parameters may be estimated by solving Equations (5.49)–(5.51) simultaneously using the Newton–Raphson method.

5.4.3 Method of Moments

Moments of the Hal-IB distribution can be derived in two ways. The first is the usual procedure by integration. In the second method, the moments of the Hal-IB distribution can be obtained from the moments of the Hal-B distribution by noting that the rth noncentral moment of the Hal-IB distribution is equivalent to the rth noncentral moment of the Hal-B distribution. The moments of the Hal-B distribution have been derived in the preceding chapter and they can be used for deriving the moments of the Hal-IB distribution. Thus, the rth noncentral moment for the Hal-IB distribution can be expressed as

$$M_r = \frac{m^r ef_{\nu-\frac{r}{2}}(\alpha)}{ef_\nu(\alpha)}. \qquad (5.52)$$

The moments exist for $r \leq 2\nu$ because the index of the exponential factorial function must be positive. The first moment or the mean is obtained from Equation (5.52) as

$$M_1 = \mu = \frac{mef_{\nu-\frac{1}{2}}(\alpha)}{ef_\nu(\alpha)}. \qquad (5.53)$$

Based on the noncentral moments, the central moments can be expressed as

$$M_2^\mu = \frac{m^2}{ef_\nu^2(\alpha)}\left(ef_{\nu-1}(\alpha)ef_\nu(\alpha) - ef_{\nu-\frac{1}{2}}^2(\alpha)\right) \tag{5.54}$$

$$M_3^\mu = \frac{m^3}{ef_\nu^3(\alpha)}\left(ef_{\nu-\frac{3}{2}}(\alpha)ef_\nu^2(\alpha) - 3ef_{(\nu-1)}(\alpha)ef_\nu(\alpha)ef_{\nu-\frac{1}{2}}(\alpha) + 2ef_{\nu-\frac{1}{2}}^3(\alpha)\right)$$

$$\tag{5.55}$$

$$M_4^\mu = \frac{m^4}{ef_\nu^4(\alpha)}\left(\begin{array}{c} ef_{\nu-2}(\alpha)ef_\nu^3(\alpha) - 4ef_{\nu-\frac{3}{2}}(\alpha)ef_{\nu-\frac{1}{2}}(\alpha)ef_\nu^2(\alpha) \\ + 6ef_{(\nu-1)}(\alpha)ef_{\nu-\frac{1}{2}}^2(\alpha)ef_\nu(\alpha) - 3ef_{\nu-\frac{1}{2}}^4(\alpha) \end{array}\right) \tag{5.56}$$

$$C_\nu = \frac{\sqrt{ef_{\nu-1}(\alpha)ef_\nu(\alpha) - ef_{\nu-\frac{1}{2}}^2(\alpha)}}{ef_{\nu-\frac{1}{2}}(\alpha)} \tag{5.57}$$

$$C_s = \frac{ef_{\nu-\frac{3}{2}}(\alpha)ef_\nu^2(\alpha) - 3ef_{\nu-1}(\alpha)ef_{\nu-\frac{1}{2}}(\alpha)ef_\nu(\alpha) + 2ef_{\nu-\frac{1}{2}}^3(\alpha)}{\left(ef_{\nu-1}(\alpha)ef_\nu(\alpha) - ef_{\nu-\frac{1}{2}}^2(\alpha)\right)^{\frac{3}{2}}} \tag{5.58}$$

$$C_g = \frac{ef_{\nu-2}(\alpha)ef_\nu^3(\alpha) - 4ef_{\nu-\frac{3}{2}}(\alpha)ef_{\nu-\frac{1}{2}}(\alpha)ef_\nu^2(\alpha) + 6ef_{\nu-1}(\alpha)ef_{\nu-\frac{1}{2}}^2(\alpha)ef_\nu(\alpha) - 3ef_{\nu-\frac{1}{2}}^4(\alpha)}{\left(ef_{(\nu-1)}(\alpha)ef_{\nu(\alpha)} - ef_{\nu-\frac{1}{2}}^2(\alpha)\right)^2}. \tag{5.59}$$

Previously, we stated that the inverse gamma distribution is the limiting distribution of the Hal-IB distribution. Using the coefficient of variation and coefficient of skewness, Figure 5.1 plots the moment ratio diagram for the Hal-IB distribution. As shown in Figure 5.1, the inverse gamma distribution is the limiting distribution of the Hal-IB distribution. The Hal-IB distribution cannot be applied if $[C_\nu, C_s]$ falls below the inverse gamma distribution.

Furthermore, the parameters of the Hal-IB distribution can be estimated analytically through the sample moments (Perreault et al., 1999) as

$$\nu = \frac{E(X)E(X^{-1})\left(E(X^{-3})E(x^{-1}) - (E(X^{-2}))^2\right) - \mathrm{var}(X^{-1})(E(X^{-1}))^2}{2\left(1 - E(x)E(X^{-1})\right)\left((E(X^{-2}))^2 - E(X^{-3})E(X^{-1})\right) - (\mathrm{var}(X^{-1}))^2} \tag{5.60}$$

$$m^2 = \frac{2\nu\left(1 - E(X)E(X^{-1})\right) + E(X)E(X^{-1})}{2\mathrm{var}(X^{-1})} \tag{5.61}$$

$$\alpha = \frac{2v\big(E(X^{-1}) - E(X^{-2})E(X)\big) + E(X^{-2})E(X)}{m\mathrm{var}(X^{-1})}. \tag{5.62}$$

5.4.4 Maximum Likelihood Estimation Method

From Equation (5.1), the log-likelihood function of the Hal-IB distribution can be written as

$$\log L = n\ln 2 + 2vn \ln m - n \ln ef_v(\alpha) - (2v+1) \sum_{i=1}^{n} \ln x_i - \sum_{i=1}^{n} \left(\frac{m}{x_i}\right)^2 + \alpha \sum_{i=1}^{n} \frac{m}{x_i}. \tag{5.63}$$

In Equation (5.63), n represents the sample size. Taking the derivative of Equation (5.63) with respect to v, m, and α and setting them equal to zero, we obtain

$$\frac{\partial \log L}{\partial v} = 2n \ln m - \frac{n}{ef_v(\alpha)} \frac{\partial ef_v(\alpha)}{\partial v} - 2 \sum_{i=1}^{n} \ln x_i = 0 \Rightarrow \ln m - \frac{1}{2ef_v(\alpha)} \frac{\partial ef_v(\alpha)}{\partial v} = \overline{\ln x} \tag{5.64}$$

$$\frac{\partial \log L}{\partial m} = \frac{2vn}{m} - 2m \sum_{i=1}^{n} x_i^{-2} + \alpha \sum_{i=1}^{n} x_i^{-1} = 0 \Rightarrow m^2 \overline{\frac{1}{x^2}} = v + \frac{\alpha m}{2} \overline{\frac{1}{x}} \tag{5.65}$$

$$\frac{\partial \log L}{\partial \alpha} = -\frac{n ef_{v+\frac{1}{2}}(\alpha)}{ef_v(\alpha)} + m \sum_{i=1}^{n} \frac{1}{x_i} = 0 \Rightarrow \frac{1}{m} \frac{ef_{v+\frac{1}{2}}(\alpha)}{ef_v(\alpha)} = \overline{\frac{1}{x}}. \tag{5.66}$$

Solving Equations (5.64)–(5.66), the parameters can then be estimated using the maximum likelihood estimation (MLE) method.

5.5 Application

5.5.1 Simulating Hal-IB Distributed Random Variables with Fixed Parameters

Similar to the discussion of Halphen type A (Hal-A) and Hal-B distributions, the acceptance-rejection method is applied to random number data simulation. In the case of the Hal-IB distribution, its limiting distribution is the inverse gamma distribution. There is a relation between the Hal-B and Hal-IB distributions; that is, if $X \sim$ Hal-B $(x; \alpha, v, m)$ then $Y = 1/X \sim$ Hal-IB

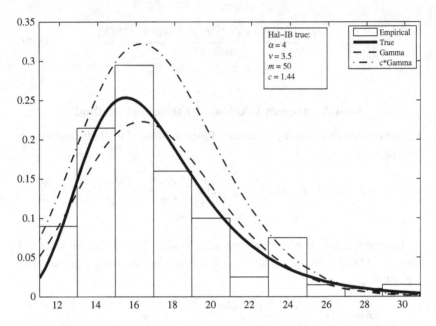

Figure 5.2 Hal-IB distributed random variable with acceptance-rejection method.

$(y; \alpha, v, m^{-1})$. Thus, the simulation procedure applied for the Hal-B distribution may also be applied for the Hal-IB distribution. Figure 5.2 illustrates the random variable simulated for the Hal-IB distribution with the acceptance-rejection method. The figure shows that the Hal-IB distributed random variable may be successfully simulated using the same acceptance-rejection method as for the Hal-B distribution.

5.5.2 Parameter Estimation Using the Simulated Random Variables

In this section, using the simulated Hal-IB distributed random variable from the true population with parameters of $\alpha = 4$, $v = 3.5$, and $m = 50$, the parameters for the sample are estimated using the method of moments (MOM), MLE method, and entropy method as listed in Table 5.1. Using the relation of Lagrange multipliers to the population parameters, that is, Equations (5.20)–(5.22), Table 5.1 shows that the parameters estimated with all three methods are close to each other. Table 5.2 lists the goodness-of-fit (GoF) results. The GoF study indicates the generated random variables are indeed from the true population with the parameters of $\alpha = 4$, $v = 3.5$, and $m = 50$.

Table 5.1. *Parameters estimated for the generated sample from true population ($\alpha = 4$, $v = 3.5$, and $m = 50$).*

Method	α	v	m	
MOM	3.67	3.67	48.91	
MLE	5.16	1.82	49.91	
Entropy	λ_0	λ_1	λ_2	λ_3
	-5.31	4.88	-253.003	2,490
Equations (5.20)–(5.22)	$\alpha = 5.07$,	$v = 1.94$,	$m = 49.90$	

Table 5.2. *GoF study for the generated sample with true population.*

Method		α (4)	v (3.5)	m (50)
MOM	95% confidence interval	[1.69, 6.34]	[1.17, 5.24]	[43.68, 57.35]
	GoF	$D_n = 0.044$, $P = 0.57$		
MLE	95% confidence interval	[2.77, 5.40]	[1.54, 4.97]	[42.96,54.46]
	GoF	$D_n = 0.044$, $P = 0.47$		
Entropy		λ_1 (8)	λ_2 (-200)	(2,500)
		$\lambda_0 = \lambda(\lambda_1, \lambda_2, \lambda_3) = 17.37$		
	95% confidence interval	[3.44, 11.00]	[-272.39, -153.33]	[2,430.4, 2,589.1]
	GoF	$D_n = 0.044$, $P = 0.55$		

Figure 5.3a–c compare the true population, empirical distribution, and fitted distribution as well as 95% confidence bounds.

5.5.3 Peak Flow

Previously, we have shown that the acceptance-rejection method may be applied to simulate the Hal-IB distributed random variables and all three methods may be applied for parameter estimation. In this section, we again apply the peak flow data from USGS09239500 as an example. Applying MOM, MLE, and entropy methods, Table 5.3 lists the parameters estimated. Table 5.4 lists the GoF results. Table 5.4 indicates that (1) the confidence bound is very tight for the parameters estimated for all three methods; (2) all three methods yield similar test statistics and P-value; (3) the test statistics are significantly higher while the P-values are lower than those obtained for Hal-A and Hal-B distributions; and (4) the P-value is around 0.15 for the fitted Hal-IB distribution with the use of the entropy method. Figure 5.4 plots the PDF of the fitted distribution versus the empirical frequency distribution. The figure again shows very similar performance among the three estimation methods. Moreover, a significant difference is shown on the cumulative probability distribution function (CDF) plot that explains the low P-value obtained from the GoF study.

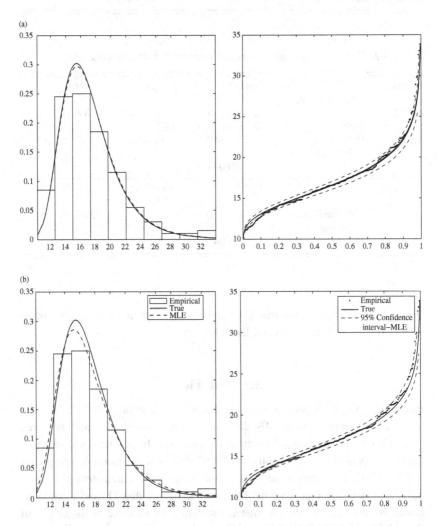

Figure 5.3 Comparison of the true population and fitted distributions, and the 95% confidence bounds: (a) MOM; (b) MLE; and (c) entropy.

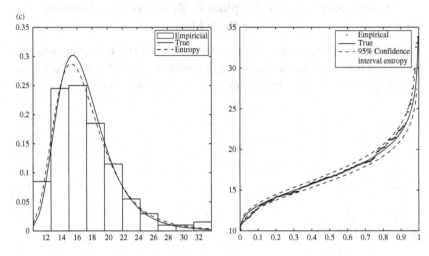

Figure 5.3 (*cont.*)

Table 5.3. *Parameters estimated: peak flow.*

Method	α	ν	m	
MOM	−6.5848	3.7756	80.2970	
MLE	−6.5845	3.7756	80.2970	
Entropy	λ_0	λ_1	λ_2	λ_3
	−37.09	8.5513	528.7426	6,447.6489
Equations (5.20)–(5.22)	$\alpha = -6.5848$,	$\nu = 3.776$,	$m = 80.2973$	

Table 5.4. *Confidence intervals and GoF results: peak flow.*

Method	Statistics	α	ν	m
MOM	Confidence interval	[−6.5860, −6.5841]	[3.7756, 3.7757]	[74.6953, 80.2971]
	GoF	$D_n = 0.116$	$P = 0.047$	
MLE	Confidence interval	[−6.589, −6.586]	[3.7756, 3.7757]	[75.741, 86.003]
	GoF	$D_n = 0.1074$	$P = 0.007$	
		λ_1	λ_2	λ_3
Entropy	Confidence interval	[8.5512, 8.5514]	[528.7411, 528.7889]	[6,447.6048, 6,447.6553]
	GoF	$D_n = 0.109$	$P = 0.152$	

5.5.4 Comparison of Halphen A, B, and IB Distributions for Peak Flow

In Chapters 3 and 4, we have investigated Hal-A and Hal-B distributions and found that both these distributions may be applied to model the peak flow data. In this section, we will further compare all three distributions. Figure 5.5

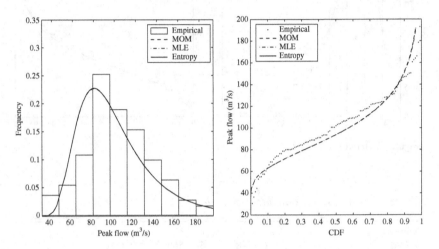

Figure 5.4 Fitted Hal-IB probability density function versus empirical frequency distribution: peak flow.

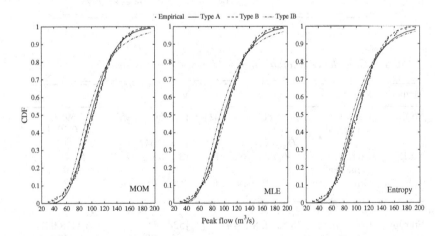

Figure 5.5 Parametric CDF versus empirical CDF: peak flow.

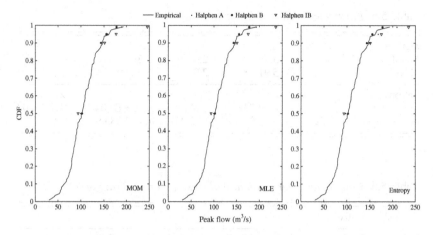

Figure 5.6 Comparison of quantiles computed for Halphen A, B, and IB distributions: peak flow.

depicts the comparison of empirical CDF with the parametric CDF for different parameter estimation methods. It shows that Hal-A and Hal-B distributions yield better performance than the Hal-IB distribution and this is in agreement with the GoF study results. Figure 5.6 compares the quantiles (i.e., 50%, 90%, 95%, and 99%) computed from different distributions for different parameter estimation methods. It confirms that Hal-A and Hal-B distributions may properly model peak flow while the Hal-IB distribution may overestimate it.

5.5.5 Annual Rainfall Amount

In the previous section, we found that the Hal-IB distribution may not be applied to model the peak flow data of USGS09239500. In this section, we choose the annual rainfall data at U330058 to test the performance of the Hal-IB distribution. Table 5.5. lists the parameters estimated for the annual rainfall data with the confidence intervals and GoF results listed in Table 5.6.

The GoF results in Table 5.6 show that the Hal-IB distribution may be applied to model the annual rainfall data at U330058. Figure 5.7a–c visually confirm the applicability of the Hal-IB distribution for the annual rainfall dataset.

Table 5.5. *Parameters estimated: annual rainfall.*

Method	α	ν	m	
MOM	5.885	2.933	3,362.8	
MLE	5.914	2.932	3,383.9	
Entropy	λ_0	λ_1	λ_2	λ_3
	−32.732	6.882	−2.00E+04	1.15E+07
Equations (5.20)–(5.22)	$\alpha = 5.914$,	$\nu = 2.941$,	$m = 3,383.9$	

Table 5.6. *Confidence intervals and GoF results: annual rainfall.*

Method	Statistics	α	ν	m
MOM	Confidence interval	[4.068, 5.885]	[2.933, 4.792]	[2,998.6, 3,362.8]
	GoF	$D_n = 0.088$	$P = 0.49$	
MLE	Confidence interval	[1.982, 7.704]	[0.12, 9.914]	[3,073.1, 4,127.4]
	GoF	$D_n = 0.082$	$P = 0.45$	
		λ_1	λ_2	λ_3
Entropy	Confidence interval	[5.408, 8.221]	[−2.005E04, −2.0013E04]	[1.142E07, 1.148E07]
	GoF	$D_n = 0.083$	$P = 0.52$	

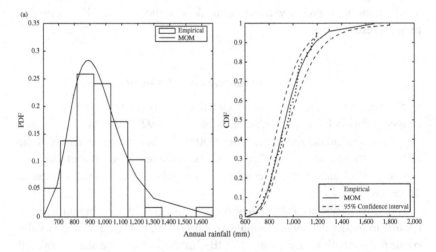

Figure 5.7 Comparison of the empirical distribution, fitted distribution, and 95% confidence bounds for the annual rainfall: (a) MOM; (b) MLE; and (c) entropy.

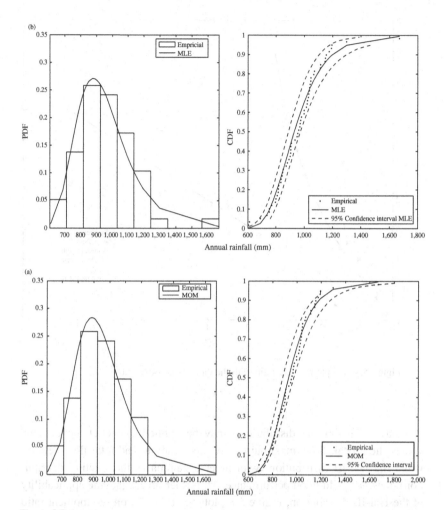

Figure 5.7 (*cont.*)

5.6 Conclusion

Similar to the discussion of Hal-A and Hal-B distributions, this chapter discusses the Hal-IB distribution, its parameter estimation, and random variable simulation. With the use of synthetic data and peak flow as examples, it is determined that the Hal-IB distribution may not be applied to model the peak flow dataset. With gamma and inverse gamma distributions as the limiting distributions, it is shown that Hal-A and Hal-B distributions (with

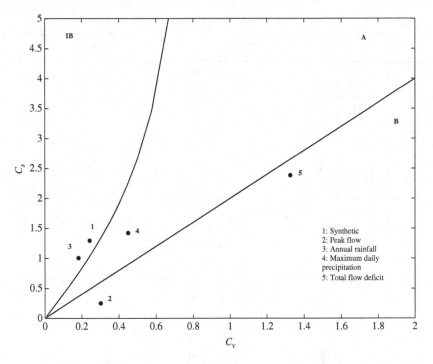

Figure 5.8 Sample moment ratio with the datasets investigated.

gamma as the limiting distribution) may be applied to model the extreme values (here peak flow, maximum daily precipitation, and total flow deficit), while the Hal-IB distribution (with inverse gamma as the limiting distribution) does not yield good performance. To further investigate the applicability of the Hal-IB distribution, Figure 5.8 plots the (Cv, Cs) on the moment ratio diagram plot and confirms that the Hal-IB distribution may be applied to model the annual rainfall at U330058 but not to the peak flow, maximum daily precipitation, and total flow deficit datasets. Figure 5.8 clearly shows that (1) sample [Cv, Cs] of the synthetic data and of the annual rainfall at U330058 are within the Cv–Cs domain for the Hal-IB distribution; (2) sample [Cv, Cs] of the peak flow and of the total flow deficit are within the Cv–Cs domain of the Hal-A distribution; (3) sample [Cv, Cs] of the total flow deficit is within the Cv–Cs domain for the Hal-A distribution. To this end, it is recommended to investigate the sample [Cv, Cs] with the limiting gamma (inverse gamma) distributions to choose between the Halphen A, B, and IB distribution candidates.

References

Perreault, L., Bobée, B., and Rasmussen, P. (1999). Halphen distribution system. I: Mathematical and statistical properties. *Journal of Hydrologic Engineering*, Vol. 4, No. 3, pp. 189–199.

Shannon, C.E. (1948). A mathematical theory of communication. *Bell System Technical Journal*, Vol. 27, No. 3, pp. 379–423.

Singh, V.P. (2011). Hydrologic synthesis using entropy theory: Review. *Journal of Hydrologic Engineering*, Vol. 16, No. 5, pp. 421–433.

Singh, V.P. (2013). *Entropy Theory and Its Application in Environmental and Water Engineering*. Wiley-Blackwell, Hoboken, NJ.

6

Three-Parameter Generalized
Gamma Distribution

6.1 Introduction

Two-parameter gamma distribution is widely applied in environmental and water engineering. However, it has limited flexibility and versatility. Hence, its generalizations as three-parameter and four-parameter gamma distributions have been proposed and applied. The generalized gamma (GG) distributions have been employed for modeling flood peaks, volumes, and mean values. The gamma distribution has also been used in rainfall-runoff modeling (see Lienhard, 1964, 1972; Lienhard and Meyer 1967; Singh, 1988, 2013). In Russia, China, and Eastern Europe, the three-parameter generalized gamma (TPGG) distribution is also referred to as the Kritsky–Menkel distribution (Klemes, 1989). Bobée and Ashkar (1991) discussed the gamma family and its applications in hydrology. This chapter derives the TPGG distribution using the entropy theory and derives its parameters using this theory (Singh, 1998) as well as methods of maximum likelihood estimation (MLE) (Natural Environment Research Council, 1999), moments (Chebana et al., 2010), probability weighted moments (Landwehr et al., 1979), L-moments (Hosking, 1990), and entropy method (Papalexiou and Koutsoyiannis, 2012). The distribution is also applied to real data.

6.2 Characteristics of GG Distribution

Let X be a random variable and x be its specific value. The probability density function (PDF) of the TPGG distribution of X can be expressed as

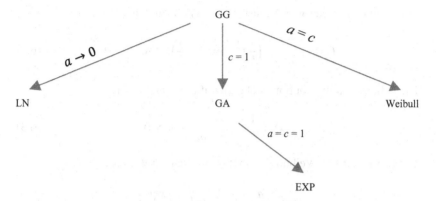

Figure 6.1 Tree diagram of the GG and its special distributions.

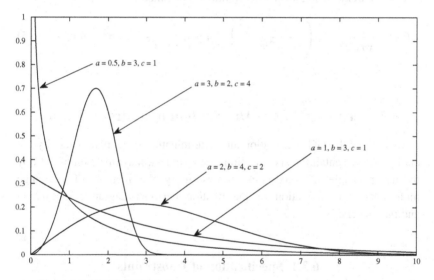

Figure 6.2 TPGG distribution.

$$f(x) = \frac{c}{b\Gamma\left(\frac{a}{c}\right)} \left(\frac{x}{b}\right)^{a-1} \exp\left(-\left(\frac{x}{b}\right)^c\right); \quad a, b, c > 0, \ x > 0, \qquad (6.1)$$

where $\Gamma(\cdot)$ is the gamma function, a and c are the shape parameters, and b is the scale parameter. Figure 6.1 shows the tree diagram of the GG and its special distributions and Figure 6.2 shows that the GG distribution can take on different shapes for different values of parameters.

The two-parameter gamma distribution is obtained if $c = 1$:

$$f(x) = \frac{1}{b\Gamma(a)} \left(\frac{x}{b}\right)^{a-1} \exp\left(-\frac{x}{b}\right); \quad a, \ b > 0. \tag{6.2}$$

The exponential distribution is obtained if $a = 1$ and $c = 1$:

$$f(x) = \frac{1}{b} \exp\left(-\frac{x}{b}\right); \quad b > 0. \tag{6.3}$$

The two-parameter Weibull distribution is obtained if $a = c$:

$$f(x) = \frac{a}{b} \left(\frac{x}{b}\right)^{a-1} \exp\left(-\left(\frac{x}{b}\right)^{a}\right). \tag{6.4}$$

The two-parameter lognormal distribution is obtained if $c \to 0$:

$$f(x) = \frac{1}{x\sigma\sqrt{2\pi}} \exp\left(-\frac{(\ln x - \mu)^2}{2\sigma^2}\right); \quad \mu = \lim_{c \to 0} \frac{ab^c - c}{c^2}; \quad \sigma^2 = \lim_{c \to 0} \frac{b^c}{c^2}. \tag{6.5}$$

6.3 Derivation of GG Distribution by Entropy Theory

Derivation of the GG distribution and determination of its parameters by the entropy theory entail the specification of constraints, maximization of entropy by the principle of maximum entropy using the method of Lagrange multipliers, and derivation of the relation between Lagrange multipliers and parameters.

6.3.1 Specification of Constraints

The Shannon entropy $H(x)$ for random variable X (Shannon, 1948) can be expressed as

$$H(x) = -\int_0^\infty f(x) \ln f(x) dx. \tag{6.6}$$

In order to derive the GG distribution, we first determine the constraints that the distribution must satisfy. Substituting Equation (6.1) in Equation (6.6), we obtain

$$H(x) = -\int_0^\infty f(x) \ln\left(\frac{c}{b\Gamma\left(\frac{a}{c}\right)} \left(\frac{x}{b}\right)^{a-1} \exp\left(-\left(\frac{x}{b}\right)^c\right)\right) dx. \quad (6.7)$$

Equation (6.7) yields

$$H(x) = -\ln\left(\frac{c}{b^a \Gamma\left(\frac{a}{c}\right)}\right) - (a-1)E(\ln x) + \frac{E(x^c)}{b^c}, \quad (6.8)$$

where $E(\cdot)$ is the expectation of $(\cdot)$. Thus, the constraints can be obtained from Equation (6.8) as

$$\int_0^\infty f(x)dx = 1 = C_0 \quad (6.9)$$

$$\int_0^\infty f(x)\ln x\, dx = E(\ln x) = C_1 \quad (6.10)$$

$$\int_0^\infty x^c f(x)dx = E(x^c) = C_2. \quad (6.11)$$

6.3.2 Entropy Maximizing

Entropy can be maximized using the method of Lagrange multipliers for which the Lagrangian function L can be constructed as

$$L = -\int_0^\infty f(x)\ln f(x)dx - (\lambda_0 - 1)\left(\int_0^\infty f(x)dx - C_0\right) - \lambda_1\left(\int_0^\infty f(x)\ln x\, dx - C_1\right)$$

$$- \lambda_2\left(\int_0^\infty x^c f(x)dx - C_2\right), \quad (6.12)$$

where λ_0, λ_1, and λ_2 are the unknown Lagrange multipliers that will be determined in terms of constraints given by Equations (6.9)–(6.11).

6.3.3 Derivation of GG Distribution

Differentiating Equation (6.12) with respect to $f(x)$ and equating the derivative to zero, we obtain

$$\frac{\partial L}{\partial f(x)} = -\ln f(x) - 1 - (\lambda_0 - 1) - \lambda_1 \ln x - \lambda_2 x^c = 0. \qquad (6.13)$$

Equation (6.13) yields

$$f(x) = \exp(-\lambda_0 - \lambda_1 \ln x - \lambda_2 x^c). \qquad (6.14)$$

Equation (6.14) is the entropy-based PDF of the TPGG distribution. Equation (6.14) can be written as

$$f(x) = \exp(-\lambda_0) x^{-\lambda_1} \exp(-\lambda_2 x^c). \qquad (6.15)$$

Comparing Equation (6.15) with Equation (6.1), it observed that

$$\lambda_0 = \ln\left(\frac{b^a \Gamma\left(\frac{a}{c}\right)}{c}\right), \quad \lambda_1 = 1 - a, \quad \lambda_2 = b^{-c}. \qquad (6.16)$$

6.3.4 Relation between Lagrange Multipliers and Parameters

Substitution of Equation (6.15) in Equation (6.9) yields

$$\int_0^\infty \exp(-\lambda_0) x^{-\lambda_1} \exp(-\lambda_2 x^c) dx = C_0 = 1. \qquad (6.17)$$

Equation (6.17) can be expressed with the use of partition function $Z(\lambda_1, \lambda_2) = \exp(\lambda_0)$ as

$$Z(\lambda_1, \lambda_2) = \exp(\lambda_0) = \int_0^\infty x^{-\lambda_1} \exp(-\lambda_2 x^c) dx. \qquad (6.18)$$

Let $y = \lambda_2 x^c$. We can solve Equation (6.18) as

$$Z(\lambda_1, \lambda_2) = \exp(\lambda_0) = \frac{1}{c} \lambda_2^{\frac{\lambda_1 - 1}{c}} \Gamma\left(\frac{1 - \lambda_1}{c}\right). \qquad (6.19)$$

In Equation (6.19), we have $c > 0$, $\lambda_1 < 1$, and $\lambda_2 > 0$. Substituting Equation (6.19) in Equation (6.15), we obtain

$$f(x) = \frac{c\lambda_2^{\frac{1-\lambda_1}{c}}}{\Gamma\left(\frac{1-\lambda_1}{c}\right)} x^{-\lambda_1} \exp(-\lambda_2 x^c).$$

(6.20)

6.4 Methods of Parameter Estimation

The GG distribution parameters are estimated by the regular entropy method, parameter space expansion method, method of moments (MOM), and MLE method. Each of these methods is now discussed.

6.4.1 Regular Entropy Method

This method involves the derivation of the relation between Lagrange multipliers and constraints and the relation between distribution parameters and constraints.

6.4.1.1 Relation between Lagrange Multipliers and Constraints

The Lagrange multiplier λ_0 is a function of λ_1 and λ_2 as expressed by Equations (6.18) and (6.19). Therefore, two sets of equations can be obtained by differentiating these equations. From Equation (6.18) we have

$$\lambda_0 = \ln \int_0^\infty \exp(-\lambda_1 \ln x - \lambda_2 x^c) dx.$$

(6.21)

Likewise, Equation (6.19) can be rewritten as

$$\lambda_0 = -\ln c + \frac{\lambda_1 - 1}{c} \ln \lambda_2 + \ln \Gamma\left(\frac{1-\lambda_1}{c}\right).$$

(6.22)

Differentiation of Equation (6.21) with respect to λ_1 and λ_2 yields

$$\frac{\partial \lambda_0}{\partial \lambda_1} = -\frac{\int_0^\infty \ln x \exp(-\lambda_1 \ln x - \lambda_2 x^c) dx}{\int_0^\infty \exp(-\lambda_1 \ln x - \lambda_2 x^c) dx} = -E(\ln x)$$

(6.23)

$$\frac{\partial \lambda_0}{\partial \lambda_2} = -\frac{\int_0^\infty x^c \exp(-\lambda_1 \ln x - \lambda_2 x^c) dx}{\int_0^\infty \exp(-\lambda_1 \ln x - \lambda_2 x^c) dx} = -E(x^c).$$

(6.24)

The differentiation of Equation (6.22) with respect to λ_1 and λ_2 yields

$$\frac{\partial \lambda_0}{\partial \lambda_1} = \frac{1}{c}\ln \lambda_2 - \frac{1}{c}\psi\left(\frac{1-\lambda_1}{c}\right) \tag{6.25}$$

$$\frac{\partial \lambda_0}{\partial \lambda_2} = \frac{\lambda_1 - 1}{c\lambda_2}. \tag{6.26}$$

In Equation (6.25), $\psi(\cdot)$ is a digamma function defined as $\psi(x) = (d\ln\Gamma(x))/dx$.

Equating Equations (6.23) and (6.24) to Equations (6.25) and (6.26) correspondingly, we obtain

$$\frac{\partial \lambda_0}{\partial \lambda_1} = \frac{1}{c}\ln \lambda_2 - \frac{1}{c}\psi\left(\frac{1-\lambda_1}{c}\right) = -E(\ln x) \tag{6.27}$$

$$\frac{\partial \lambda_0}{\partial \lambda_2} = \frac{\lambda_1 - 1}{c\lambda_2} = -E(x^c). \tag{6.28}$$

Since the GG distribution has three parameters, Equations (6.27) and (6.28) are not sufficient for calculating all the parameters. Therefore, one additional equation is needed, which can be obtained by differentiating Equation (6.27) again as

$$\frac{\partial^2 \lambda_0}{\partial \lambda_1^2} = -\frac{1}{c}\frac{\partial\psi\left(\frac{1-\lambda_1}{c}\right)}{\partial\lambda_1} = \frac{1}{c^2}\psi^{(1)}\left(\frac{(1-\lambda_1)}{c}\right) = \text{var}(\ln x), \tag{6.29}$$

where $\psi^{(1)}(\cdot)$ is a trigamma function as $\psi^{(1)}(x) = (d^2\ln\Gamma(x))/dx^2$.

6.4.1.2 Relation between Parameters and Constraints

Substituting $\lambda_1 = 1 - a$ and $\lambda_2 = b^{-c}$, that is, Equation (6.16), in Equations (6.27)–(6.29), we can establish the relation between the population parameters and the constraints as

$$\frac{\partial \lambda_0}{\partial \lambda_1} = \frac{1}{c}\ln \lambda_2 - \frac{1}{c}\psi\left(\frac{1-\lambda_1}{c}\right) = \frac{1}{c}\ln b^{-c} - \frac{1}{c}\psi\left(\frac{a}{c}\right) \Rightarrow \ln b + \frac{1}{c}\psi\left(\frac{a}{c}\right) = E(\ln x) \tag{6.30}$$

$$\frac{\partial \lambda_0}{\partial \lambda_2} = \frac{\lambda_1 - 1}{c\lambda_2} \Rightarrow \frac{a}{cb^{-c}} = E(x^c) \tag{6.31}$$

$$\frac{\partial^2 \lambda_0}{\partial \lambda_1^2} = \frac{1}{c^2} \psi^{(1)} \left(\frac{(1 - \lambda_1)}{c} \right) \Rightarrow \frac{1}{c^2} \psi^{(1)} \left(\frac{a}{c} \right) = \text{var}(\ln x). \qquad (6.32)$$

6.4.2 Parameter Space Expansion Method

The GG distribution parameters can be estimated using the parameter space expansion method (Singh et al., 1986; Singh, 1998).

6.4.2.1 Specification of Constraints

Using the parameter space expansion method, the constraints for the GG distribution can be defined as

$$\int_0^\infty \ln \left(\left(\frac{x}{b} \right)^{a-1} \right) f(x) dx = E \left(\ln \left(\left(\frac{x}{b} \right)^{a-1} \right) \right) = C_1 \qquad (6.33)$$

$$\int_0^\infty \left(\frac{x}{b} \right)^c f(x) dx = E \left(\left(\frac{x}{b} \right)^c \right) = C_2. \qquad (6.34)$$

6.4.2.2 Derivation of Entropy Function

Following the regular procedure, the entropy-based PDF of the GG distribution becomes

$$f(x) = \exp \left(- \lambda_0 - \lambda_1 \ln \left(\left(\frac{x}{b} \right)^{a-1} \right) - \lambda_2 \left(\frac{x}{b} \right)^c \right), \qquad (6.35)$$

where λ_0, λ_1, and λ_2 are the unknown Lagrange multipliers that need to be determined. Inserting Equation (6.35) in Equation (6.9), we have

$$\exp(\lambda_0) = \int_0^\infty \left(\frac{x}{b} \right)^{-\lambda_1(a-1)} \exp \left(- \lambda_2 \left(\frac{x}{b} \right)^c \right) dx. \qquad (6.36)$$

Let $y = \lambda_2 (x/b)^c$. Then Equation (6.36) may be solved as

$$\exp(\lambda_0) = \frac{b}{c} \lambda_2^{\frac{\lambda_1(a-1)-1}{c}} \Gamma \left(\frac{1 - \lambda_1(a-1)}{c} \right), \qquad (6.37)$$

where $a, b, c > 0$, $\lambda_2 > 0$, and $\lambda_1(a - 1) < 1$.

Substituting Equation (6.37) in Equation (6.35), we have

$$f(x) = \frac{c\lambda_2^{\frac{1-\lambda_1(a-1)}{c}}}{b\Gamma\left(\frac{1-\lambda_1(a-1)}{c}\right)} \left(\frac{x}{b}\right)^{-\lambda_1(a-1)} \exp\left(-\lambda_2\left(\frac{x}{b}\right)^c\right). \tag{6.38}$$

Substituting Equation (6.38) in Equation (6.6), the entropy is expressed as

$$H(x) = \frac{c\lambda_2^{\frac{1-\lambda_1(a-1)}{c}}}{b\Gamma\left(\frac{1-\lambda_1(a-1)}{c}\right)} - \lambda_1 E\left[\ln\left(\frac{x}{b}\right)^{a-1}\right] - \lambda_2 E\left[\left(\frac{x}{b}\right)^c\right]. \tag{6.39}$$

6.4.2.3 Relation between Parameters and Constraints

Differentiating Equation (6.39) with respect to the Lagrange multipliers and parameters a, b, and c, and equating each derivative to zero, we get

$$\frac{\partial H}{\partial \lambda_1} = \frac{(a-1)\lambda_2^{\frac{1-\lambda_1(a-1)}{c}}\left(\psi\left(\frac{1-\lambda_1(a-1)}{c}\right) - \ln\lambda_2\right)}{b\Gamma\left(\frac{1-\lambda_1(a-1)}{c}\right)} - E\left(\ln\left(\frac{x}{b}\right)^{a-1}\right) = 0$$

$$\tag{6.40}$$

$$\frac{\partial H}{\partial \lambda_2} = \frac{\lambda_2^{\frac{1-\lambda_1(a-1)}{c}}(1-\lambda_1(a-1))}{b\lambda_2\Gamma\left(\frac{1-\lambda_1(a-1)}{c}\right)} - E\left(\left(\frac{x}{b}\right)^c\right) = 0 \tag{6.41}$$

$$\frac{\partial H}{\partial a} = \frac{\lambda_1\lambda_2^{\frac{1-\lambda_1(a-1)}{c}}\left(\psi\left(\frac{1-\lambda_1(a-1)}{c}\right) - \ln\lambda_2\right)}{b\Gamma\left(\frac{1-\lambda_1(a-1)}{c}\right)} - \lambda_1 E\left(\ln\left(\frac{x}{b}\right)\right) = 0 \tag{6.42}$$

$$\frac{\partial H}{\partial b} = -\frac{c\lambda_2^{\frac{1-\lambda_1(a-1)}{c}}}{b^2\Gamma\left(\frac{1-\lambda_1(a-1)}{c}\right)} + \frac{\lambda_1(a-1)}{b} + \frac{c\lambda_2}{b}E\left(\left(\frac{x}{b}\right)^c\right) = 0 \tag{6.43}$$

$$\frac{\partial H}{\partial c} = \frac{c+(1-\lambda_1(a-1))\left[\psi\left(\frac{(1-\lambda_1(a-1))}{c}\right) - \ln\lambda_2\right]}{bc\lambda_2^{\frac{\lambda_1(a-1)-1}{c}}\Gamma\left(\frac{1-\lambda_1(a-1)}{c}\right)} - \lambda_2 E\left(\left(\frac{x}{b}\right)^c\ln\left(\frac{x}{b}\right)\right) = 0.$$

$$\tag{6.44}$$

From $\lambda_1 = 1 - a$ we have $d\lambda_1 = -da$. Equation (6.42) may be expressed through Equation (6.40) as

$$\frac{\partial H}{\partial a} = \frac{\partial H}{\partial \lambda_1} \frac{d\lambda_1}{da} = -\frac{\partial H}{\partial \lambda_1} = -\frac{(a-1)\lambda_2^{\frac{1-\lambda_1(a-1)}{c}}\left(\psi\left(\frac{1-\lambda_1(a-1)}{c}\right) - \ln\lambda_2\right)}{b\Gamma\left(\frac{1-\lambda_1(a-1)}{c}\right)} + E\left(\ln\left(\frac{x}{b}\right)^{a-1}\right)$$

$$\Rightarrow \frac{\partial H}{\partial a} = (1-a)\frac{(a-1)\lambda_2^{\frac{1-\lambda_1(a-1)}{c}}\left(\psi\left(\frac{1-\lambda_1(a-1)}{c}\right) - \ln\lambda_2\right)}{b\Gamma\left(\frac{1-\lambda_1(a-1)}{c}\right)} - (1-a)E\left(\ln\left(\frac{x}{b}\right)\right).$$

$$(6.45)$$

Substituting equation $\lambda_1 = 1 - a$ in Equation (6.42), we obtain exactly the same equation as Equation (6.45). Finally, substituting $\lambda_1 = 1 - a$ and $\lambda_2 = b^{-c}$ in Equation (6.45) and setting to zero, we have

$$\frac{(a-1)b^{-1-(a-1)^2}\left(\psi\left(\frac{1+(a-1)^2}{c}\right) + c\ln b\right)}{b\Gamma\left(\frac{1+(a-1)^2}{c}\right)} = E\left(\ln\left(\frac{x}{b}\right)\right). \qquad (6.46)$$

From $\lambda_2 = b^{-c}$, we have $\partial\lambda_2 = -cb^{-c-1}\partial b$. Equation (6.43) may be expressed through Equation (6.33) as

$$\frac{\partial H}{\partial b} = \frac{\partial H}{\partial \lambda_2} \frac{\partial\lambda_2}{\partial b} = -\frac{cb^{-c}}{b}\left[\frac{\lambda_2^{\frac{1-\lambda_1(a-1)}{c}}(1-\lambda_1(a-1))}{b\lambda_2\Gamma\left(\frac{1-\lambda_1(a-1)}{c}\right)} - E\left(\left(\frac{x}{b}\right)^c\right)\right]. \qquad (6.47)$$

Equating Equation (6.47) to Equation (6.43) yields

$$-\frac{cb^{-c}}{b}\left[\frac{\lambda_2^{\frac{1-\lambda_1(a-1)}{c}}(1-\lambda_1(a-1))}{b\lambda_2\Gamma\left(\frac{1-\lambda_1(a-1)}{c}\right)} - E\left(\left(\frac{x}{b}\right)^c\right)\right]$$

$$(6.48)$$

$$= -\frac{c\lambda_2^{\frac{1-\lambda_1(a-1)}{c}}}{b^2\Gamma\left(\frac{1-\lambda_1(a-1)}{c}\right)} + \frac{\lambda_1(a-1)}{b} + \frac{c\lambda_2}{b}E\left(\left(\frac{x}{b}\right)^c\right).$$

Substituting $\lambda_1 = 1 - a$ and $\lambda_2 = b^{-c}$ in both sides of Equation (6.48), we have

$$-\frac{cb^{-(a-1)^2-1}\left(1+(a-1)^2\right)}{b^2\Gamma\left(\dfrac{1+(a+1)^2}{c}\right)}+\frac{cb^{-c}}{b}E\left(\left(\frac{x}{b}\right)^c\right)=-\frac{cb^{-(a-1)^2-1}}{b^2\Gamma\left(\dfrac{1+(a-1)^2}{c}\right)}$$

$$-\frac{(a-1)^2}{b}+\frac{cb^{-c}}{b}E\left(\left(\frac{x}{b}\right)^c\right)$$

$$\Rightarrow -\frac{cb^{-(a-1)^2-1}\left(1+(a-1)^2\right)}{b^2\Gamma\left(\dfrac{1+(a-1)^2}{c}\right)}=-\frac{cb^{-(a-1)^2-1}}{b^2\Gamma\left(\dfrac{1+(a-1)^2}{c}\right)}-\frac{(a-1)^2}{b}.$$

$$(6.49)$$

Equation (6.49) can be further simplified as

$$cb^{-(a-1)^2-1}=b\Gamma\left(\frac{1+(a-1)^2}{c}\right).\qquad(6.50)$$

Furthermore, substituting $\lambda_1 = 1 - a$ and $\lambda_2 = b^{-c}$ in Equation (6.44), we have

$$\frac{c+\left(1+(a-1)^2\right)\left[\psi\left(\dfrac{\left(1+(a-1)^2\right)}{c}\right)+c\ln b\right]}{cb^{(a-1)^2+2}\Gamma\left(\dfrac{1+(a-1)^2}{c}\right)}-b^{-c}E\left(\left(\frac{x}{b}\right)^c\ln\left(\frac{x}{b}\right)\right)=0.$$

$$(6.51)$$

Now we reduce five equations to three equations by using the relation between Lagrange parameters and distribution parameters. The parameters may then be solved by simultaneously solving Equations (6.46), (6.50), and (6.51).

6.4.3 MOM

For the TPGG distribution, its noncentral moments may be computed as

$$\mu_r=\int\limits_0^\infty x^r\frac{c}{b\Gamma\left(\dfrac{a}{c}\right)}\left(\frac{x}{b}\right)^{a-1}\exp\left(-\left(\frac{x}{b}\right)^c\right)dx.\qquad(6.52)$$

Let $y = (x/b)^c$. With some simple algebraic manipulation, Equation (6.52) may be expressed as follows:

$$\mu_r = \frac{1}{\Gamma\left(\frac{a}{c}\right)} \int_0^\infty \left(by^{\frac{1}{c}}\right)^r y^{\frac{a-1}{c}} \exp(-y) y^{\frac{1}{c}-1} dy = \frac{b^{r+1}}{c} \int_0^\infty y^{\frac{r+a}{c}-1} \exp(-y) dy$$

$$\Rightarrow \mu_r = \frac{b^r \Gamma\left(\frac{a+r}{c}\right)}{\Gamma\left(\frac{a}{c}\right)}. \tag{6.53}$$

Now the parameters may be estimated by solving the following system of equations:

$$\frac{b\Gamma\left(\frac{a+1}{c}\right)}{\Gamma\left(\frac{a}{c}\right)} = \bar{x} \tag{6.54}$$

$$\frac{b^2\Gamma\left(\frac{a+2}{c}\right)}{\Gamma\left(\frac{a}{c}\right)} = \overline{x^2} \tag{6.55}$$

$$\frac{b^3\Gamma\left(\frac{a+3}{c}\right)}{\Gamma\left(\frac{a}{c}\right)} = \overline{x^3}. \tag{6.56}$$

The moment diagram may be expressed with the kurtosis and skewness as follows:

$$\mu_4 = E(X^4) - 4E(X^3)\mu + 6E(X^2)\mu^2 - 4E(X^3)\mu^3 + \mu^4$$

$$= b^4 \frac{\Gamma\left(\frac{a+4}{c}\right)\Gamma^3\left(\frac{a}{c}\right) - 4\Gamma\left(\frac{a+3}{c}\right)\Gamma\left(\frac{a+1}{c}\right)\Gamma^2\left(\frac{a}{c}\right) + 6\Gamma\left(\frac{a+2}{c}\right)\Gamma^2\left(\frac{a+1}{c}\right)\Gamma\left(\frac{a}{c}\right) - 3\Gamma^4\left(\frac{a+1}{c}\right)}{\Gamma^4\left(\frac{a}{c}\right)}$$

$$\tag{6.57}$$

$$\mu_3 = E(X^3) - 3E(X^2)\mu + 2\mu^2 = b^3 \frac{\Gamma\left(\frac{a+3}{c}\right)\Gamma^2\left(\frac{a}{c}\right) - 3\Gamma\left(\frac{a+2}{c}\right)\Gamma\left(\frac{a}{c}\right) + 2\Gamma^3\left(\frac{a+1}{c}\right)}{\Gamma^3\left(\frac{a}{c}\right)}$$

$$\tag{6.58}$$

$$\mu_2 = E(X^2) - \mu^2 = b^2 \frac{\Gamma\left(\frac{a+2}{c}\right)\Gamma\left(\frac{a}{c}\right) - \Gamma^2\left(\frac{a+1}{c}\right)}{\Gamma^2\left(\frac{a}{c}\right)}. \tag{6.59}$$

Let γ_1 and γ_2 represent the skewness and kurtosis, which may then be expressed as

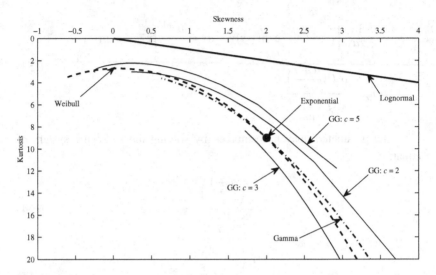

Figure 6.3 Moment diagram of GG distribution and its special cases.

$$\gamma_1 = \frac{\mu_3}{\mu_2^{\frac{3}{2}}} = \frac{\Gamma\left(\frac{a+3}{c}\right)\Gamma^2\left(\frac{a}{c}\right) - 3\Gamma\left(\frac{a+2}{c}\right)\Gamma\left(\frac{a}{c}\right) + 2\Gamma^3\left(\frac{a+1}{c}\right)}{\left(\Gamma\left(\frac{a+2}{c}\right)\Gamma\left(\frac{a}{c}\right) - \Gamma^2\left(\frac{a+1}{c}\right)\right)^{\frac{3}{2}}} \qquad (6.60)$$

$$\gamma_2 = \frac{\mu_4}{\mu_2^2}$$

$$= \frac{\Gamma\left(\frac{a+4}{c}\right)\Gamma^3\left(\frac{a}{c}\right) - 4\Gamma\left(\frac{a+3}{c}\right)\Gamma\left(\frac{a+1}{c}\right)\Gamma^2\left(\frac{a}{c}\right) + 6\Gamma\left(\frac{a+2}{c}\right)\Gamma^2\left(\frac{a+1}{c}\right)\Gamma\left(\frac{a}{c}\right) - 3\Gamma^4\left(\frac{a+1}{c}\right)}{\left(\Gamma\left(\frac{a+2}{c}\right)\Gamma\left(\frac{a}{c}\right) - \Gamma^2\left(\frac{a+1}{c}\right)\right)^2}.$$

$$(6.61)$$

Figure 6.3 plots the moment diagram for the GG distributions as well as its special cases, including the exponential, two-parameter gamma, and two-parameter Weibull distributions.

6.4.4 MLE Method

From the PDF, that is, Equation (6.1), the likelihood function for the TPGG can be expressed as

$$L = \prod_{i=1}^{n} \frac{c}{b\Gamma\left(\frac{a}{c}\right)} \left(\frac{x_i}{b}\right)^{a-1} \exp\left(-\left(\frac{x_i}{b}\right)^c\right), \qquad (6.62)$$

where n is the sample size.

Then, the corresponding log-likelihood function can be expressed as

$$
\ln L = n \ln\left(\frac{c}{b\Gamma\left(\frac{a}{c}\right)}\right) + (a-1)\sum_{i=1}^{n}\ln\left(\frac{x_i}{b}\right) - \sum_{i=1}^{n}\left(\frac{x_i}{b}\right)^c
$$

$$
= n \ln c - na \ln b - n \ln\left(\Gamma\left(\frac{a}{c}\right)\right) + (a-1)\sum_{i=1}^{n}\ln x_i - \sum_{i=1}^{n}\left(\frac{x_i}{b}\right)^c.
$$

$$(6.63)$$

Taking the derivative of $\ln L$ with respect to parameters a, b, and c and setting the differential equation equal to zero, we have

$$
\frac{\partial \ln L}{\partial a} = -n \ln b - \frac{n\psi\left(\frac{a}{c}\right)}{c} + \sum_{i=1}^{n}\ln x_i = 0
$$

$$(6.64)$$

$$
\frac{\partial \ln L}{\partial b} = \frac{c}{b^c}\sum_{i=1}^{n}x_i^c - na = 0
$$

$$(6.65)$$

$$
\frac{\partial \ln L}{\partial c} = \frac{n}{c} + \frac{an}{c^2}\psi\left(\frac{a}{c}\right) - \sum_{i=1}^{n}\left(\frac{x_i}{b}\right)^c \ln\left(\frac{x_i}{b}\right) = 0.
$$

$$(6.66)$$

Now the parameters may be estimated by solving the above system of equations.

As discussed in the previous chapters, the confidence interval may be constructed from Fisher information for the parameters estimated with MLE. The second derivatives of the sample likelihood function (i.e., the components of the observed Hessian matrix) are given as

$$
\frac{\partial^2 \ln L}{\partial a^2} = -\frac{n\psi^{(1)}\left(\frac{a}{c}\right)}{c^2}
$$

$$(6.67)$$

$$
\frac{\partial^2 \ln L}{\partial a \partial b} = -\frac{n}{b}
$$

$$(6.68)$$

$$
\frac{\partial^2 \ln L}{\partial a \partial c} = n\frac{c\psi\left(\frac{a}{c}\right) + a\psi^{(1)}\left(\frac{a}{c}\right)}{c^3}
$$

$$(6.69)$$

$$
\frac{\partial^2 \ln L}{\partial b^2} = -\frac{c^2}{b^{c+1}}\sum_{i=1}^{n}x_i^c
$$

$$(6.70)$$

$$
\frac{\partial^2 \ln L}{\partial b \partial c} = \frac{1}{b^c}\sum_{i=1}^{n}(x_i^c - x_i^c \ln b + cx_i^c \ln x_i)
$$

$$(6.71)$$

$$\frac{\partial^2 \ln L}{\partial c^2} = -\sum_{i=1}^{n} \left(\frac{x_i}{b}\right)^c \ln^2\left(\frac{x_i}{b}\right) - \frac{n}{c^2} - \frac{a^2 n \psi^{(1)}\left(\frac{a}{c}\right)}{c^4} - \frac{2an\psi\left(\frac{a}{c}\right)}{c^3}. \quad (6.72)$$

Furthermore, for parameters estimated using MOM and entropy methods, including MLE, the parametric bootstrap method may be applied to construct the confidence interval with the general procedure listed in the Appendix.

6.5 Application

6.5.1 Synthetic Data

In this section, the synthetic data generated from the TPGG distribution are applied first to illustrate the estimation methods, including MLE, MOM, and entropy. Table 6.1 lists the synthetic data generated from the known

Table 6.1. *Synthetic data generated from TPGG distribution with parameters* $a = 2$, $b = 0.5$, $c = 3$.

No.	X	No.	X	No.	X	No.	X
1	0.29	26	0.37	51	0.30	76	0.34
2	0.59	27	0.48	52	0.20	77	0.38
3	0.73	28	0.52	53	0.46	78	0.31
4	0.12	29	0.31	54	0.36	79	0.53
5	0.50	30	0.20	55	0.10	80	0.41
6	0.47	31	0.26	56	0.08	81	0.38
7	0.59	32	0.30	57	0.75	82	0.05
8	0.72	33	0.38	58	0.27	83	0.52
9	0.64	34	0.46	59	0.38	84	0.12
10	0.38	35	0.08	60	0.60	85	0.52
11	0.32	36	0.73	61	0.41	86	0.49
12	0.07	37	0.30	62	0.43	87	0.29
13	0.51	38	0.32	63	0.17	88	0.55
14	0.50	39	0.46	64	0.63	89	0.44
15	0.34	40	0.66	65	0.17	90	0.38
16	0.59	41	0.32	66	0.67	91	0.63
17	0.46	42	0.22	67	0.35	92	0.32
18	0.36	43	0.25	68	0.15	93	0.19
19	0.29	44	0.46	69	0.39	94	0.38
20	0.40	45	0.37	70	0.38	95	0.54
21	0.16	46	0.26	71	0.33	96	0.64
22	0.14	47	0.08	72	0.29	97	0.30
23	0.40	48	0.15	73	0.61	98	0.21
24	0.31	49	0.24	74	0.28	99	0.62
25	0.41	50	0.13	75	0.24	100	0.44

Table 6.2. *Parameters estimated, confidence intervals, and goodness-of-fit results: synthetic data.*

		Parameter			Goodness-of-fit	
	a	b	c		D_n	P-value
Gamma (2) initial	3.91	0.1	1			
MLE Fisher[a]	1.9	0.55	3.64			
	[1.42, 2.38]	[0.46, 0.64]	[2.02, 5.27]			
	[1.50, 3.36]	[0.25, 0.63]	[1.57, 6.71]		0.0696	0.158
MOM	2.01	0.5	3			
	[1.42, 2.43]	[0.44, 0.55]	[2.99, 3.02]		0.0696	0.21
	l0	11	12	C		
Entropy	−2.22	−1.05	7.91	3.02		
		[−2.68, −0.46]	[6.08, 29.29]	[1.31, 7.09]	0.696	0.21

[a] This is the confidence interval. The first one is from Fisher's information, and the second one is from simulation.

population. Using the aforementioned three estimation methods, Table 6.2 lists the parameters estimated. To apply the MLE, the initial parameters obtained from the gamma distribution are set as the initial parameters. From Table 6.2, it is seen that the parameters estimated from the different methods are not far from the true parameters. Additionally, the goodness-of-fit study on the known population parameters indicates that the synthetic data are properly generated and can be used to represent the true population. Figures 6.3–6.5 compare the fitted distribution with the true population distribution. The plots show that the MOM and MLE method result in very similar performances for frequency analysis, and the entropy method results in the best overall performance that is closest to the frequency distribution of the true population.

6.5.2 Peak Flow

The peak flow data from USGS09239500 are applied to evaluate the performances of different estimation methods. Applying the three estimation methods discussed earlier, Table 6.3 lists the parameters estimated as well as the corresponding confidence interval and goodness-of-fit. Table 6.3 indicates that (1) the parameters estimated using MOM and MLE are quite different from that fitted with two-parameter gamma distribution that was used to initiate the parameter estimation, that is, $a = 10.1$ and $b = 10.19$ and (2) it is appropriate to apply any of the three methods for parameter estimation to study the peak

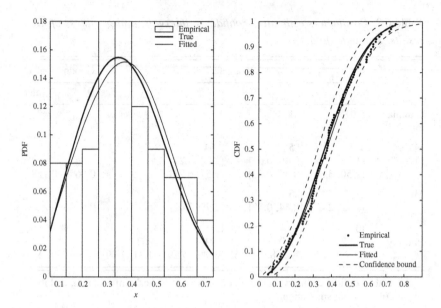

Figure 6.4 Comparison of fitted frequency and distribution function to those of true population distribution (MLE).

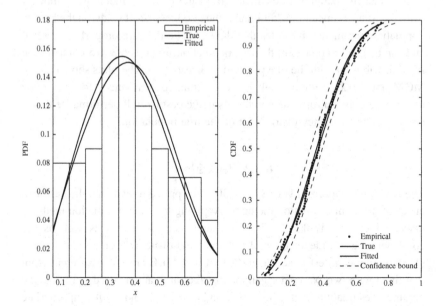

Figure 6.5 Comparison of fitted frequency and distribution function to those of true population distribution (MOM).

Table 6.3. *Parameters estimated, corresponding confidence intervals, and goodness-of-fit study: peak flow.*

	Parameter				Goodness-of-fit	
	a	b	c		D_n	P-value
Gamma (2) initial	10.10	10.19	1			
MLE	4.980 [3.287, 10.859]	79.470 [12.496, 123.173]	2.360 [1.058, 4.458]		0.054	0.441
MOM	4.985 [3.30, 6.40]	79.420 [44.12, 102.04]	2.357 [1.07, 2.89]		0.055	0.479
Entropy	$l0$ 20.970	$l1$ −3.977 [−4.69, −3.97]	$l2$ 3.32E−05 [3.32E−05, 3.86E−05]	C 2.357 [2.35, 2.38]	0.054	0.72

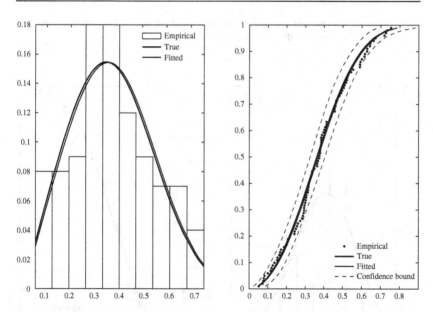

Figure 6.6 Comparison of fitted frequency and distribution function to those of true population (entropy).

flow data. Figures 6.7–6.9 compare empirical frequencies and fitted frequencies, empirical CDF and fitted parametric CDF, and the constructed 95% confidence interval. These plots show that all three methods yield very similar performances.

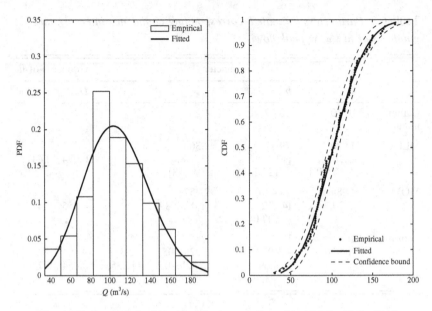

Figure 6.7 Comparison of empirical frequency and empirical CDF with the fitted frequency and fitted CDF: peak flow (MLE).

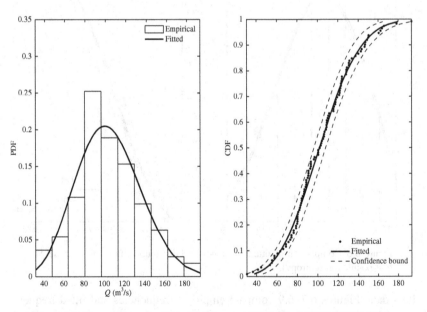

Figure 6.8 Comparison of empirical frequency and empirical CDF with the fitted frequency and fitted CDF: peak flow (MOM).

Table 6.4. *Quantiles computed for different probability values: peak flow.*

| Distribution | Method | Probability | | | | |
		0.5	0.8	0.9	0.95	0.99
Two-parameter gamma	MLE	99.61	128.80	146.08	161.42	192.91
Three-parameter gamma	MOM	101.73	129.15	143.91	156.26	179.69
	MLE	101.72	129.16	143.93	156.29	179.74
	Entropy	101.72	129.16	143.94	156.29	179.74

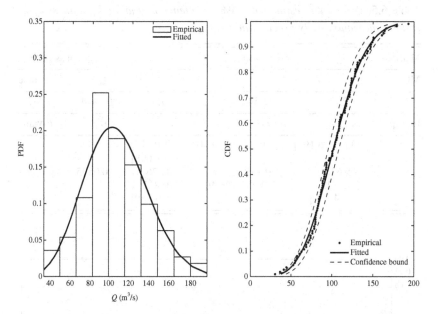

Figure 6.9 Comparison of empirical frequency and empirical CDF with the fitted frequency and fitted CDF: peak flow (entropy).

To further compare the three-parameter with the two-parameter gamma distribution, the quantiles are computed and listed in Table 6.4 with the comparison shown in Figure 6.10. Table 6.4 indicates that the quantiles obtained using the parameters estimated with MLE are very close for the two- and three-parameter gamma distributions for the 50, 80, 90, and 95 percentile with the maximum absolute difference observed at 99 percentile (about 7% of relative difference). Additionally, the quantiles estimated with the two-parameter gamma distribution are observed higher than those estimated from the GG distribution.

6.5.3 Maximum Daily Precipitation

The maximum daily precipitation (Brenham, Texas; GHCND: USC00411048) is applied to evaluate the application of the TPGG distribution. In this case, MLE and entropy methods are applied for parameter estimation. Table 6.5 lists the parameters estimated, the corresponding 95% confidence bound, and the goodness-of-fit results. The goodness-of-fit test results indicate that the TPGG distribution may be applied to model the maximum daily precipitation. Figures 6.11 and 6.12 compare the empirical distribution with the fitted

Table 6.5. *Parameters estimated, 95% confidence bound, and goodness-of-fit results: maximum daily precipitation.*

Method	Parameter			Goodness-of-fit	
MLE	a	b	c	D_n	P-value
	5.22	22.344	1.071	0.0474	0.767
Entropy	λ_1	λ_2	c	D_n	P-value
	[3.651, 6.193]	[17375, 26.434]	[1.026, 1.098]		
	−5.646 [−5.722, −4.324]	0.139 [0.004, 0.179]	0.876 [0.811, 0.989]	0.0535	0.645

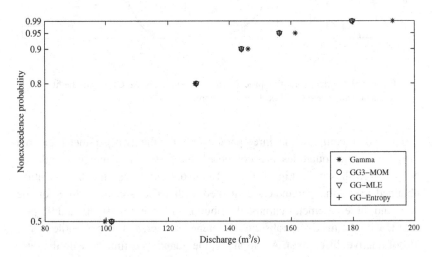

Figure 6.10 Comparison of quantiles obtained with fitted gamma distribution and generalized gamma distribution: peak flow.

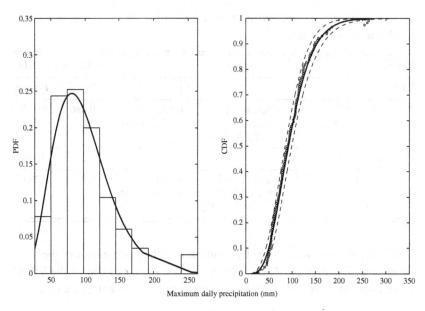

Figure 6.11 Comparison of empirical distribution with the fitted TPGG distribution: maximum daily precipitation (MLE).

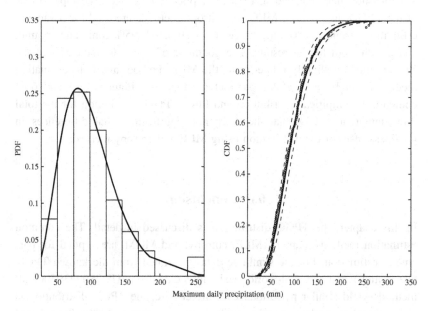

Figure 6.12 Comparison of empirical distribution with the fitted TPGG distribution: maximum daily precipitation (entropy).

Table 6.6. *Parameters estimated, 95% confidence bound, and goodness-of-fit results: total flow deficit.*

Method	Parameter			Goodness-of-fit	
MLE	a	b	c	D_n	P-value
	0.692 [0.554, 0.786]	8.95E+04 [8.816E+04, 9.370E+04]	0.951 [0.723, 1.091]	0.086	0.062
Entropy	λ_1	λ_2	c	D_n	P-value
	0.308 [0.171, 0.489]	2.041E−05 [1.402E−05, 2.091E−05]	0.951 [0.936, 0.985]	0.078	0.293

TPGG distribution, which visually confirms that the performance of the maximum entropy-based TPGG distribution is very similar to that of the TPGG distribution fitted with MLE for the maximum daily precipitation dataset.

6.5.4 Total Flow Deficit

The total flow deficit at Tilden, Texas, is applied to investigate the applicability of the TPGG distribution. MLE and entropy methods are used for parameter estimation. Table 6.6 lists the parameters estimated, 95% confidence bound, and goodness-of-fit test results. The goodness-of-fit results indicate that the entropy method slightly outperforms the MLE method, and both estimation methods may be applied for parameter estimation. Figures 6.13 and 6.14 compare the empirical distribution and fitted TPGG distribution for the total flow deficit and show that there are not significant visual differences in goodness-of-fit of the distribution using MLE and entropy methods.

6.6 Conclusion

In this chapter, the TPGG distribution is discussed in detail. The common estimation methods, namely, MLE, entropy, and MOM, are applied for parameter estimation. For the synthetic data (generated from the known TPGG distribution) and the real-world peak flow data from USGS09239500, all methods yield similar performances. Furthermore, the TPGG distribution is also applied to evaluate maximum daily precipitation and total flow deficit with the use of the estimation methods of MLE and entropy. It can be

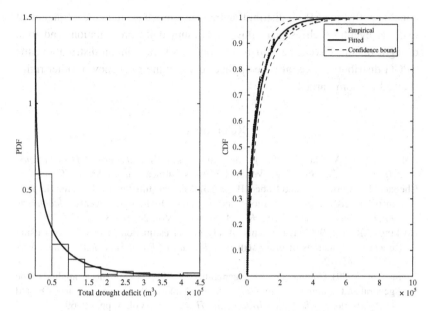

Figure 6.13 Comparison of empirical distribution with the fitted TPGG distribution: total flow deficit (MLE).

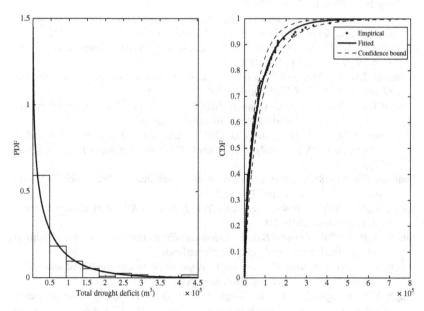

Figure 6.14 Comparison of empirical distribution with the fitted TPGG distribution: total flow deficit (entropy).

concluded that the TPGG distribution may be applied to study the frequency of extreme events, such as peak flow, maximum daily precipitation, and total drought deficit. Given the statistical property of the gamma distribution, the TPGG distribution may also be applied to study the frequency of other right-skewed random variables.

References

Bobée, B. and Ashkar, F. (1991). *The Gamma Family and Derived Distributions: Application Sin Hydrology.* Water Resources Publications, Littleton, CO.

Chebana, F., Adlouni, S., and Bobée, B. (2010). Mixed estimation methods for Halphen distributions with applications in extreme hydrologic events. *Stochastic Environmental Research and Risk Assessment*, Vol. 24, No. 3, pp. 359–376.

Hosking, J.R.M. (1990). L-moments: Analysis and estimation of distributions using linear combinations of order statistics. *Journal of Royal Statistical Society*, Vol. 52, pp. 105–124.

Klemes, V. (1989). Comment on "The generalized method of moments as applied to the generalized gamma distribution," by F. Ashkar, B. Bobée, Leroux, D. and D. Morisette, *Stochastic Hydrology and Hydraulics*, Vol. 3, pp. 68–69.

Landwehr, J.M., Matalas, N.C., and Wallis, J.R. (1979). Probability weighted moments compared with some traditional techniques in estimating Gumbel Parameters and quantiles. *Water Resources Research*, Vol. 15, No. 5, pp. 1055–1064. doi:10.1029/WR015i005p01055.

Lienhard, J.H. (1964). A statistical mechanical prediction of the dimensionless unit hydrograph. *Journal of the Geophysical Research*, Vol. 69, pp. 5231–5238.

Lienhard, J.H. (1972). Prediction of the dimensionless unit hydrograph. *Nordic Hydrology*, Vol. 3, pp. 107–109.

Lienhard, J.H. and Meyer, P.L. (1967). A physical basis for the gamma distribution. *Quarterly of Applied Mathematics*, Vol. 25, No. 3, pp. 330–334.

Natural Environment Research Council (NERC). (1999). Flood studies report, in five volumes. Institute of Hydrology, Wallingford, England.

Papalexiou, S.M. and Koutsoyiannis, D. (2012). Entropy based derivation of probability distributions: A case study to daily rainfall. *Advances in Water Resources*, Vol. 45, pp. 51–57.

Shannon, C.E. (1948). A mathematical theory of communication. *Bell System Technical Journal*, Vol. 27, No. 3, pp. 379–423.

Singh, V.P. (1988). *Hydrologic Systems: Vol. 1 Rainfall-Runoff Modeling.* Prentice Hall, Englewood Cliffs, NJ.

Singh, V.P. (1998). *Entropy-Based Parameter Estimation in Hydrology.* Kluwer Academic Publishers, Dordrecht, the Netherlands.

Singh, V.P. (2013). *Entropy Theory and Its Application in Environmental and Water Engineering.* Wiley-Blackwell, Hoboken, NJ.

Singh, V.P., Rajagopal, A.K., and Singh, K. (1986). Derivation of some frequency distributions using the principle of maximum entropy (POME). *Advances in Water Resources*, Vol. 9, No. 2, pp. 91–106.

7

Generalized Beta Lomax Distribution

7.1 Introduction

The beta Lomax distribution is a four-parameter frequency distribution that can be obtained from a general class of distributions defined by Eugene et al. (2002) and is discussed by Mahmoud and Abd El Ghafour (2013). This distribution can be obtained by transforming the beta distribution. Let Y be a beta-distributed random variable; then its probability density function (PDF) $f(y)$ can be given as

$$f(y) = \frac{1}{B(p,q)} y^{p-1}(1-y)^{q-1}; \quad y \in [0,1], \tag{7.1}$$

where p and q are the shape parameters and $B(p, q)$ is the beta function defined as

$$B(p,q) = \frac{\Gamma(p)\Gamma(q)}{\Gamma(p+q)} = \int\limits_0^1 y^{p-1}(1-y)^{q-1}dy; \quad p,q > 0, y \in [0,1]. \tag{7.2}$$

Now, let variable X be the monotone transformation of Y and defined as

$$x = b\left(y^{-\frac{1}{c}} - 1\right), \quad b > 0, c > 0. \tag{7.3}$$

Then, the PDF of X can be expressed as

$$f(x) = f(y(x))\left|\frac{dy}{dx}\right|. \tag{7.4}$$

From Equation (7.3), we have

$$y = \left(\frac{x}{b}+1\right)^{-c}, \quad \frac{dy}{dx} = -\frac{c}{b}\left(\frac{x}{b}+1\right)^{-c-1}dx. \tag{7.5}$$

Substituting Equation (7.5) in Equation (7.4) we have

$$f(x) = \frac{c}{bB(p,q)} \left(\frac{x}{b}+1\right)^{-cp-1} \left(1-\left(1+\frac{x}{b}\right)^{-c}\right)^{q-1} \; ; \quad b,c,p,q > 0, x \in (0, +\infty),$$

(7.6)

where b is the scale parameter; c, p, and q are the shape parameters.

It should be noted that when y is 1, x is 0 and when y tends to 0, x tends to $+\infty$. Thus, y varies from 0 to 1. Figure 7.1 plots the PDF of the beta Lomax distribution with different fixed parameters. As shown in Figure 7.1, the beta Lomax distribution is positively skewed, which may be applied for frequency analysis in water resources and environmental engineering.

From the four-parameter beta Lomax distribution, the following special cases may be derived:

Three-parameter Lomax distribution: If $q = 1$, Equation (7.6) reduces to

$$f(x) = \frac{c}{bB(p,1)} \left(1+\frac{x}{b}\right)^{-cp-1} = \frac{cp}{b}\left(1+\frac{x}{b}\right)^{-(cp+1)}.$$

(7.7)

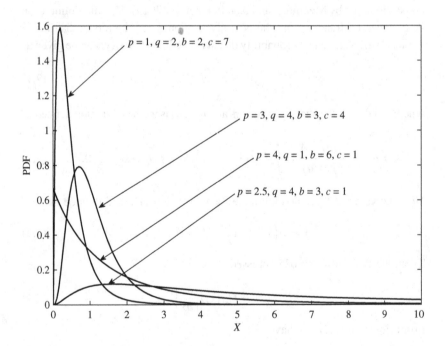

Figure 7.1 Shapes of beta Lomax PDF with fixed parameters.

Two-parameter Lomax distribution: In Equation (7.7) let $\alpha = cp$; the two-parameter Lomax distribution is then obtained as

$$f(x) = \frac{\alpha}{b}\left(1 + \frac{x}{b}\right)^{-(\alpha+1)}.$$

(7.8)

It is worth to note that the two-parameter Lomax distribution is also called Pareto II distribution without the location parameter.

7.2 Generating Differential Equation

Taking the logarithm of Equation (7.6), we obtain

$$\ln(f(x)) = \ln\left(\frac{c}{bB(p,q)}\right) - (cp+1)\ln\left(1 + \frac{x}{b}\right) + (q-1)\ln\left(1 - \left(1 + \frac{x}{b}\right)^{-c}\right).$$

(7.9)

Differentiating Equation (7.9) with respect to x, we have

$$\frac{1}{f(x)}\frac{df}{dx} = -(cp+1)\frac{1}{1+\frac{x}{b}}\left(\frac{1}{b}\right) + (q-1)\frac{1}{1-\left(1+\frac{x}{b}\right)^{-c}}\left(\frac{c}{b}\left(1+\frac{x}{b}\right)^{-c-1}\right).$$

(7.10)

Equation (7.10) can be further simplified as

$$\frac{1}{f}\frac{df}{dx} = -\frac{1}{b}\left(\frac{cp+1}{1+\frac{x}{b}} - \frac{c(q-1)}{\left(1+\frac{x}{b}\right)^{c+1} - \left(1+\frac{x}{b}\right)}\right).$$

(7.11)

7.3 Derivation Using Entropy Theory

The Shannon entropy can be defined as

$$H(x) = -\int f(x)\ln f(x)dx.$$

(7.12)

Substitution of Equation (7.6) in Equation (7.12) yields

$$H(x) = -\int f(x)\left(\ln\left(\frac{c}{bB(p,q)}\right) - (cp+1)\ln\left(1 + \frac{x}{b}\right) + (q-1)\ln\left(1 - \left(1 + \frac{x}{b}\right)^{-c}\right)\right)dx.$$

(7.13)

From Equation (7.12), the following constraints can be defined:

$$\int f(x)dx = 1 \tag{7.14}$$

$$\int f(x)\ln\left(1 + \frac{x}{b}\right)dx = C_1 \tag{7.15}$$

$$\int f(x)\ln\left(1 - \left(1 + \frac{x}{b}\right)^{-c}\right)dx = C_2. \tag{7.16}$$

For determining $f(x)$, entropy is maximized subject to the constraints given by Equations (7.14)–(7.16). To that end, we construct the Lagrangian function L as

$$
\begin{aligned}
L = &-\int f(x)\ln f(x)dx - (\lambda_0 - 1)\left(\int f(x)dx - 1\right) \\
&- \lambda_1 \left(\int f(x)\ln\left(1 + \frac{x}{b}\right)dx - E\left(\ln\left(1 + \frac{x}{b}\right)\right)\right) \\
&- \lambda_2 \left(\int f(x)\ln\left(1 - \left(1 + \frac{x}{b}\right)^{-c}\right)dx - E\left(\ln\left(1 - \left(1 + \frac{x}{b}\right)^{-c}\right)\right)\right),
\end{aligned}
\tag{7.17}
$$

where λ_0, λ_1, and λ_2 are Lagrange multipliers. Differentiating Equation (7.17) with $f(x)$, we get the entropy-based PDF:

$$f(x) = \exp\left(-\lambda_0 - \lambda_1 \ln\left(1 + \frac{x}{b}\right) - \lambda_2 \ln\left(1 - \left(1 + \frac{x}{b}\right)^{-c}\right)\right). \tag{7.18}$$

Now the objective is to determine the Lagrange multipliers and in turn the distribution parameters, which entails two steps. First, the Lagrange multipliers are expressed in terms of distribution parameters, and second, the parameters are expressed in terms of the specified constraints.

Substituting Equation (7.18) in Equation (7.14), we get

$$\exp(\lambda_0) = \int \exp\left(-\lambda_1 \ln\left(1 + \frac{x}{b}\right) - \lambda_2 \ln\left(1 - \left(1 + \frac{x}{b}\right)^{-c}\right)\right)dx. \tag{7.19}$$

Equation (7.19) can also be written as

$$\exp(\lambda_0) = \int \left(1 + \frac{x}{b}\right)^{-\lambda_1}\left(1 - \left(1 + \frac{x}{b}\right)^{-c}\right)^{\lambda_2}dx. \tag{7.20}$$

Equation (7.20) can be solved analytically by setting $z = (1 + (x/b))^{-c}$ as

$$\exp(\lambda_0) = \frac{b}{c} \int_0^1 z^{\frac{\lambda_1}{c}-\frac{1}{c}-1}(1-z)^{-\lambda_2}dz = \frac{b}{c}B\left(\frac{\lambda_1}{c}-\frac{1}{c}, 1-\lambda_2\right), \qquad (7.21)$$

where $B(\cdot,\cdot)$ is a beta function. Inserting Equation (7.21) in Equation (7.18), we obtain

$$f(x) = \frac{c}{bB\left(\frac{\lambda_1}{c}-\frac{1}{c}, -\lambda_2+1\right)}\left(1+\frac{x}{b}\right)^{-\lambda_1}\left(1-\left(1+\frac{x}{b}\right)^{-c}\right)^{-\lambda_2}. \qquad (7.22)$$

Comparing Equation (7.22) with Equation (7.6), it is observed that

$$\lambda_1 = cp + 1; \quad \lambda_2 = 1 - q. \qquad (7.23)$$

7.4 Parameter Estimation

The distribution parameters are estimated using four methods: (1) regular entropy method, (2) parameter space expansion method, (3) maximum likelihood estimation (MLE) method, and (4) method of moments.

7.4.1 Regular Entropy Method

The zeroth Lagrange multiplier can be written from Equation (7.21) as

$$\lambda_0 = \ln b - \ln c + \ln \Gamma\left(\frac{\lambda_1 - 1}{c}\right) + \ln \Gamma(1-\lambda_2) - \ln\left(\frac{\lambda_1 - 1}{c}+1-\lambda_2\right). \qquad (7.24)$$

From Equation (7.24), it is shown that $\lambda_1 > 1$ and $\lambda_2 < 1$.

Differentiating Equation (7.24) with respect to λ_1 and λ_2 separately, the result is

$$\frac{\partial \lambda_0}{\partial \lambda_1} = \frac{1}{c}\psi\left(\frac{\lambda_1 - 1}{c}\right) - \frac{1}{c}\psi\left(\frac{\lambda_1 - 1}{c}+1-\lambda_2\right) \qquad (7.25)$$

$$\frac{\partial \lambda_0}{\partial \lambda_2} = -\psi(1-\lambda_2) + \psi\left(\frac{(\lambda_1 - 1)}{c}+1-\lambda_2\right), \qquad (7.26)$$

where $\psi(x) = (d\ln\Gamma(x))/dx$ is a digamma function.

From Equation (7.19), differentiating λ_0 with respect to λ_1 and λ_2 separately, the result is

$$\frac{\partial \lambda_0}{\partial \lambda_1} = -\frac{\int\limits_0^\infty \ln\left(1 + \frac{x}{b}\right) \exp\left(-\lambda_1 \ln\left(1 + \frac{x}{b}\right) - \lambda_2 \ln\left(1 - \left(1 + \frac{x}{b}\right)^{-c}\right)\right) dx}{\int\limits_0^\infty \exp\left(-\lambda_1 \ln\left(1 + \frac{x}{b}\right) - \lambda_2 \ln\left(1 - \left(1 + \frac{x}{b}\right)^{-c}\right)\right) dx}$$

$$= -E\left(\ln\left(1 + \frac{x}{b}\right)\right)$$

$$(7.27)$$

$$\frac{\partial \lambda_0}{\partial \lambda_2} = -\frac{\int\limits_0^\infty \ln\left(1 - \left(1 + \frac{x}{b}\right)^{-c}\right) \exp\left(-\lambda_1 \ln\left(1 + \frac{x}{b}\right) - \lambda_2 \ln\left(1 - \left(1 + \frac{x}{b}\right)^{-c}\right)\right) dx}{\int\limits_0^\infty \exp\left(-\lambda_1 \ln\left(1 + \frac{x}{b}\right) - \lambda_2 \ln\left(1 - \left(1 + \frac{x}{b}\right)^{-c}\right)\right) dx}$$

$$= -E\left(\ln\left(1 - \left(1 + \frac{x}{b}\right)^{-c}\right)\right).$$

$$(7.28)$$

Equating Equation (7.25) to Equation (7.27) and Equation (7.26) to Equation (7.28), we obtain two equations for determining distribution parameters. However, the beta Lomax distribution has four parameters, so two additional equations are needed. These are obtained as follows.

Taking the second derivatives of the zeroth Lagrange multiplier with respect to the first and second Lagrange multipliers, that is, $\partial^2 \lambda_0 / \partial \lambda_1^2$ and $\partial^2 \lambda_0 / \partial \lambda_2^2$, from Equations (7.25) and (7.26), we have

$$\frac{\partial^2 \lambda_0}{\partial \lambda_1^2} = \frac{1}{c^2} \psi'\left(\frac{\lambda_1 - 1}{c}\right) - \frac{1}{c^2} \psi'\left(\frac{\lambda_1 - 1}{c} + 1 - \lambda_2\right) \qquad (7.29)$$

$$\frac{\partial^2 \lambda_0}{\partial \lambda_2^2} = \psi'(1 - \lambda_2) - \psi'\left(\frac{\lambda_1 - 1}{c} + 1 - \lambda_2\right), \qquad (7.30)$$

where $\psi'(x) = \frac{d\psi(x)}{d(x)} = \frac{d^2 \ln \Gamma(x)}{dx^2}$.

Similarly, from Equations (7.27) and (7.28), we have

$$\frac{\partial^2 \lambda_0}{\partial \lambda_1^2} = \text{var}\left(\ln\left(1 + \frac{x}{b}\right)\right) \qquad (7.31)$$

$$\frac{\partial^2 \lambda_0}{\partial \lambda_2^2} = \text{var}\left(\ln\left(1 - \left(1 + \frac{x}{b}\right)^{-c}\right)\right). \qquad (7.32)$$

Equating Equations (7.29) and (7.30) to Equations (7.31) and (7.32) correspondingly, we get two additional equations.

7.4.2 Parameter Space Expansion Method

In this method, constraints are redefined as Equation (7.13) and

$$\int_0^\infty (cq+1)\ln\left(1+\left(\frac{x}{b}\right)\right)f(x)dx = E\left((cq+1)\ln\left(1+\frac{x}{b}\right)\right) = C_1 \quad (7.33)$$

$$\int_0^\infty (p-1)\ln\left(1-\left(1+\frac{x}{b}\right)^{-c}\right)f(x)dx = E\left((p-1)\ln\left(1-\left(1+\frac{x}{b}\right)^{-c}\right)\right) = C_2. \quad (7.34)$$

Following the same procedure as before, we get the entropy-based PDF:

$$f(x) = \exp\left(-\lambda_0 - \lambda_1(cp+1)\ln\left(1+\frac{x}{b}\right) - \lambda_2(q-1)\ln\left(1-\left(1+\frac{x}{b}\right)^{-c}\right)\right). \quad (7.35)$$

Substituting Equation (7.35) in Equation (7.14), we get

$$\exp(\lambda_0) = \int_0^\infty \exp\left(-\lambda_1(cp+1)\ln\left(1+\frac{x}{b}\right) - \lambda_2(q-1)\ln\left(1-\left(1+\frac{x}{b}\right)^{-c}\right)\right)dx. \quad (7.36)$$

Equation (7.36) can be rewritten as

$$\exp(\lambda_0) = \int_0^\infty \left(1+\frac{x}{b}\right)^{-\lambda_1(cp+1)}\left(1-\left(1+\frac{x}{b}\right)^{-c}\right)^{-\lambda_2(q-1)}dx. \quad (7.37)$$

Equation (7.37) can be solved by letting $y = (1+(x/b))^{-c}$ as

$$\exp(\lambda_0) = \frac{b}{c}B\left(\frac{\lambda_1(cp+1)-1}{c}, -\lambda_2(q-1)+1\right), \quad (7.38)$$

where $B(\cdot,\cdot)$ is a beta function.

Inserting Equation (7.38) in Equation (7.35) we obtain

$$f(x) = \frac{c}{bB\left(\frac{\lambda_1(cp+1)-1}{c}, -\lambda_2(q-1)+1\right)}\left(1+\frac{x}{b}\right)^{-\lambda_1(cp+1)}\left(1-\left(1+\frac{x}{b}\right)^{-c}\right)^{-\lambda_2(q-1)}. \quad (7.39)$$

Taking the logarithm of Equation (7.39),

$$\ln f(x) = \ln\left(\frac{c}{bB\left(\frac{\lambda_1(cp+1)-1}{c}, -\lambda_2(q-1)+1\right)}\right) - \lambda_1(cp+1)\ln\left(1+\frac{x}{b}\right)$$
$$- \lambda_2(q-1)\ln\left(1-\left(1+\frac{x}{b}\right)^{-c}\right). \quad (7.40)$$

Thus, entropy of the distribution can be written as

$$H(f) = \ln \left(\frac{c}{bB\left(\frac{\lambda_1(cp+1)-1}{c}, -\lambda_2(q-1)+1 \right)} \right) - \lambda_1(cp+1)E\left(\ln\left(1+\frac{x}{b}\right) \right)$$

$$- \lambda_2(q-1)E\left(\ln\left(1 - \left(1+\frac{x}{b}\right)^{-c}\right) \right).$$

$$(7.41)$$

Differentiating Equation (7.41) with respect to Lagrange multipliers and distribution parameters, we obtain

$$\frac{\partial H}{\partial \lambda_1} = 0$$

$$= \frac{cp+1}{c}\left(\psi\left(\frac{\lambda_1(cp+1)-1}{c} - \lambda_2(q-1)+1 \right) - \psi\left(\frac{\lambda_1(cp+1)-1}{c} \right) \right)$$

$$-(cp+1)E\left[\ln\left(1+\frac{x}{b}\right) \right]$$

$$\Rightarrow E\left(\ln\left(1+\frac{x}{b}\right) \right)$$

$$= \frac{1}{c}\left(\psi\left(\frac{\lambda_1(cp+1)-1}{c} - \lambda_2(q-1)+1 \right) - \psi\left(\frac{\lambda_1(cp+1)-1}{c} \right) \right)$$

$$(7.42)$$

$$\frac{\partial H}{\partial \lambda_2} = 0$$

$$= (q-1)\left(\psi\left(-\lambda_2(q-1)+1 \right) - \psi\left(\frac{\lambda_1(cp+1)-1}{c} - \lambda_2(q-1)+1 \right) \right)$$

$$-(q-1)E\left(\ln\left(1 - \left(1+\frac{x}{b}\right)^{-c}\right) \right)$$

$$\Rightarrow E\left[\ln\left(1 - \left(1+\frac{x}{b}\right)^{c}\right) \right]$$

$$= \left(\psi\left(-\lambda_2(q-1)+1 \right) - \psi\left(\frac{\lambda_1(cp+1)-1}{c} - \lambda_2(q-1)+1 \right) \right)$$

$$(7.43)$$

$$\frac{\partial H}{\partial p} = 0$$

$$= \lambda_1 \left(\psi \left(\frac{\lambda_1(cp+1)-1}{c} - \lambda_2(q-1) + 1 \right) - \psi \left(\frac{\lambda_1(cp+1)-1}{c} \right) \right)$$

$$- c\lambda_1 E \left[\ln \left(1 + \frac{x}{b} \right) \right]$$

$$\Rightarrow E \left[\ln \left(1 + \frac{x}{b} \right) \right] = \frac{1}{c} \left(\psi \left(\frac{\lambda_1(cp+1)-1}{c} - \lambda_2(q-1) + 1 \right) - \psi \left(\frac{\lambda_1(cp+1)-1}{c} \right) \right)$$

$$(7.44)$$

$$\frac{\partial H}{\partial q} = 0$$

$$= \lambda_2 \left(\psi \left(-\lambda_2(q-1) + 1 \right) - \psi \left(\frac{\lambda_1(cp+1)-1}{c} - \lambda_2(q-1) + 1 \right) \right)$$

$$- \lambda_2 E \left[\ln \left(1 - \left(1 + \frac{x}{b} \right)^{-c} \right) \right] \Rightarrow E \left[\ln \left(1 - \left(1 + \frac{x}{b} \right)^{-c} \right) \right]$$

$$= \left(\psi \left(-\lambda_2(q-1) + 1 \right) - \psi \left(\frac{\lambda_1(cp+1)-1}{c} - \lambda_2(q-1) + 1 \right) \right)$$

$$(7.45)$$

$$\frac{\partial H}{\partial b} = -\frac{1}{b} + \lambda_1(cp+1)E\left(\frac{x}{b(b+x)} \right) + \lambda_2(q-1)E\left(\frac{cx}{b^2} \left(1 + \frac{x}{b} \right)^{-c-1} \right) = 0$$

$$(7.46)$$

$$\frac{\partial H}{\partial c} = 0 = \frac{1}{c} - \lambda_1 p E \left(\ln \left(1 + \frac{x}{b} \right) \right) + \lambda_2(q-1)E\left(\frac{\left(1 + \frac{x}{b} \right)^{-c} \ln \left(1 + \frac{x}{b} \right)}{1 - \left(1 + \frac{x}{b} \right)^{-c}} \right).$$

$$(7.47)$$

Comparing Equation (7.42) with Equation (7.44), we have $\lambda_1 = cp + 1$. Comparing Equation (7.43) with Equation (7.45), we have $\lambda_2 = q - 1$. To this end, Equations (7.44) and (7.45) can be rewritten as

$$E\left[\ln \left(1 + \frac{x}{b} \right) \right] = \frac{1}{c} \left(\psi \left(\frac{(cp+1)^2-1}{c} - (q-1)^2 + 1 \right) - \psi \left(\frac{(cp+1)^2-1}{c} \right) \right)$$

$$(7.48)$$

$$E\left[\ln \left(1 - \left(1 + \left(\frac{x}{b} \right)^{-c} \right) \right) \right] = \left(\psi \left(-(q-1)^2 + 1 \right) - \psi \left(\frac{(cp+1)^2-1}{c} - (q-1)^2 + 1 \right) \right).$$

$$(7.49)$$

Similarly substituting $\lambda_1 = cp + 1$ and $\lambda_2 = q - 1$ in Equations (7.46) and (7.47), we have

$$-\frac{1}{b} + (cp+1)^2 E\left(\frac{x}{b(b+x)}\right) + (q-1)^2 E\left(\frac{cx}{b^2}\left(1+\frac{x}{b}\right)^{-c-1}\right) = 0 \quad (7.50)$$

$$\frac{1}{c} - p(cp+1)E\left(\ln\left(1+\frac{x}{b}\right)\right) + (q-1)^2 E\left(\frac{\left(1+\frac{x}{b}\right)^{-c}\ln\left(1+\frac{x}{b}\right)}{1-\left(1+\frac{x}{b}\right)^{-c}}\right) = 0.$$

$$(7.51)$$

To this end, we have reduced to four equations with four parameters. Now, the parameters may be estimated by solving the system of equations numerically.

7.4.3 MLE Method

From Equation (7.6), the log-likelihood function of the beta Lomax distribution may be expressed as

$$\ln L = n\ln c - n\ln b - n\ln B(p,q) - (cp+1)\sum_{i=1}^{n}\ln\left(1+\frac{x_i}{b}\right)$$
$$+(q-1)\sum_{i=1}^{n}\ln\left(1-\left(1+\frac{x_i}{b}\right)^{-c}\right). \quad (7.52)$$

Taking the partial derivative of Equation (7.52) with respect to parameters p, q, b, and c, we have

$$\frac{\partial \ln L}{\partial p} = -n\psi(p) + n\psi(p+q) + c\sum_{i=1}^{n}\ln\left(1+\frac{x_i}{b}\right) \quad (7.53)$$

$$\frac{\partial \ln L}{\partial q} = -n\psi(q) + n\psi(p+q) + \sum_{i=1}^{n}\ln\left(1-\left(1+\frac{x_i}{b}\right)^{-c}\right) \quad (7.54)$$

$$\frac{\partial \ln L}{\partial b} = -\frac{n}{b} - (cp+1)\sum_{i=1}^{n}\frac{x_i}{b^2+bx_i} \quad (7.55)$$

$$\frac{\partial \ln L}{\partial c} = \frac{n}{c} - p\sum_{i=1}^{n}\ln\left(1+\frac{x_i}{b}\right) - (q-1)\sum_{i=1}^{n}\frac{1}{\left(1+\frac{x_i}{b}\right)^{c}-1}\ln\left(1+\frac{x_i}{b}\right). \quad (7.56)$$

Equating Equations (7.53)–(7.56) to zero, we can estimate the parameters of the beta Lomax distribution numerically by solving the system of equations.

For the parameters estimated with MLE, the confidence bound may be constructed using the observed information matrix through the observed Hessian matrix. The second derivative of the log-likelihood function with respect to parameters (i.e., the elements of the observed Hessian matrix) can be given as

$$\frac{\partial^2 \ln L}{\partial p^2} = -n\psi^{(1)}(p) + n\psi^{(1)}(p+q) \tag{7.57}$$

$$\frac{\partial^2 \ln L}{\partial p \partial q} = n\psi^{(1)}(p+q) \tag{7.58}$$

$$\frac{\partial^2 \ln L}{\partial p \partial b} = -\frac{c}{b^2} \sum_{i=1}^{n} \frac{x_i}{1 + \frac{x_i}{b}} \tag{7.59}$$

$$\frac{\partial^2 \ln L}{\partial p \partial c} = -\sum_{i=1}^{n} \ln\left(1 + \frac{x_i}{b}\right) \tag{7.60}$$

$$\frac{\partial^2 \ln L}{\partial q^2} = n\psi^{(1)}(p+q) - n\psi^{(1)}(q) \tag{7.61}$$

$$\frac{\partial^2 \ln L}{\partial q \partial b} = -\frac{c}{b^2} \sum_{i=1}^{n} \frac{x_i}{\left(1 + \frac{x_i}{b}\right)^{c+1} - \left(1 + \frac{x_i}{b}\right)} \tag{7.62}$$

$$\frac{\partial^2 \ln L}{\partial q \partial c} = \sum_{i=1}^{n} \frac{\ln\left(1 + \frac{x_i}{b}\right)}{\left(\frac{x_i}{b} + 1\right)^c - 1} \tag{7.63}$$

$$\frac{\partial^2 \ln L}{\partial b^2} = \frac{n}{b^2} + (cp+1) \sum_{i=1}^{n} \frac{x_i(x_i + 2b)}{(b^2 + bx_i)^2} \tag{7.64}$$

$$\frac{\partial^2 \ln L}{\partial b \partial c} = -\frac{p}{b} \sum_{i=1}^{n} \frac{x_i}{b + x_i} \tag{7.65}$$

$$\frac{\partial^2 \ln L}{\partial c^2} = -\frac{n}{c^2} + (q-1) \sum_{i=1}^{n} \frac{\left(1 + \frac{x_i}{b}\right)^c \left(\ln\left(\frac{x_i}{b} + 1\right)\right)^2}{\left(\left(1 + \frac{x_i}{b}\right)^c - 1\right)^2}. \tag{7.66}$$

Then, the confidence bound for the parameters estimated with MLE can be approximated by diagonal components of the inverse of the negative observed Hessian matrix constructed using Equations (7.57)–(7.66). More specifically, the critical value of the 95% bound is 1.96, which is obtained from the standard normal distribution.

7.4.4 Method of Moments

The rth noncentral moment for the beta Lomax distribution can be expressed as

$$E(X^r) = \int_0^\infty x^r \frac{c}{bB(p,q)} \left(\frac{x}{b}+1\right)^{-cp-1} \left(1 - \left(1+\frac{x}{b}\right)^{-c}\right)^{q-1} dx. \quad (7.67)$$

It is obvious that there is no closed form solution for Equation (7.67) to easily apply the method of moment estimation, but numerically the moment equations can be solved. However, if $q = 1$, the four-parameter beta Lomax distribution is reduced to the three-parameter beta Lomax distribution, and Equation (7.6) can be rewritten as

$$f(x; b, c, p) = \frac{c}{bB(p,1)} \left(\frac{x}{b}+1\right)^{-cp-1}. \quad (7.68)$$

Equation (7.67) can be rewritten as

$$E(X^r) = \int_0^\infty \frac{cp}{b} x^r \left(\frac{x}{b}+1\right)^{-cp-1} dx. \quad (7.69)$$

Equation (7.69) can be solved as

$$E(X^r) = cpb^r \int_0^\infty \left(\frac{x}{b}\right)^r \left(1+\frac{x}{b}\right)^{-(cp+1)} d\left(\frac{x}{b}\right) = cbp^r B(r+1, cp-r); \quad \exists cp > r.$$

$$(7.70)$$

If $cp > 3$, then the first three moments are computed as

$$E(X) = cpbB(2, cp - 1) = \frac{b}{cp - 1} \quad (7.71)$$

$$E(X^2) = cpb^2 B(3, cp - 2) = \frac{2b^2}{(cp - 1)(cp - 2)} \quad (7.72)$$

$$E(X^3) = cpb^3 B(4, cp - 3) = \frac{6b^3}{(cp - 1)(cp - 2)(cp - 3)}. \quad (7.73)$$

Solving the system of equations by equating the population moments to the sample moments numerically, we can estimate the parameters of the three-parameter beta Lomax distribution.

7.5 Application

7.5.1 Synthetic Data

Previously, we have shown that with the transformation, $y = (1 + (x/b))^{-c}$ follows the beta distribution with parameters (p, q). In this section, we first

Table 7.1. *Random variable simulated from beta Lomax distribution with parameters* ($p = 3.5$, $q = 2$, $b = 6$, *and* $c = 2.5$).

No.	X	No.	X	No.	X	No.	X
1	1.97	26	3.18	51	1.31	76	1.35
2	2.16	27	1.70	52	0.16	77	0.80
3	1.03	28	1.53	53	0.79	78	1.32
4	5.78	29	1.24	54	0.21	79	0.25
5	0.42	30	2.04	55	2.37	80	2.47
6	2.26	31	0.84	56	2.23	81	1.02
7	2.95	32	0.49	57	1.59	82	1.49
8	0.37	33	1.06	58	0.98	83	0.41
9	0.73	34	1.01	59	1.31	84	0.65
10	1.90	35	2.53	60	2.79	85	1.13
11	0.92	36	1.40	61	0.99	86	1.29
12	0.98	37	0.35	62	1.03	87	0.85
13	1.30	38	1.60	63	1.20	88	1.47
14	3.77	39	2.34	64	1.45	89	1.36
15	0.59	40	1.28	65	0.90	90	0.58
16	0.09	41	0.97	66	1.51	91	0.28
17	2.24	42	4.44	67	2.37	92	1.80
18	0.95	43	0.79	68	1.14	93	0.26
19	0.48	44	0.98	69	1.93	94	0.52
20	0.63	45	2.09	70	0.22	95	0.53
21	0.39	46	2.29	71	0.50	96	0.64
22	0.74	47	0.79	72	2.79	97	2.39
23	0.85	48	1.33	73	1.54	98	0.81
24	6.36	49	1.11	74	0.66	99	0.29
25	1.20	50	1.44	75	1.17	100	2.07

simulate the random variables from the true population to evaluate the different parameter estimations methods, including entropy and MLE methods. Table 7.1 lists the random variable data generated from the true beta Lomax population with the parameters of $p = 3.5$, $q = 2$, $b = 6$, and $c = 2.5$. Table 7.2 lists the parameters estimated with the use of MLE and entropy methods for the simulated random variable data as well as the corresponding confidence intervals and goodness-of-fit measures with the use of the parametric bootstrap approach. The goodness-of-fit results indicate that the random variables are properly sampled from the true population. Figure 7.2a and b compare the true distribution with the empirical and fitted probability functions for the simulated random samples, indicating that the frequency distribution computed with the parameters estimated using the MLE and entropy methods reaches the agreement with the true frequency distribution.

7.5.2 Peak Flow

Previously, we have shown the application of MLE and entropy methods to the simulated random variable data from the true beta Lomax population. In what follows, we will evaluate the application with the use of peak flow data from

Table 7.2. *Parameters estimated using MLE and entropy methods, corresponding confidence intervals, and goodness-of-fit results: synthetic data.*

Method	Parameter and confidence interval				Goodness-of-fit	
MLE	p	q	b	c	D_n	P
	3.78 [2.39, 22.38]	2.56 [1.39, 2.61]	5.5 [3.27, 18.71]	2.67 [0.82, 10.46]	0.08	0.15
Entropy	λ_0	λ_1	λ_2	b	c	
	−3.75	13.21 [6.89, 15.7]	−1.45 [−6.73, −0.56]	6.5 [2.80, 12.87]	1.36 [0,3.2]	0.08	0.20
	Equivalent: $p = 8.9$, $q = 2.45$, $b = 6.5$, $c = 1.36$					

(a)

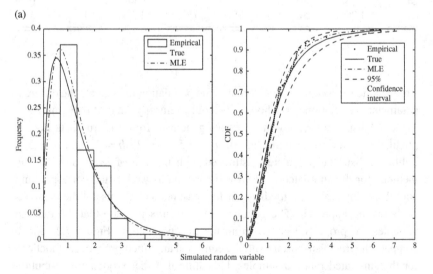

Figure 7.2 Comparison of fitted frequency distribution to frequency distribution of the true population: (a) MLE and (b) entropy.

(b)

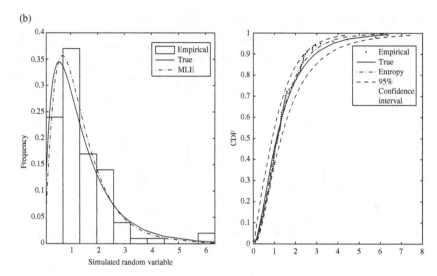

Figure 7.2 (*cont.*)

Table 7.3. *Parameters estimated using MLE and entropy methods,*
corresponding confidence intervals, and goodness-of-fit results: peak flow.

Method	Parameter and confidence interval				Goodness-of-fit	
MLE	p	q	b	c	D_n	P
	115.96 [34.68, 117.99]	10.11 [6.75, 12.32]	3.15E+05 [2.43E+05, 3.15E+05]	256.9 [224.14, 350.00]	0.059	0.44
Entropy	λ_0	λ_1	λ_2	b	c	
	−26.76	2.98E+04 [2.62E+04, 3.13E+04]	−8.54 [−10.59, −5.47]	3.34E+05 [2.45E+05, 4.06E+05]	255.54 [141.52, 261.49]	0.064 0.33
	Equivalent: $p = 116.16$, $q = 9.54$, $b = 3.34E+05$, $c = 255.54$					

gaging station USGS09239500. Using the MLE method, Table 7.3 lists the
parameters estimated where the initial parameters are generated using the
genetic algorithm. Then, applying the relationship between Lagrange
multipliers and the population parameters, that is, Equation (7.23), Table 7.3
also lists the estimated parameters (including multipliers and embedded parameters) with the initial estimates deduced from the parameters estimated
using MLE. The goodness-of-fit results in Table 7.3 indicate both MLE and

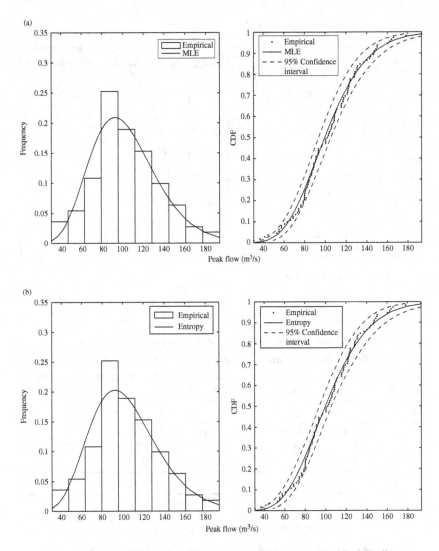

Figure 7.3 Comparison of fitted beta Lomax distribution with empirical distribution for peak flow data: (a) MLE and (b) entropy.

entropy estimation methods may be applied for parameter estimation and the beta Lomax distribution may be properly applied to analyze the peak flow dataset. Figure 7.3a and b compare the fitted parametric (MLE- and entropy-estimated) with the empirical frequency distribution. The plots visually show the adequacy of the beta Lomax distribution. Figure 7.4 plots the quantiles computed with the parameters estimated using the MLE and entropy methods

Table 7.4. *Quantiles computed from fitted and empirical distributions for peak flow (m³/s).*

| Probability | Beta Lomax distribution | | Empirical distribution |
	MLE	Entropy	
0.5	99.59	99.77	101.94
0.9	146.10	148.30	146.85
0.95	161.49	164.43	146.85
0.99	193.10	197.65	191.35

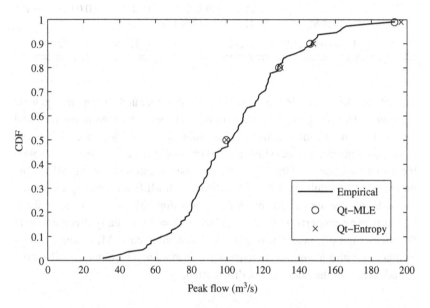

Figure 7.4 Comparison of the quantiles computed from MLE- and entropy-estimated population with observed peak flow data.

(with probability = {0.5,0.8,0.9,0.99}) and compares with the empirical distribution. Figure 7.4 shows that the computed quantiles are very similar for both methods as shown in Table 7.4.

7.5.3 Maximum Daily Precipitation

In this section, the maximum daily precipitation at Brenham, Texas (THCND: USC00411048), is analyzed. MLE and entropy methods are applied for parameter estimation. Using the MLE method, Table 7.5 lists the parameters

Table 7.5. *Parameters estimated, corresponding confidence intervals, and goodness-of-fit results: maximum daily precipitation.*

Method	Parameter and confidence interval				Goodness-of-fit	
MLE	p	Q	b	c	D_n	P
	43.838 [34.194, 54.960]	12.253 [8.061, 29.223]	72.149 [21.162, 123.12]	0.302 [0.254, 0.387]	0.044	0.811
Entropy	λ_1	λ_2	b	c		
	14.234 [8.431, 19.969]	−11.253 [−15.744, −7.52]	72.149 [49.846, 112.52]	0.302 [0.082, 1.77]	0.044	0.801
Equation (7.23): $p = 43.82$, $q = 12.253$, $b = 72.149$, $c = 0302$						

estimated where the initial parameters are generated using the genetic algorithm. Then, applying the relationship between Lagrange multipliers and the population parameters, that is, Equation (7.23), Table 7.5 also lists the estimated parameters (including multipliers and embedded parameters) with the initial estimates deduced from the parameters estimated using MLE. The goodness-of-fit results in Table 7.5 indicate both MLE and entropy estimation methods may be applied for parameter estimation and the beta Lomax distribution may be properly applied to analyze the maximum daily precipitation at Brenham. Figure 7.5a and b compare the fitted parametric (MLE- and entropy-estimated) with the empirical frequency distribution. The plots visually show the adequacy of the beta Lomax distribution.

7.5.4 Total Flow Deficit

The total flow deficit data at Tilden, Texas, is applied to evaluate the applicability of the beta Lomax distribution to model extreme events. The special case of the beta Lomax distribution (Pareto II distribution, i.e., Equation (7.8)) is applied to set up the initial parameters as $p = 1$, $q = 1$, $b = 9.59 \times 10^4$, and $c = 2.42$. With these initial parameter values, the MLE and entropy methods are then applied to fit the distribution. Table 7.6 lists the parameters estimated, 95% confidence bounds, and goodness-of-fit test results. The goodness-of-fit results in Table 7.6 show that (1) there is minimal difference in the computed Kolmogorov–Smirnov test statistic values; (2) the equivalent population parameters computed from the entropy estimates converge to the MLE estimates;

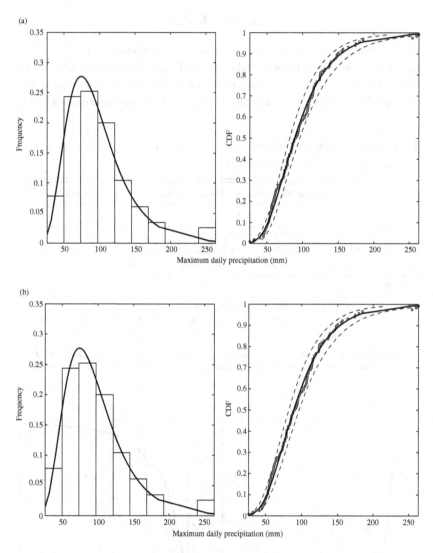

Figure 7.5 Comparison of fitted beta Lomax distribution with empirical distribution for maximum daily precipitation: (a) MLE and (b) entropy.

and (3) the beta Lomax distribution may be applied to properly model the total flow deficit dataset. Figure 7.6a and b compare the empirical distribution with the fitted beta Lomax distribution and visually confirm that the performances are very similar for the beta Lomax distributions fitted with MLE and entropy methods for the total flow deficit.

7.6 Conclusion

In this chapter, the four-parameter beta Lomax distribution is discussed. The common estimation methods, including MLE and entropy, are applied for parameter estimation. The method of moments may not be easily applied to estimate the parameters for the four-parameter beta Lomax distribution due to

Table 7.6. *Parameters estimated, corresponding confidence intervals, and goodness-of-fit results: total flow deficit.*

Method	Parameter and confidence interval				Goodness-of-fit	
MLE	p	q	b	c	D_n	P
	1.125 [0.26, 6.93]	0.953 [0.74, 1.23]	9.54E+04 [6.75E+04, 2.08E+05]	2.077 [0.31, 6.14]	0.069	0.236
Entropy	λ_1	λ_2	b	C		
	3.337 [2.43, 8.85]	0.047 [−8.51, 0.43]	9.54E+04 [4.10E+04, 425E+05]	2.077 [2.7E−09, 1E+03]	0.069	0.217

Equation (7.23) equivalent: $p = 1.125$, $q = 0.953$, $b = 9.542E+04$, $c = 2.077$

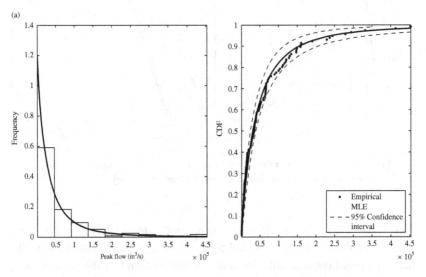

Figure 7.6 Comparison of fitted beta Lomax distribution with empirical distribution for total flow deficit: (a) MLE and (b) entropy.

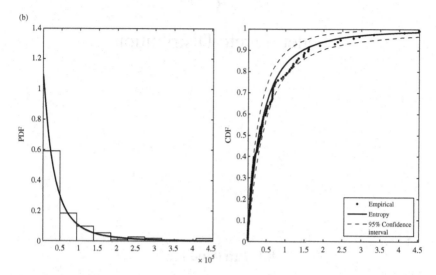

Figure 7.6 (*cont.*)

the nonexistence of closed form solutions. With the synthetic data (generated from four-parameter beta Lomax distribution) and the real-world data, namely, peak flow, maximum daily precipitation, and total flow deficit, the two methods yield very similar performances. It can be concluded that one may choose any method (i.e., MLE or entropy) to estimate the parameters of the four-parameter beta Lomax distribution.

References

Eugene, N., Lee, C., and Famoye, F. (2002). Beta-normal distribution and its applications. *Communications in Statistics – Theory and Method*, Vol. 31, pp. 497–512.

Mahmoud, M.R. and Abd El Ghafour, A.S. (2013). Shannon entropy for generalized Feller–Pareto (GFP) family and order statistics of GFP subfamilies. *Applied Mathematical Sciences*, Vol. 7, No. 65, pp. 3247–3253.

8

Feller–Pareto Distribution

8.1 Introduction

The five-parameter Feller–Pareto (FP) distribution is a general distribution that includes several distributions as special cases, such as a hierarchy of Pareto distributions, generalized Pareto distribution, Burr distribution, inverse Burr distribution, and transformed beta distribution. Thus, this distribution can be considered to represent a family, called the Feller–Pareto family. Similar to the four-parameter beta Lomax distribution, the five-parameter FP distribution may also be derived by a transformation of the beta distribution. Let Y be a beta-distributed random variable with the probability density function (PDF) as

$$f(y) = \frac{1}{B(p,q)} y^{p-1} (1-y)^{q-1}; \quad p,q > 0, y \in (0,1), \tag{8.1}$$

where $B(\cdot)$ is the beta function and p and q are the shape parameters.
Let

$$y = \left(1 + \left(\frac{x-c}{b}\right)^{\frac{1}{a}}\right)^{-1}; \quad a,b > 0, c \in (-\infty, \infty), x \in [c, \infty).$$

We have

$$dy = -\frac{1}{ab}\left(\frac{x-c}{b}\right)^{\frac{1}{a}-1}\left(1 + \left(\frac{x-c}{b}\right)^{\frac{1}{a}}\right)^{-2} dx; \quad x = b(y^{-1} - 1)^a + c. \tag{8.2}$$

Substitution of Equation (8.2) in Equation (8.1) yields

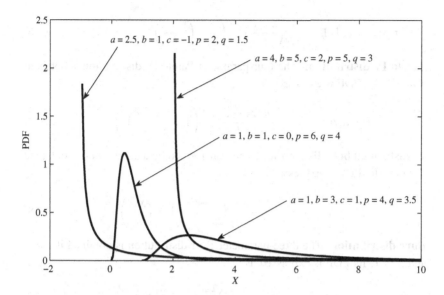

Figure 8.1 Five-parameter FP distribution.

$$f(x) = f(y(x)) \left| \frac{dy}{dx} \right| = \frac{1}{abB(p,q)} \left(\frac{x-c}{b} \right)^{\frac{q}{a}-1} \left(1 + \left(\frac{x-c}{b} \right)^{\frac{1}{a}} \right)^{-(p+q)}, \quad (8.3)$$

which is the PDF of the FP distribution. Here c is the location parameter; b is the scale parameter; and a, p, and q are the shape parameters. Figure 8.1 shows the distribution shapes for different parameters. As shown in the figure, the five-parameter FP distribution is right skewed.

8.2 Special Cases of Five-Parameter FP Distribution

Pareto I distribution: The two-parameter Pareto I distribution may be obtained if $a = 1$, $b = c$, and $q = 1$ and can be expressed as

$$f(x; 1, b, b, p, 1) = pb^p x^{-(p+1)}; \quad x \geq b. \quad (8.4)$$

Pareto II distribution: The three-parameter Pareto II distribution may be obtained if $a = 1$ and $q = 1$ and can be expressed as

$$f(x; 1, b, c, p, 1) = \frac{p}{b} \left(1 + \frac{x-c}{b} \right)^{-(p+1)}; \quad x \geq c. \quad (8.5)$$

Pareto III distribution: The three-parameter Pareto III distribution can be obtained if $p = q = 1$ and can be expressed as

$$f(x; a, b, c, 1, 1) = \frac{1}{ab} \left(\frac{x-c}{b}\right)^{\frac{1}{a}-1} \left(1 + \left(\frac{x-c}{b}\right)^{\frac{1}{a}}\right)^{-2}; \quad x \geq c. \qquad (8.6)$$

Pareto IV distribution: The four-parameter Pareto IV distribution is obtained if $q = 1$. Its PDF is expressed as

$$f(x; a, b, c, p, 1) = \frac{p}{ab} \left(\frac{x-c}{b}\right)^{\frac{1}{a}-1} \left(1 + \left(\frac{x-c}{b}\right)^{\frac{1}{a}}\right)^{-(p+1)}. \qquad (8.7)$$

Transformed beta distribution: The transformed beta distribution is obtained if $c = 0$. Its PDF is expressed as

$$f(x; a, b, 0, p, q) = \frac{ab}{B(p, q)} \left(\frac{x}{b}\right)^{aq-1} \left(1 + \left(\frac{x}{b}\right)^{\frac{1}{a}}\right)^{-(p+q)}. \qquad (8.8)$$

Burr distribution: The three-parameter Burr distribution is obtained if $c = 0$ and $q = 1$. Its PDF is expressed as

$$f(x; a, b, 0, p, 1) = \frac{p}{ab} \left(\frac{x}{b}\right)^{\frac{1}{a}-1} \left(1 + \left(\frac{x}{b}\right)^{\frac{1}{a}}\right)^{-(p+1)}. \qquad (8.9)$$

Log-logistic distribution: The two-parameter log-logistic distribution is obtained if $c = 0$ and $p = q = 1$. Its PDF is expressed as

$$f(x; a, b, 0, 1, 1) = \frac{1}{ab} \left(\frac{x}{b}\right)^{\frac{1}{a}-1} \left(1 + \left(\frac{x}{b}\right)^{\frac{1}{a}}\right)^{-2}. \qquad (8.10)$$

Paralogistic distribution: The two-parameter paralogistic distribution is obtained if $c = 0$ and $a = p$. Its PDF is expressed as

$$f(x; a, b, 0, a, 1) = \frac{1}{b} \left(\frac{x}{b}\right)^{\frac{1}{a}-1} \left(1 + \left(\frac{x}{b}\right)^{\frac{1}{a}}\right)^{-a-1}. \qquad (8.11)$$

Generalized Pareto distribution: The three-parameter generalized Pareto distribution is obtained if $c = 0$ and $a = 1$. Its PDF is expressed as

$$f(x; 1, b, 0, p, q) = \frac{1}{bB(p, q)} \left(\frac{x}{b}\right)^{q-1} \left(1 + \left(\frac{x}{b}\right)\right)^{-(p+q)}. \qquad (8.12)$$

Scaled F distribution: The two-parameter scaled F distribution is obtained if $c = 0$ and $a = b = 1$.

$$f(x; 1, 1, 0, p, q) = \frac{1}{B(p, q)} (x)^{q-1} (1 + x)^{-(p+q)}. \qquad (8.13)$$

Inverse Pareto distribution: The two-parameter inverse Pareto distribution is obtained if $c = 0$ and $a = p = 1$. Its PDF is expressed as

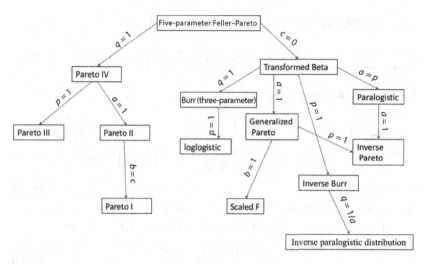

Figure 8.2 Five-parameter FP distribution and its special cases.

$$f(x; 1, b, 0, 1, q) = \frac{q}{b} \left(\frac{x}{b}\right)^{q-1} \left(1 + \frac{x}{b}\right)^{-(q+1)}. \tag{8.14}$$

Inverse Burr distribution: The three-parameter inverse Burr distribution is obtained if $c = 0$ and $p = 1$. Its PDF is expressed as

$$f(x; a, b, 0, 1, q) = \frac{q}{ab} \left(\frac{x}{b}\right)^{\frac{q}{a}-1} \left(1 + \left(\frac{x}{b}\right)^{\frac{1}{a}}\right)^{-(q+1)}. \tag{8.15}$$

Inverse paralogistic distribution: The two-parameter inverse paralogistic distribution is obtained if $c = 0$, $p = 1$, and $q = 1/a$. Its PDF is expressed as

$$f\left(x; a, b, 0, 1, \frac{1}{a}\right) = \frac{1}{a^2 b} \left(\frac{x}{b}\right)^{\frac{1}{a^2}-1} \left(1 + \left(\frac{x-c}{b}\right)^{\frac{1}{a}}\right)^{-\left(\frac{1}{a}+1\right)}. \tag{8.16}$$

Figure 8.2 illustrates the five-parameter FP distribution and its special cases in a tree diagram.

8.3 Generating Differential Equation

Taking the logarithm of Equation (8.3), we obtain

$$\ln f(x) = \ln\left(\frac{1}{abB(p, q)}\right) + \left(\frac{q}{a} - 1\right) \ln\left(\frac{x-c}{b}\right) - (p+q) \ln\left(1 + \left(\frac{x-c}{b}\right)^{\frac{1}{a}}\right). \tag{8.17}$$

Differentiating Equation (8.17) with respect to x, we get

$$\frac{1}{f(x)}\frac{df(x)}{dx} = \frac{q-a}{ab}\left(\frac{x-c}{b}\right)^{-1} - \frac{p+q}{ab}\left(\frac{x-c}{b}\right)^{\frac{1}{a}-1}\left(1+\left(\frac{x-c}{b}\right)^{\frac{1}{a}}\right)^{-1}. \quad (8.18)$$

Equation (8.18) can be simplified as

$$\frac{1}{f(x)}\frac{df(x)}{dx} = \frac{(q-a)-(p+a)\left(\frac{x-c}{b}\right)^{\frac{1}{a}}}{ab\left(\frac{x-c}{b}\right)\left(1+\left(\frac{x-c}{b}\right)^{\frac{1}{a}}\right)} = \frac{(q-a)-(p+a)\left(\frac{x-c}{b}\right)^{\frac{1}{a}}}{ab\left(\frac{x-c}{b}\right)+ab\left(\frac{x-c}{b}\right)^{\frac{1}{a}+1}}.$$

$$(8.19)$$

In Equation (8.19), let $a_0 = (q-a)$, $a_1 = -(a+p)$, $b_0 = ab$, $x_* = (x-c)/b$, $c_* = 1/a$, $c_1 = (1/a)+1$; then Equation (8.19) can be rewritten as

$$\frac{1}{f(x)}\frac{df(x)}{dx} = \frac{a_0 + a_1 x_*^{c_*}}{b_0(x_* + x_*^{c_1})}. \quad (8.20)$$

Now, the mode of five-parameter FP distribution can be determined by setting Equation (8.18) or (8.20) to zero, yielding

$$\frac{x-c}{b} = \left(\frac{q-a}{q+a}\right)^a \quad \text{or} \quad a_0 + a_1 x_*^{c_*} = 0. \quad (8.21)$$

Solving Equation (8.21) we can obtain the mode as

$$x = c + b\left(\frac{q-a}{q+a}\right)^a. \quad (8.22)$$

Equations (8.21) and (8.22) reveal another condition for the five-parameter FP distribution, that is, $q \geq a$.

8.4 Derivation of FP Distribution Using Entropy Theory

The Shannon entropy of the five-parameter FP distribution can be written as

$$H(f) = -\int_c^\infty f(x)\ln f(x)dx. \quad (8.23)$$

Substituting Equation (8.3) in Equation (8.23), we get

$$H(f) = -\int_c^\infty f(x)\ln\left(\frac{1}{abB(p,q)}\right) + \left(\frac{q}{a}-1\right)\ln\left(\frac{x-c}{b}\right) - (p+q)\ln\left(1+\left(\frac{x-c}{b}\right)^{\frac{1}{a}}\right)dx.$$

$$(8.24)$$

Equation (8.24) leads to the following constraints:

$$C_0 = \int_c^\infty f(x)dx = 1 \tag{8.25}$$

$$C_1 = \int_c^\infty \ln\left(\frac{x-c}{b}\right)f(x) = E\left(\ln\left(\frac{x-c}{b}\right)\right) \tag{8.26}$$

$$C_2 = \int_c^\infty \ln\left(1 + \left(\frac{x-c}{b}\right)^{\frac{1}{a}}\right)f(x)dx = E\left(\ln\left(1 + \left(\frac{x-c}{b}\right)^{\frac{1}{a}}\right)\right). \tag{8.27}$$

The method of Lagrange multipliers is then applied to derive $f(x)$ subject to the constraints defined by Equations (8.25)–(8.27). For the method of Lagrange multipliers, the Lagrangian function L can be constructed as

$$
\begin{aligned}
L = &-\int_c^\infty f(x)\ln f(x)dx - (\lambda_0 - 1)\left(\int_c^\infty f(x)dx - 1\right) \\
&- \lambda_1\left(\int_c^\infty \ln\left(\frac{x-c}{b}\right)f(x)dx - E\left(\ln\left(\frac{x-c}{b}\right)\right)\right) \\
&- \lambda_2\left(\int_c^\infty \ln\left(1 + \left(\frac{x-c}{b}\right)^{\frac{1}{a}}\right)f(x)dx - E\left(\ln\left(1 + \left(\frac{x-c}{b}\right)^{\frac{1}{a}}\right)\right)\right),
\end{aligned}
\tag{8.28}
$$

where λ_0, λ_1, and λ_2 are the Lagrange multipliers. Differentiating Equation (8.28) with respect to $f(x)$, we get

$$f(x) = \exp\left(-\lambda_0 - \lambda_1 \ln\left(\frac{x-c}{b}\right) - \lambda_2 \ln\left(1 + \left(\frac{x-c}{b}\right)^{\frac{1}{a}}\right)\right). \tag{8.29}$$

Equation (8.29) is the entropy-based PDF of the FP distribution.

Substituting Equation (8.29) in Equation (8.25), we get

$$
\begin{aligned}
\exp(\lambda_0) &= \int_c^\infty \exp\left(-\lambda_1 \ln\left(\frac{x-c}{b}\right) - \lambda_2 \ln\left(1 + \left(\frac{x-c}{b}\right)^{\frac{1}{a}}\right)\right)dx \\
&= \int_c^\infty \left(\frac{x-c}{b}\right)^{-\lambda_1}\left(1 + \left(\frac{x-c}{b}\right)^{\frac{1}{a}}\right)^{-\lambda_2} dx.
\end{aligned}
\tag{8.30}
$$

In Equation (8.30), let $y = ((x-c)/b)^{1/a}$; we have

$$x = c + by^a, \quad dx = aby^{a-1}dy. \tag{8.31}$$

Substituting Equation (8.31) in Equation (8.30), we obtain

$$\exp(\lambda_0) = ab \int_0^\infty y^{-a\lambda_1+a-1}(1+y)^{-\lambda_2}dy = abB(a - a\lambda_1, \lambda_2 + a\lambda_1 - a). \tag{8.32}$$

In Equation (8.32) we have $\lambda_1 < 1$; $\lambda_2 > a(1 - \lambda_1)$. Furthermore, Equation (8.32) can be rewritten as

$$\begin{aligned}\lambda_0 &= \ln(ab) + \ln B(a - a\lambda_1, \lambda_2 + a\lambda_1 - a) \\ &= \ln(ab) + \ln \Gamma(a - a\lambda_1) + \ln \Gamma(\lambda_2 + a\lambda_1 - a) - \ln \Gamma(\lambda_2).\end{aligned} \tag{8.33}$$

Substituting Equation (8.32) in Equation (8.29), the entropy-based five-parameter FP distribution can be rewritten as

$$\begin{aligned}f(x) &= \frac{1}{abB(a - a\lambda_1, \lambda_2 + a\lambda_1 - a)} \exp\left(-\lambda_1 \ln\left(\frac{x-c}{b}\right) - \lambda_2 \ln\left(1 + \left(\frac{x-c}{b}\right)^{\frac{1}{a}}\right)\right) \\ &= \frac{1}{abB(a - a\lambda_1, \lambda_2 + a\lambda_1 - a)} \left(\frac{x-c}{b}\right)^{-\lambda_1}\left(1 + \left(\frac{x-c}{b}\right)^{\frac{1}{a}}\right)^{-\lambda_2}.\end{aligned} \tag{8.34}$$

Comparing Equation (8.34) with Equation (8.3), the relationship of Lagrange multipliers and population parameters may be expressed as

$$\lambda_1 = 1 - \frac{q}{a} = \frac{a-q}{a}; \quad \lambda_2 = p + q. \tag{8.35}$$

8.5 Parameter Estimation

8.5.1 Ordinary Entropy Method

Differentiating Equation (8.33) with respect to λ_1 and λ_2, we get

$$\frac{\partial \lambda_0}{\partial \lambda_1} = -a\psi(a - a\lambda_1) + a\psi(\lambda_2 + a\lambda_1 - a) \tag{8.36}$$

$$\frac{\partial \lambda_0}{\partial \lambda_2} = \psi(\lambda_2 + a\lambda_1 - a) - \psi(\lambda_2). \tag{8.37}$$

Differentiating Equation (8.30) with respect to λ_1 and λ_2, we get

$$\frac{\partial \lambda_0}{\partial \lambda_1} = -\frac{\int_c^\infty \ln\left(\frac{x-c}{b}\right)\left[\left(\frac{x-c}{b}\right)^{-\lambda_1}\left(1+\left(\frac{x-c}{b}\right)^{\frac{1}{a}}\right)^{-\lambda_2}\right]dx}{\exp(\lambda_0)} \tag{8.38}$$

$$= -\int_c^\infty f(x)\ln\left(\frac{x-c}{b}\right)dx = -E\left(\ln\left(\frac{x-c}{b}\right)\right)$$

$$\frac{\partial \lambda_0}{\partial \lambda_2} = -\frac{\int_c^\infty \ln\left(1+\left(\frac{x-c}{b}\right)^{\frac{1}{a}}\right)\left[\left(\frac{x-c}{b}\right)^{-\lambda_1}\left(1+\left(\frac{x-c}{b}\right)^{\frac{1}{a}}\right)^{-\lambda_2}\right]dx}{\exp(\lambda_0)} \tag{8.39}$$

$$= -\int_c^\infty f(x)\ln\left(1+\left(\frac{x-c}{b}\right)^{\frac{1}{a}}\right)dx = -E\left(\ln\left(1+\left(\frac{x-c}{b}\right)^{\frac{1}{a}}\right)\right).$$

Equating Equation (8.38) to Equation (8.36) and Equation (8.39) to Equation (8.37), we get

$$\frac{\partial \lambda_0}{\partial \lambda_1} = -a\psi(a-a\lambda_1) + a\psi(\lambda_2 + a\lambda_1 - a) = -E\left(\ln\left(\frac{x-c}{b}\right)\right) \tag{8.40}$$

$$\frac{\partial \lambda_0}{\partial \lambda_2} = \psi(\lambda_2 + a\lambda_1 - a) - \psi(\lambda_2) = -E\left(\ln\left(1+\left(\frac{x-c}{b}\right)^{\frac{1}{a}}\right)\right). \tag{8.41}$$

Equations (8.40) and (8.41) are parameter estimation equations, but the FP distribution has five parameters so three additional equations are needed. These are obtained by differentiating Equations (8.40) and (8.41) as follows:

$$\frac{\partial^2 \lambda_0}{\partial \lambda_1^2} = a^2\psi'(a-a\lambda_1) + a^2\psi'(\lambda_2 + a\lambda_1 - a) = \text{var}\left(\ln\left(\frac{x-c}{b}\right)\right) \tag{8.42}$$

$$\frac{\partial^2 \lambda_0}{\partial \lambda_2^2} = \psi'(\lambda_2 + a\lambda_1 - a) - \psi'(\lambda_2) = \text{var}\left(\ln\left(1+\left(\frac{x-c}{b}\right)^{\frac{1}{a}}\right)\right) \tag{8.43}$$

$$\frac{\partial^2 \lambda_0}{\partial \lambda_1 \partial \lambda_2} = a\psi'(\lambda_2 + a\lambda_1 - a) = \text{cov}\left(\ln\left(\frac{x-c}{b}\right), \ln\left(1+\left(\frac{x-c}{b}\right)^{\frac{1}{a}}\right)\right). \tag{8.44}$$

To this end, the Lagrange multipliers (λ_1, λ_2) and the embedded distribution parameters (a, b, c) can be estimated by solving the system of equations (8.40)–(8.44) numerically for the entropy-based five-parameter FP distribution.

8.5.2 Parameter Space Expansion Method

To apply the parameter space expansion method, the constraints $[C_1, C_2]$ are given as

$$C_1 = \int_c^\infty \left(\frac{q}{a} - 1\right) \ln\left(\frac{x-c}{b}\right) f(x)dx = E\left[\left(\frac{q}{a} - 1\right) \ln\left(\frac{x-c}{b}\right)\right] \qquad (8.45)$$

$$C_2 = \int_c^\infty (p+q) \ln\left(1 + \left(\frac{x-c}{b}\right)^{\frac{1}{a}}\right) f(x)dx = E\left[(p+q) \ln\left(1 + \left(\frac{x-c}{b}\right)^{\frac{1}{a}}\right)\right].$$

$$(8.46)$$

Following the same procedure as in the ordinary entropy method, the Lagrangian function L can be constructed as

$$L = -\int_c^\infty f(x) \ln f(x)dx - (\lambda_0 - 1)\left[\int_c^\infty f(x)dx - 1\right]$$

$$- \lambda_1 \left[\int_c^\infty \left(\frac{q}{a} - 1\right) \ln\left(\frac{x-c}{b}\right) f(x)dx - E\left(\left(\frac{q}{a} - 1\right)\left(\ln\left(\frac{x-c}{b}\right)\right)\right)\right]$$

$$- \lambda_2 \left[\int_c^\infty (p+q) \ln\left(1 + \left(\frac{x-c}{b}\right)^{\frac{1}{a}}\right) f(x)dx - E\left((p+q) \ln\left(1 + \left(\frac{x-c}{b}\right)^{\frac{1}{a}}\right)\right)\right],$$

$$(8.47)$$

where λ_0, λ_1, and λ_2 are the Lagrange multipliers. Differentiating Equation (8.47) with respect to $f(x)$, we get

$$f(x) = \exp\left(-\lambda_0 - \lambda_1\left(\frac{q}{a} - 1\right) \ln\left(\frac{x-c}{b}\right) - \lambda_2(p+q) \ln\left(1 + \left(\frac{x-c}{b}\right)^{\frac{1}{a}}\right)\right).$$

$$(8.48)$$

Equation (8.48) is the entropy-based PDF of the FP distribution.

Substituting Equation (8.48) in constraint C_0 in Equation (8.25), we get

$$\exp(\lambda_0) = \int_c^\infty \left(\frac{x-c}{b}\right)^{-\lambda_1\left(\frac{q}{a}-1\right)} \left(1 + \left(\frac{x-c}{b}\right)^{\frac{1}{a}}\right)^{-\lambda_2(p+q)} dx. \qquad (8.49)$$

Again, let $y = ((x-c)/b)^{1/a}$. Substituting Equation (8.31) in Equation (8.49), we have

$$\exp(\lambda_0) = ab \int_0^\infty y^{-\lambda_1(q-a)+a-1}(1+y)^{-\lambda_2(p+q)}dy. \tag{8.50}$$

Equation (8.50) may be expressed through the beta function as

$$\exp(\lambda_0) = abB\big(-\lambda_1(q-a)+a, \lambda_2(p+q)+\lambda_1(q-a)-a\big). \tag{8.51}$$

Equation (8.51) can be rewritten as

$$\lambda_0 = \ln(ab) + \ln\Gamma\big(-\lambda_1(q-a)+a\big) + \ln\Gamma\big(\lambda_2(p+q)+\lambda_1(q-a)-a\big)$$
$$- \ln\Gamma\big(\lambda_2(p+q)\big). \tag{8.52}$$

Substitution of Equation (8.51) in Equation (8.48) leads to

$$f(x) = \frac{1}{abB\big(-\lambda_1(q-a)+a, \lambda_2(p+q)+\lambda_1(q-a)-a\big)}$$
$$\times \left(\frac{x-c}{b}\right)^{-\lambda_1\left(\frac{q}{a}-1\right)}\left(1+\left(\frac{x-c}{b}\right)^{\frac{1}{a}}\right)^{-\lambda_2(p+q)}. \tag{8.53}$$

Taking the logarithms of Equation (8.53), we get

$$\ln f(x) = -\ln(ab) - \ln\Gamma\big(-\lambda_1(q-a)+a\big) - \ln\Gamma\big(\lambda_2(p+q)+\lambda_1(q-a)-a\big)$$
$$+ \ln\Gamma\big(\lambda_2(p+q)\big) - \lambda_1(q-a)\ln\left(\frac{x-c}{b}\right) - \lambda_2(p+q)\ln\left(1+\left(\frac{x-c}{b}\right)^{\frac{1}{a}}\right). \tag{8.54}$$

The corresponding entropy is then given as

$$H = \ln(ab) + \ln\Gamma\big(-\lambda_1(q-a)+a\big) + \ln\Gamma\big(\lambda_2(p+q)+\lambda_1(q-a)-a\big)$$
$$- \ln\Gamma\big(\lambda_2(p+q)\big) - E\left(\left(\frac{x-c}{b}\right)^{-\lambda_1(q-a)}\right) - E\left(\left(1+\left(\frac{x-c}{b}\right)^{\frac{1}{a}}\right)^{-\lambda_2(p+q)}\right). \tag{8.55}$$

Differentiating Equation (8.55) with respect to Lagrange multipliers and distribution parameters, we get

$$\frac{\partial H}{\partial \lambda_1} = -(q-a)\psi\big(-\lambda_1(q-a)+a\big) + (q-a)\psi\big(\lambda_2(p+q)+\lambda_1(q-a)-a\big)$$

$$-E\left(-(q-a)\left(\left(\frac{x-c}{b}\right)\right)^{-\lambda_1(q-a)} \ln\left(\frac{x-c}{b}\right)\right)$$

$$\Rightarrow (q-a)E\left[\left(\frac{x-c}{b}\right)^{-\lambda_1(q-a)} \ln\left(\frac{x-c}{b}\right)\right]$$

$$= (q-a)\left[\psi\big(-\lambda_1(q-a)+a\big) - \psi\big(\lambda_2(p+q)+\lambda_1(q-a)-a\big)\right]$$

$$(8.56)$$

$$\frac{\partial H}{\partial \lambda_2} = (p+q)\psi\big(\lambda_2(p+q)+\lambda_1(q-a)-a\big) - (p+q)\psi\big(\lambda_2(p+q)\big)$$

$$-E\left(-(p+q)\left(1+\left(\frac{x-c}{b}\right)^{\frac{1}{a}}\right)^{-\lambda_2(p+q)} \ln\left(1+\left(\frac{x-c}{b}\right)^{\frac{1}{a}}\right)\right)$$

$$\Rightarrow (p+q)E\left[\left(1+\left(\frac{x-c}{b}\right)^{\frac{1}{a}}\right)^{-\lambda_2(p+q)} \ln\left(1+\left(\frac{x-c}{b}\right)^{\frac{1}{a}}\right)\right]$$

$$(8.57)$$

$$= (p+q)\left[\psi\big(\lambda_2(p+q)\big) - \psi\big(-\lambda_1(q-a)+a\big)\right]$$

$$\frac{\partial H}{\partial p} = \lambda_2\psi\big(\lambda_2(p+q)+\lambda_1(q-a)-a\big) - \lambda_2\psi\big(\lambda_2(p+q)\big)$$

$$-E\left(-\lambda_2\left(1+\left(\frac{x-c}{b}\right)^{\frac{1}{a}}\right)^{-\lambda_2(p+q)} \ln\left(1+\left(\frac{x-c}{b}\right)^{\frac{1}{a}}\right)\right)$$

$$\Rightarrow \lambda_2 E\left[\left(1+\left(\frac{x-c}{b}\right)^{\frac{1}{a}}\right)^{-\lambda_2(p+q)} \ln\left(1+\left(\frac{x-c}{b}\right)^{\frac{1}{a}}\right)\right]$$

$$(8.58)$$

$$= \lambda_2\left(\psi\big(\lambda_2(p+q)\big) - \psi\big(\lambda_2(p+q)+\lambda_1(q-a)-a\big)\right)$$

$$\frac{\partial H}{\partial q} = -\lambda_1\psi\big(-\lambda_1(q-a)+a\big) + (\lambda_1+\lambda_2)\psi\big(\lambda_2(p+q)+\lambda_1(q-a)-a\big) - \lambda_2\psi\big(\lambda_2(p+q)\big)$$

$$-E\left(-\lambda_1\left(\frac{x-c}{b}\right)^{-\lambda_1(q-a)} \ln\left(\frac{x-c}{b}\right)\right) - E\left(-\lambda_2\left(1+\left(\frac{x-c}{b}\right)^{\frac{1}{a}}\right)^{-\lambda_2(p+q)} \ln\left(1+\left(\frac{x-c}{b}\right)^{\frac{1}{a}}\right)\right).$$

$$(8.59)$$

Substituting Equation (8.58) in Equation (8.59), we have

$$\frac{\partial H}{\partial q} = -\lambda_1\psi\big(-\lambda_1(q-a)+a\big) + (\lambda_1+\lambda_2)\psi\big(\lambda_2(p+q)+\lambda_1(q-a)-a\big) - \lambda_2\psi\big(\lambda_2(p+q)\big)$$

$$+\lambda_1 E\left(\left(\frac{x-c}{b}\right)^{-\lambda_1(q-a)} \ln\left(\frac{x-c}{b}\right)\right) + \lambda_2\left(\psi\big(\lambda_2(p+q)\big) - \psi\big(\lambda_2(p+q)+\lambda_1(q-a)-a\big)\right)$$

$$\Rightarrow \lambda_1 E\left(\left(\frac{x-c}{b}\right)^{-\lambda_1(q-a)} \ln\left(\frac{x-c}{b}\right)\right) = \lambda_1\left(\psi\big(-\lambda_1(q-a)+a\big)\right.$$

$$\left. -\psi\big(\lambda_2(p+q)+\lambda_1(q-a)-a\big)\right)$$

$$(8.60)$$

$$\frac{\partial H}{\partial a} = \frac{1}{a} + (\lambda_1 + 1)\psi\left(-\lambda_1(q-a) + a\right) - (\lambda_1 + 1)\psi\left(\lambda_2(p+q) + \lambda_1(q-a) - a\right)$$

$$-E\left(\lambda_1\left(\frac{x-c}{b}\right)^{-\lambda_1(q-a)}\ln\left(\frac{x-c}{b}\right)\right)$$

$$-E\left(\frac{\lambda_2(p+q)}{a^2}\left(1+\left(\frac{x-c}{b}\right)^{\frac{1}{a}}\right)^{-\lambda_2(p+q)-1}\left(\frac{x-c}{b}\right)^{\frac{1}{a}}\ln\left(\frac{x-c}{b}\right)\right)$$

$$\tag{8.61}$$

$$\frac{\partial H}{\partial b} = \frac{1}{b} - E\left(\frac{\lambda_1(q-a)}{b}\left(\frac{x-c}{b}\right)^{-\lambda_1(q-a)}\right)$$

$$-E\left(\frac{\lambda_2(p+q)}{ab}\left(1+\left(\frac{x-c}{b}\right)^{\frac{1}{a}}\right)^{-\lambda_2(p+q)-1}\left(\frac{x-c}{b}\right)^{\frac{1}{a}}\right) \tag{8.62}$$

$$\frac{\partial H}{\partial c} = -E\left(\frac{\lambda_1(q-a)}{b}\left(\frac{x-c}{b}\right)^{-\lambda_1(q-a)-1}\right)$$

$$-E\left(\frac{\lambda_2(p+q)}{ab}\left(1+\left(\frac{x-c}{b}\right)^{\frac{1}{a}}\right)^{-\lambda_2(p+q)-1}\left(\frac{x-c}{b}\right)^{\frac{1}{a}-1}\right). \tag{8.63}$$

Comparing Equation (8.56) with Equation (8.60), we have $\lambda_1 = q - a$. Comparing Equation (8.57) with Equation (8.58), we have $\lambda_2 = p + q$. Now Equations (8.58) and (8.60) can be rewritten as

$$\frac{\partial H}{\partial p} = \frac{\partial H}{\partial \lambda_2}\frac{\partial \lambda_2}{\partial p} \Rightarrow E\left[\left(1+\left(\frac{x-c}{b}\right)^{\frac{1}{a}}\right)^{-(p+q)^2}\ln\left(1+\left(\frac{x-c}{b}\right)^{\frac{1}{a}}\right)\right] \tag{8.64}$$

$$= \psi\left((p+q)^2\right) - \psi\left((p+q)^2 + (q-a)^2 - a\right)$$

$$\frac{\partial H}{\partial q} = \frac{\partial H}{\partial \lambda_1}\frac{\partial \lambda_1}{\partial q} \Rightarrow E\left[\left(\frac{x-c}{b}\right)^{-(q-a)^2}\ln\left(\frac{x-c}{b}\right)\right] \tag{8.65}$$

$$= \psi\left(-(q-a)^2 + a\right) - \psi\left((p+q)^2 + (q-a)^2 - a\right).$$

Similarly, Equations (8.61)–(8.63) can be rewritten as

$$E\left((q-a)\left(\frac{x-c}{b}\right)^{-(q-a)^2}\ln\left(\frac{x-c}{b}\right)\right)$$

$$+E\left(\frac{(p+q)^2}{a^2}\left(1+\left(\frac{x-c}{b}\right)^{\frac{1}{a}}\right)^{-(p+q)^2-1}\left(\frac{x-c}{b}\right)^{\frac{1}{a}}\ln\left(\frac{x-c}{b}\right)\right)$$

$$= \frac{1}{a} + (q-a+1)\left[\psi\left(-(q-a)^2 + a\right) - \psi\left((p+q)^2 + (q-a)^2 - a\right)\right] \tag{8.66}$$

$$E\left(\frac{(q-a)^2}{b}\left(\frac{x-c}{b}\right)^{-(q-a)^2}\right) - E\left(\frac{(p+q)^2}{ab}\left(1+\left(\frac{x-c}{b}\right)^{\frac{1}{a}}\right)^{-(p+q)^2-1}\left(\frac{x-c}{b}\right)^{\frac{1}{a}}\right) = \frac{1}{b}$$

(8.67)

$$E\left(\frac{(q-a)^2}{b}\left(\frac{x-c}{b}\right)^{-(q-a)^2-1}\right) + E\left(\frac{(p+q)^2}{ab}\left(1+\left(\frac{x-c}{b}\right)^{\frac{1}{a}}\right)^{-(p+q)^2-1}\left(\frac{x-c}{b}\right)^{\frac{1}{a}-1}\right) = 0.$$

(8.68)

To this end, we have reduced seven equations to five equations with five parameters. And the parameters may be estimated by solving Equations (8.64)–(8.68) simultaneously.

8.5.3 Maximum Likelihood Estimation Method

From the PDF, that is, Equation (8.3), the likelihood function for the five-parameter FP distribution can be expressed as

$$L = \prod_{i=1}^{n} \frac{1}{abB(p,q)}\left(\frac{x_i-c}{b}\right)^{\frac{q}{a}-1}\left(1+\left(\frac{x_i-c}{b}\right)^{\frac{1}{a}}\right)^{-(p+q)},$$

(8.69)

where n is the sample size.

Then, the corresponding log-likelihood function can be expressed as

$$\ln L = -n\ln\left(abB(p,q)\right) + \left(\frac{q}{a}-1\right)\sum_{i=1}^{n}\ln\left(\frac{x_i-c}{b}\right) - (p+q)\sum_{i=1}^{n}\ln\left(1+\left(\frac{x_i-c}{b}\right)^{\frac{1}{a}}\right).$$

(8.70)

Taking the derivative of $\ln L$ with respect to the distribution parameters and setting the partial differential equation equal to zero, we have

$$\frac{\partial \ln L}{\partial a} = -\frac{n}{a} - \frac{q}{a^2}\sum_{i=1}^{n}\ln\left(\frac{x_i-c}{b}\right) + \frac{(p+q)}{a^2}\sum_{i=1}^{n}\left(1+\left(\frac{x_i-c}{b}\right)^{\frac{1}{a}}\right)^{-1}\left(\frac{x_i-c}{b}\right)^{\frac{1}{a}}\ln\left(\frac{x_i-c}{b}\right) = 0$$

(8.71)

$$\frac{\partial \ln L}{\partial b} = -\frac{n}{b} - \frac{n(q-a)}{ab} + \frac{(p+q)}{ab}\sum_{i=1}^{n}\left(1+\left(\frac{x_i-c}{b}\right)^{\frac{1}{a}}\right)^{-1}\left(\frac{x_i-c}{b}\right)^{\frac{1}{a}} = 0$$ (8.72)

$$\frac{\partial \ln L}{\partial c} = -\frac{q-a}{ab}\sum_{i=1}^{n}\left(\frac{x_i-c}{b}\right)^{-1} + \frac{(p+q)}{ab}\sum_{i=1}^{n}\left(1+\left(\frac{x_i-c}{b}\right)^{\frac{1}{a}}\right)^{-1}\left(\frac{x_i-c}{b}\right)^{\frac{1}{a}-1} = 0$$

(8.73)

$$\frac{\partial \ln L}{\partial p} = -\psi(p) + \psi(p+q) - \sum_{i=1}^{n}\ln\left(1+\left(\frac{x_i-c}{b}\right)^{\frac{1}{a}}\right) = 0$$

(8.74)

$$\frac{\partial \ln L}{\partial q} = -\psi(q) + \psi(p+q) + \frac{1}{a}\sum_{i=1}^{n}\ln\left(\frac{x_i - c}{b}\right) - \sum_{i=1}^{n}\ln\left(1 + \left(\frac{x_i - c}{b}\right)^{\frac{1}{a}}\right).$$

(8.75)

Then the parameters may be estimated by solving Equations (8.71)–(8.75) numerically.

8.5.4 Method of Moments

For the five-parameter FP distribution, its noncentral moments may be computed as

$$E(X^r) = \int_c^{\infty} x^r \frac{1}{abB(p,q)}\left(\frac{x-c}{b}\right)^{\frac{q}{a}-1}\left(1 + \left(\frac{x-c}{b}\right)^{\frac{1}{a}}\right)^{-(p+q)} dx.$$

(8.76)

In Equation (8.76), let $y = ((x-c)/b)^{1/a}$. Then, we have $x = c + by^a$ and $dx = aby^{a-1}dy$; Equation (8.76) may be rewritten as

$$E(X^r) = E(c + bY^a) = \frac{1}{abB(p,q)}\int_0^{\infty}(c + by^a)^r y^{q-a}(1+y)^{-(p+q)}aby^{a-1}dy$$

$$= \frac{1}{B(p,q)}\int_0^{\infty}(c + by^a)^r y^{q-1}(1+y)^{-(p+q)}dy$$

$$= \frac{1}{B(p,q)}\int_0^{\infty}\sum_{k=0}^{r}\binom{r}{k}(by^a)^k c^{r-k}y^{q-1}(1+y)^{-(p+q)}dy.$$

(8.77)

To estimate the parameters, the first five moments need to exist to apply the method of moments (MOM), which can be expressed as

$$E(X) = \frac{1}{B(p,q)}\int_0^{\infty}(c + by^a)y^{q-1}(1+y)^{-(p+q)}dy$$

$$= \frac{c}{B(p,q)}\int_0^{\infty}y^{q-1}(1+y)^{-(p+q)}dy + \frac{b}{B(p,q)}\int_0^{\infty}y^{q+a-1}(1+y)^{-(p+q)}dy$$

$$= c + \frac{bB(q+a,p-a)}{B(p,q)}$$

(8.78)

$$E(X^2) = \frac{1}{B(p,q)} \int_0^\infty (c + by^a)^2 y^{q-1} (1+y)^{-(p+q)} dy$$

$$= c^2 + \frac{2cbB(q+a, p-a)}{B(p,q)} + \frac{b^2 B(q+2a, p-2a)}{B(p,q)} \tag{8.79}$$

$$E(X^3) = \frac{1}{B(p,q)} \int_0^\infty (c + by^a)^3 y^{q-1} (1+y)^{-(p+q)} dy$$

$$= c^3 + \frac{3bc^2 B(q+a, p-a)}{B(p,q)} + \frac{3b^2 cB(q+2a, p-2a)}{B(p,q)} + \frac{b^3 B(q+3a, p-3a)}{B(p,q)} \tag{8.80}$$

$$E(X^4) = \frac{1}{B(p,q)} \int_0^\infty (c + by^a)^4 y^{q-1} (1+y)^{-(p+q)} dy$$

$$= c^4 + \frac{4bc^3 B(q+a, p-a)}{B(p,q)} + \frac{6b^2 c^2 B(q+2a, p-2a)}{B(p,q)} \tag{8.81}$$

$$+ \frac{4b^3 cB(q+3a, p-3a)}{B(p,q)} + \frac{b^4 B(q+4a, p-4a)}{B(p,q)}$$

$$E(X^5) = \frac{1}{B(p,q)} \int_0^\infty (c + by^a)^5 y^{q-1} (1+y)^{-(p+q)} dy$$

$$= c^5 + \frac{5bc^4 B(q+a, p-a)}{B(p,q)} + \frac{10b^2 c^3 B(q+2a, p-2a)}{B(p,q)}$$

$$+ \frac{10b^3 c^2 B(q+3a, p-3a)}{B(p,q)} + \frac{5b^4 cB(q+4a, p-4a)}{B(p,q)} + \frac{b^5 B(q+5a, p-5a)}{B(p,q)}. \tag{8.82}$$

It is seen from Equation (8.82) that one more condition is needed to apply the MOM, that is, $p > 5a$.

Figure 8.3 illustrates the moment ratio diagram with different parameters.

8.6 Application

Similar to the previous chapters, the estimation methods are first applied to the random variables simulated from the true population for verification. Then the real-world data are applied to evaluate the appropriateness of the five-parameter FP distribution.

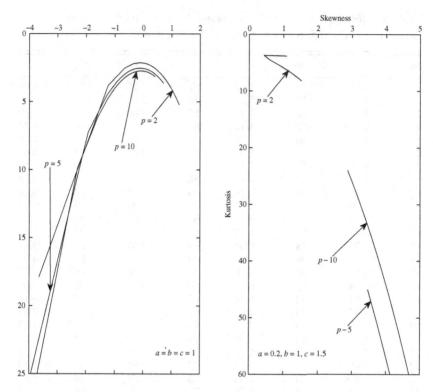

Figure 8.3 Moment ratio diagram with different parameters for q varying from 0.1 to 40.

8.6.1 Synthetic Data

The random variable data are simulated from the five-parameter FP distribution with parameters of $\{a = 0.5, b = 1.5, c = 3, p = 3.4, q = 4\}$. Early in the chapter, we have shown the beta distribution with parameters of p and q if $y = \left(1 + \left(\frac{x-c}{b}\right)^{1/a}\right)^{-1}$, that is, $Y \sim \text{Beta}(p,q)$. With this in mind, the random variable X can be simulated as follows:

Step 1: Generate uniformly distributed random variable U.
Step 2: Compute Y using $Y = \text{Beta}^{-1}(U; p, q)$.
Step 3: Compute X using $X = b(Y^{-1} - 1)^a + c$.

Table 8.1 lists the random variable simulated using the above-mentioned parameters. Table 8.2 lists the parameters estimated by maximum likelihood estimation (MLE), entropy, and MOM. To estimate the parameters with the entropy method, the initial Lagrange multipliers are set using the estimates

Table 8.1. *Random variable data simulated from true population with parameters: {a = 0.5, b = 1.5, c = 3, p = 3.4, q = 4}.*

No.	X	No.	X	No.	X	No.	X
1	4.61	26	4.62	51	5.20	76	4.06
2	5.20	27	5.29	52	3.98	77	4.06
3	5.33	28	4.07	53	3.69	78	5.30
4	5.22	29	4.44	54	6.26	79	4.24
5	4.76	30	4.82	55	4.97	80	4.77
6	3.79	31	4.01	56	4.61	81	4.62
7	4.15	32	5.33	57	4.93	82	4.50
8	4.86	33	6.75	58	4.31	83	4.34
9	6.53	34	5.78	59	5.56	84	6.79
10	5.30	35	4.61	60	3.97	85	5.65
11	5.14	36	4.93	61	5.06	86	4.20
12	4.54	37	4.75	62	4.72	87	5.59
13	4.75	38	5.04	63	4.37	88	4.90
14	4.47	39	4.41	64	5.37	89	4.07
15	6.19	40	5.30	65	5.56	90	5.11
16	4.18	41	6.01	66	4.76	91	5.05
17	4.63	42	5.08	67	5.83	92	5.19
18	4.97	43	4.15	68	4.19	93	4.01
19	3.97	44	4.93	69	7.57	94	4.89
20	4.50	45	3.81	70	4.39	95	4.03
21	4.53	46	5.02	71	3.67	96	6.72
22	10.61	47	4.84	72	4.99	97	5.19
23	5.42	48	4.45	73	4.51	98	5.62
24	5.06	49	4.83	74	3.91	99	4.32
25	4.29	50	5.32	75	3.61	100	4.87

Table 8.2. *Parameters estimated for the synthetic data listed in Table 8.1.*

Parameter	a	b	c	p	q
MLE	0.54	1.75	3.23	3.25	2.53
MOM	0.96	0.33	3	4.79	0.92

Lagrange multipliers and parameters	λ_0	λ_1	λ_2	a	b	c
Entropy	−3.17	−3.72	5.76	0.52	1.86	3.19

from MLE. The parameters estimated by MLE are also applied as the initial estimates for MOM. Figure 8.4 compares the empirical frequency distribution and the frequency distribution of the true population with the fitted frequency distribution. It is shown that MLE and entropy methods yield similar results and may be applied to model the synthetic data simulated from the five-parameter FP distribution. However, the MOM does not yield satisfactory results. Thus, we will only apply the MLE and entropy methods for the rest

Table 8.3. *95% confidence intervals and GoF study on the parameters.*

Estimation		a	b	c	p	q
MLE	Confidence interval	[0.34, 0.93]	[0.53, 2.31]	[2.76, 4.98]	[1.68, 5.81]	0.82, 4.44]
	GoF	$D = 0.07, P = 0.26$				
		a	b	c	λ_0	λ_2
Entropy	Confidence interval	[0.29, 0.63]	[0.62, 2.68]	[2.97, 3.53]	[−4.95, −0.36]	[1.86, 5.85]
	GoF	$D = 0.063, P = 0.32$				

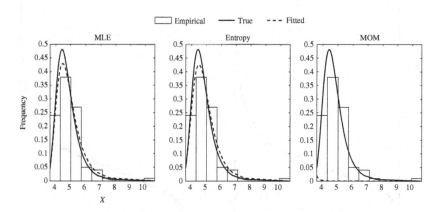

Figure 8.4 Comparison of empirical frequency distribution and frequency distribution of true population with fitted frequency distribution with different estimation techniques: synthetic data.

of the applications. Furthermore, Table 8.3 lists the confidence interval and goodness-of-fit (GoF) results for the parameters estimated using MLE and entropy methods. The GoF study indicates that that these two methods yield similar results and can be applied to model the synthetic data. Figure 8.5 compares the empirical cumulative probability distribution function (CDF) with the parametric CDF and also constructs the 95% confidence bounds for the CDF. Comparison indicates that the synthetic data simulated from the true population fall into the 95% confidence bound and further confirms the conclusion that MLE and entropy methods may be successfully applied to study the five-parameter FP distribution.

Previously, we have shown the application of MLE and entropy methods to the simulated random variable data from the true five-parameter FP population.

Table 8.4. *Parameters estimated: peak flow*

Parameter			a	b	c	p	q
MLE			1.04	36.28	−2.06	13.79	35.94
Lagrange multipliers and parameters			a	b	c	λ_1	λ_2
Entropy			29.44	0.39	301.47	−8.36	−11.85

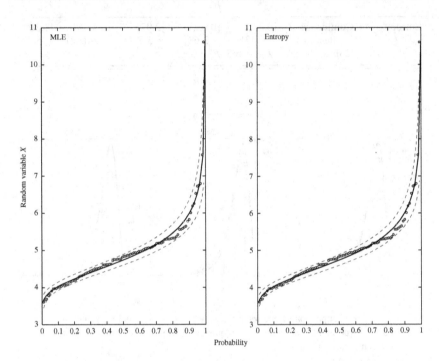

Figure 8.5 Comparison of empirical CDF with fitted parametric CDF and its 95% confidence interval: synthetic data.

In what follows, we will again evaluate the application to the real-world data using MLE and entropy parameter estimation.

8.6.2 Peak Flow

With the use of peak flow data from USGS09239500, Table 8.4 lists the parameters estimated using MLE and entropy (Lagrange multipliers and embedded population parameters) methods, and Table 8.5 lists the 95% confidence bounds for the estimated parameters as well as the GoF results. The GoF study indicates that the entropy method yields better performance. Figure 8.6

Table 8.5. *95% confidence intervals and GoF study results: peak flow.*

MLE	a	b	c	p	q
	[0.83, 1.32]	[46.52, 59.47]	[−28.57, −10.64]	[16.93, 24.47]	[37.37, 49.43]
GoF	$D_n = 0.086$, $P = 0.14$				
Entropy	a	b	c	λ_1	λ_2
	[0.15, 0.86]	[278.48, 344.36]	[−128.12, −5.18]	[−14.47, −9.27]	[13.35, 50.26]
	$Dn = 0.057$, $P = 0.402$				

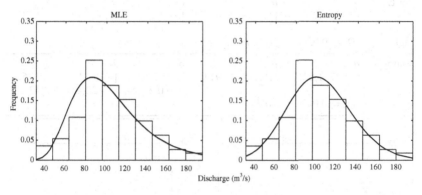

Figure 8.6 Comparison of empirical frequency distribution with fitted frequency distribution: peak flow.

compares the fitted frequency distribution to the empirical frequency distribution of peak flow. As shown in Figure 8.6, both MLE and entropy methods may be applied to estimate the parameters that are in agreement with the GoF results. Figure 8.7 compares the empirical CDF with the fitted parametric CDF as well as its 95% confidence bound. Comparison shows that the entropy method yields better performance, which is in agreement with the GoF results (i.e., lower D_n and higher P-value obtained for the entropy method).

8.6.3 Maximum Daily Precipitation

With the use of the maximum daily precipitation at Brenham, Texas (GHCND: USC00411048), Table 8.6 lists the parameters estimated using MLE and entropy estimation. Table 8.7 lists the corresponding 95% confidence bound and the GoF test results. The GoF study indicates that the MLE method yields

Table 8.6. *Parameters estimated: maximum daily precipitation.*

Parameter	a	b	c	p	q
MLE	1.05	20.187	5.942	6.229	23.877
Lagrange multipliers and parameters	a	b	c	λ_1	λ_2
Entropy	1.138	35.832	−8.507	−25.37	42.688

Table 8.7. *95% confidence intervals and GoF study results: maximum daily precipitation.*

MLE	a	b	c	p	q
	[0.398, 1.305]	[11.954, 20.468]	[3.024, 10.177]	[1.09, 7.06]	[17.231, 26.896]
GoF	$D_n = 0.053, P = 0.619$				
Entropy	a	b	c	λ_1	λ_2
	[1.067, 1.245]	[16.182, 45.914]	[−15.766, −0.287]	[−28.14, −22.48]	[40.76, 43.22]
	$D_n = 0.047, P = 0.775$				

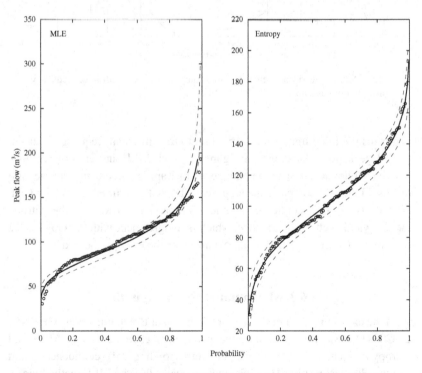

Figure 8.7 Comparison of empirical CDF with fitted parametric CDF and its 95% confidence interval: peak flow.

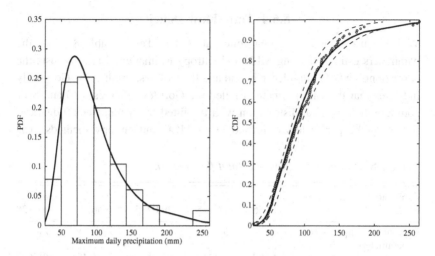

Figure 8.8 Comparison of empirical and fitted distribution (MLE): maximum daily precipitation.

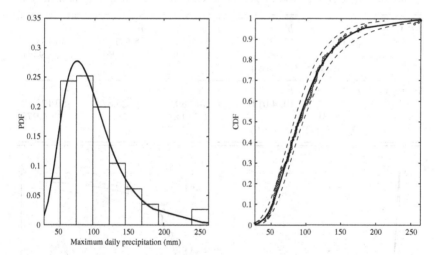

Figure 8.9 Comparison of empirical and fitted distribution (entropy): maximum daily precipitation.

better performance. Figures 8.8 and 8.9 compare the empirical distribution with the FP distributions fitted with the MLE and entropy methods, respectively. Comparison indicates that the entropy and MLE methods yield similar performances visually.

8.6.4 Total Flow Deficit

With the use of the total flow deficit at Tilden, Texas, Table 8.8 lists the parameters estimated using MLE and entropy estimation. Table 8.9 lists the corresponding 95% confidence bound and the GoF test results. The GoF study indicates that the entropy method failed the GoF test. Figures 8.10 and 8.11 compare the empirical distribution with the fitted five-parameter FP distribution with the parameters estimated using MLE and entropy methods. The

Table 8.8. *Parameters estimated: total flow deficit.*

Parameter	a	b	c	p	q
MLE	3.656	28.155	20.0735	6.666	41.128
Lagrange multipliers and parameters	a	b	c	λ_1	λ_2
Entropy	4.797	2.774	22.965	-16.481	96.648

Table 8.9. *95% confidence intervals and GoF study results: total flow deficit.*

MLE	a	b	c	p	q
	[3.379, 4.303]	[23.322, 38.705]	[16.450, 33.199]	[5.697, 9.596]	[36.135, 51.949]
GoF	$D_n = 0.085, P = 0.084$				
Entropy	a	b	C	λ_1	λ_2
	[3.229, 6.823]	(0, 4.014]	(−742.895, 1,349.8]	[−25.624, -1)	[80.122, 106.973]
	$D_n = 0.092, P = 0.013$				

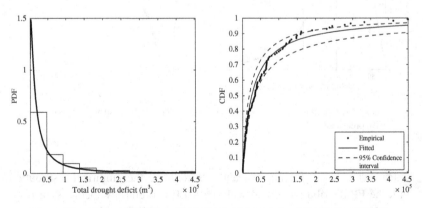

Figure 8.10 Comparison of empirical distribution with fitted distribution (MLE): total flow deficit.

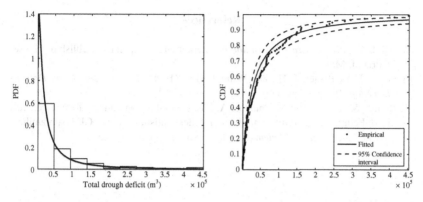

Figure 8.11 Comparison of empirical distribution with fitted distribution (entropy): total flow deficit.

comparison does not show significant differences visually, though statistically the entropy method may not be applied for this analysis.

8.7 Conclusion

For the five-parameter FP distribution, the common estimation methods, including MLE, entropy, and MOM, are applied for parameter estimation. Compared to the MOM, the MLE and entropy methods are preferred due to the ease of implementation. To employ the MOM, there is one more constraint: $p > 5a$ to guarantee the existence of the first five moments. As the results indicate, MLE and entropy methods are applied to study synthetic data simulated from the true population. Furthermore, the real-world data, namely, peak flow, maximum daily precipitation, and total flow deficit, are applied for the evaluation of the five-parameter FP distribution. MLE and entropy methods are again applied for parameter estimation. The case studies indicate that (1) the five-parameter FP distribution may be applied to evaluate peak flow and maximum daily precipitation with MLE and entropy methods; (2) compared to peak flow and maximum daily precipitation, the five-parameter FP distribution may be applied to model total flow deficit with the parameters estimated using MLE; and (3) the performance of the fitted five-parameter FP distribution to total flow deficit is inferior to the fitted five-parameter FP distribution to peak flow and maximum daily precipitation. Furthermore, though all three methods (with the existence of the first five moments) may be applied, the MLE and entropy methods are preferable for the estimation of parameters of the five-parameter FP distribution.

References

Arnold, B.C. (1983). *Pareto Distributions*. International Cooperative Publishing House, Fairland, MD.

Klugman, S.A., Panjer, H.H., and Wilmot, G.E. (1998). *Loss Model: From Data to Decision*. Wiley, New York.

Mahmoud, M.R. and Abd El Ghafour, A.S. (2013). Shannon entropy for the generalized Feller–Pareto (GFP) family and order statistics of the GFP subfamilies. *Applied Mathematical Sciences*, Vol. 7, No. 65, pp. 3247–3253.

9

Kappa Distribution

9.1 Introduction

The four-parameter kappa distribution has been applied to analyze extreme events. Since the discussion of parameter estimation by Hosking (1994), researchers have investigated different methods of parameter estimation for the kappa distribution and its application. Parida (1999) studied the application of the four-parameter kappa distribution to model Indian summer monsoon. Park and Park (2002) evaluated the four-parameter kappa distribution using maximum precipitation data. The maximum likelihood estimation (MLE) method with penalty function and the linear moments (LM or L-moments) method were applied for parameter estimation. Dupuis and Winchester (2001) examined the LM method for the kappa distribution with weekly maximum wind speeds from Tropical Pacific. Singh and Deng (2003) derived the entropy method for parameter estimation and applied to the annual maximum rainfall and peak flow data. Murshed et al. (2014) evaluated the four-parameter kappa distribution with the parameters estimated using L-moments. The objective of this chapter is to discuss the four-parameter kappa distribution, its parameter estimation with the use of the LM, MLE, method of moments (MOM), and entropy method, and its applicability with the use of annual peak flow data as an example.

9.2 Kappa Distribution and Its Characteristics

Following Hosking (1994), the kappa distribution is a four-parameter distribution. Its probability density function (PDF) is given as

233

$$f(x;a,h,k,\xi) = \frac{1}{a}\left(1 - \frac{k}{a}(x-\xi)\right)^{\frac{1}{k}-1}\left(1 - h\left(1 - \frac{k}{a}(x-\xi)\right)^{\frac{1}{k}}\right)^{\frac{1}{h}-1} ; \quad a > 0, h, k, \xi \in R.$$

(9.1)

The cumulative probability distribution function (CDF) of the four-parameter kappa distribution is given as

$$F(x;a,h,k,\xi) = \left(1 - h\left(1 - \frac{k(x-\xi)}{a}\right)^{\frac{1}{k}}\right)^{\frac{1}{h}}.$$

(9.2)

From Equation (9.2), the quantile may be expressed as

$$x(F) = \xi + \frac{a}{k}\left(1 - \left(\frac{1 - F^h}{h}\right)^k\right).$$

(9.3)

Substituting Equation (9.2) in Equation (9.1), the density function, that is, Equation (9.1), may be written as a function of the cumulative distribution function, that is, Equation (9.2), as

$$f(x;a,h,k,\xi) = \frac{1}{a}\left(1 - \frac{k}{a}(x-\xi)\right)^{\frac{1}{k}-1} F(x;a,h,k,\xi)^{1-h}.$$

(9.4)

In Equations (9.1)–(9.4), a is the scale parameter, ξ is the location parameter, and h and k are the shape parameters.

Depending on the values of parameters, the random variable may be bounded as

$$x \in \left[\xi + \frac{a(1 - h^{-k})}{k}, \xi + \frac{a}{k}\right], \quad \text{if } k > 0, h > 0$$

(9.5)

$$x \in [\xi + a\ln h, \infty), \quad \text{if } k = 0, h > 0$$

(9.6)

$$x \in \left[\xi + \frac{a(1 - h^{-k})}{k}, \infty\right), \quad \text{if } k < 0, h > 0$$

(9.7)

$$x \in \left(-\infty, \xi + \frac{a}{k}\right], \quad \text{if } k > 0, h \leq 0$$

(9.8)

$$x \in (-\infty, \infty), \quad \text{if } k = 0, h \leq 0$$

(9.9)

$$x \in \left[\xi + \frac{a}{k}, \infty\right), \quad \text{if } k < 0, h \leq 0.$$

(9.10)

Following Dupuis and Winchester (2001), and Singh and Deng (2003), some special distributions may be derived.

Three-parameter kappa distribution ($k = 0, h \neq 0$): Following Hosking (1994), the three-parameter kappa distribution is obtained with its density, cumulative distribution, and quantile functions given as

$$f(x; a, h, \xi) = \frac{1}{a} \exp\left(-\frac{x - \xi}{a}\right)\left(1 - h\exp\left(-\frac{x - \xi}{a}\right)\right)^{\frac{1}{h} - 1} \tag{9.11}$$

$$F(x; a, h, \xi) = \left(1 - h\exp\left(-\frac{x - \xi}{a}\right)\right)^{\frac{1}{h}} \tag{9.12}$$

$$x(F) = \xi - a \ln\left(\frac{1 - F^h}{h}\right). \tag{9.13}$$

Generalized extreme value (GEV) distribution: The GEV distribution is obtained if $h = 0$.

(1) If $k \neq 0$, the three-parameter GEV distribution is obtained. Its density, cumulative distribution, and quantile functions are given as

$$f(x; a, k, \xi) = \frac{1}{a}\left(1 - \frac{k}{a}(x - \xi)\right)^{\frac{1}{k} - 1} \exp\left(-\left(1 - \frac{k}{a}(x - \xi)\right)^{\frac{1}{k}}\right) \tag{9.14}$$

$$F(x; a, k, \xi) = \exp\left(-\left(1 - \frac{k}{a}(x - \xi)\right)^{\frac{1}{k}}\right). \tag{9.15}$$

Substituting Equation (9.15) in Equation (9.14), Equation (9.14) can be rewritten as

$$f(x; a, k, \xi) = \frac{1}{a}\left(1 - \frac{k}{a}(x - \xi)\right)^{\frac{1}{k} - 1} F(x; a, k, \xi). \tag{9.16}$$

In Equations (9.14)–(9.16), $x \in \left(-\infty, \xi + \frac{a}{k}\right]$ if $k > 0$ and $x \in \left[\xi + \frac{a}{k}, \infty\right)$ if $k < 0$.

$$x(F) = \xi + \frac{k}{a}\left(1 - (-\ln F)^k\right). \tag{9.17}$$

(2) If $k = 0$, the two-parameter distribution [i.e., extreme value type 1 distribution] is deduced. Its density, cumulative distribution, and quantile functions are given as

$$f(x; a, \xi) = \frac{1}{a} \exp\left(-\frac{x - \xi}{a}\right) \exp\left(-\exp\left(-\frac{x - \xi}{a}\right)\right); \quad x \in (-\infty, +\infty) \tag{9.18}$$

$$F(x; a, \xi) = \exp\left(-\exp\left(-\frac{x-\xi}{a}\right)\right) \tag{9.19}$$

$$f(x; a, \xi) = \frac{1}{a}\exp\left(-\frac{x-\xi}{a}\right)F(x; a, \xi) \tag{9.20}$$

$$x(F) = \xi - a\ln(-\ln F). \tag{9.21}$$

(3) Furthermore, if $k = 1$, the reverse two-parameter exponential distribution is deduced. Its density, cumulative distribution, and quantile functions are given as

$$f(x; a, \xi) = \frac{1}{a}\exp\left(-\left(1 - \frac{x-\xi}{a}\right)\right); \quad x \in (-\infty, a + \xi] \tag{9.22}$$

$$F(x; a, \xi) = \exp\left(-\left(1 - \frac{x-\xi}{a}\right)\right) \tag{9.23}$$

$$x(F) = \xi + a(1 + \ln F). \tag{9.24}$$

Generalized Pareto distribution: The generalized Pareto distribution is obtained if $h = 1$.

(1) If $k \neq 0$, the three-parameter generalized Pareto distribution is obtained. Its density, cumulative distribution, and quantile functions are given as

$$f(x; a, k, \xi) = \frac{1}{a}\left(1 - \frac{k}{a}(x - \xi)\right)^{\frac{1}{k}-1} \tag{9.25}$$

$$F(x; a, k, \xi) = 1 - \left(1 - \frac{k}{a}(x - \xi)\right)^{\frac{1}{k}} \tag{9.26}$$

$$x(F) = \xi + \frac{a}{k}\left(1 - (1 - F)^k\right). \tag{9.27}$$

In Equations (9.25)–(9.27), $x \in \left[\xi, \xi + \frac{a}{k}\right]$ if $k > 0$ and $x \in [\xi, \infty)$ if $k < 0$.

(2) If $k = 0$, the exponential distribution with location parameter (also called the two-parameter generalized Pareto distribution) is obtained. Its density, cumulative distribution, and quantile functions are given as

$$f(x; a, \xi) = \frac{1}{a}\exp\left(-\frac{x-\xi}{a}\right); \quad x \geq \xi \tag{9.28}$$

$$F(x; a, \xi) = 1 - \exp\left(-\frac{x-\xi}{a}\right) \tag{9.29}$$

$$x(F) = \xi - \alpha \ln(1 - F). \tag{9.30}$$

(3) Furthermore, if $k = 1$, the uniform distribution is obtained as

$$f(x; \alpha, \xi) = \frac{1}{\alpha}; \quad x \in [\xi, \xi + \alpha] \tag{9.31}$$

$$F(x; \alpha, \xi) = \frac{x - \xi}{\alpha} \tag{9.32}$$

$$x(F) = \xi + \alpha F. \tag{9.33}$$

Generalized logistic distribution: The generalized logistic distribution is obtained if $h = -1$.

(1) If $k \neq 0$, the three-parameter generalized logistic distribution is obtained. Its density, cumulative distribution, and quantile functions are expressed as

$$f(x; \alpha, k, \xi) = \frac{1}{\alpha}\left(1 - \frac{k}{\alpha}(x - \xi)\right)^{\frac{1}{k}-1}\left(1 + \left(1 - \frac{k}{\alpha}(x - \xi)\right)^{\frac{1}{k}}\right)^{-2} \tag{9.34}$$

$$F(x; \alpha, k, \xi) = \left(1 + \left(1 - \frac{k}{\alpha}(x - \xi)\right)^{\frac{1}{k}}\right)^{-1} \tag{9.35}$$

$$x(F) = \xi + \frac{\alpha}{k}\left(1 - (F^{-1} - 1)^k\right). \tag{9.36}$$

In Equations (9.34)–(9.36), $x \in \left[\xi, \xi + \frac{\alpha}{k}\right]$ if $k > 0$ and $x \in [\xi, \infty)$ if $k < 0$.

(2) If $k = 0$, the two-parameter generalized logistic distribution is obtained. Its density, cumulative distribution, and quantile functions are given as

$$f(x; \alpha, \xi) = \frac{1}{\alpha}\exp\left(-\frac{x - \xi}{\alpha}\right)\left(1 + \exp\left(-\frac{x - \xi}{\alpha}\right)\right)^{-2}; \quad x \in (-\infty, \xi] \tag{9.37}$$

$$F(x; \alpha, \xi) = \left(1 + \exp\left(-\frac{x - \xi}{\alpha}\right)\right)^{-1} \tag{9.38}$$

$$x(F) = \xi - \alpha \ln(F^{-1} - 1). \tag{9.39}$$

Mielke's three-parameter kappa distribution: Mielke's three-parameter kappa distribution is obtained by setting the parameters $c = \gamma$, $\alpha = (\gamma/\alpha'\beta)$,

$b = -(1/\alpha'\beta)$, and $h = -\alpha'$. The density and cumulative distribution functions are expressed as

$$f(x; \alpha', \beta, \gamma) = \frac{\alpha'\beta}{\gamma^\beta} x^{\beta-1} \left(\alpha' + \left(\frac{x}{\gamma} \right)^{\alpha'\beta} \right)^{-1-\frac{1}{\alpha'}} \quad ; \quad \alpha' > 0, \beta > 0, \gamma > 0, x \in (0, \infty)$$

(9.40)

$$F(x; \alpha', \beta, \gamma) = \left(\frac{x}{\gamma} \right)^\beta \left(\alpha' + \left(\frac{x}{\gamma} \right)^{\alpha'\beta} \right)^{-\frac{1}{\alpha'}}.$$

(9.41)

Additionally, the Burr type III distribution is obtained if $h < 0$ and $k < 0$, and a reverse Burr type XII distribution is obtained if $h < 0$ and $k > 0$.

Figure 9.1 depicts the shape of the kappa distribution. It shows that (1) a J-shaped curve is obtained if $k > 0$ and $k \neq h$; (2) an L-shaped curve is obtained if $k < 0$; and (3) a U-shaped curve is obtained if $k = h > 0$. Figure 9.2 plots the special cases deduced from the four-parameter kappa distribution.

9.3 Differential Equation for Kappa Distribution

Taking the logarithm of Equation (9.1), we have

$$\ln f(x) = -\ln \alpha + \left(\frac{1}{k} - 1 \right) \ln \left(1 - \frac{k}{\alpha}(x - \xi) \right)$$
$$+ \left(\frac{1}{h} - 1 \right) \ln \left(1 - h \left(1 - \frac{k}{\alpha}(x - \xi) \right)^{\frac{1}{k}} \right).$$

(9.42)

Taking the derivative of Equation (9.42) with respect to x, we have

$$\frac{1}{f}\frac{df}{dx} = \frac{(1 - h) \left(1 - \frac{k}{\alpha}(x - \xi) \right)^{\frac{1}{k}-1}}{\alpha \left(1 - h \left(1 - \frac{k}{\alpha}(x - \xi) \right)^{\frac{1}{k}} \right)} - \frac{1 - k}{\alpha \left(1 - \frac{k}{\alpha}(x - \xi) \right)}.$$

(9.43)

Substituting Equation (9.1) in Equation (9.43), we have

$$\frac{df}{dx} = \frac{k - 1}{\alpha^2} \left(1 - \frac{k}{\alpha}(x - \xi) \right)^{\frac{1}{k}-2} \left(1 - h \left(1 - \frac{k}{\alpha}(x - \xi) \right)^{\frac{1}{k}} \right)^{\frac{1}{h}-1}$$
$$- \frac{h - 1}{\alpha^2} \left(1 - \frac{k}{\alpha}(x - \xi) \right)^{\frac{2}{k}-2} \left(1 - h \left(1 - \frac{k}{\alpha}(x - \xi) \right)^{\frac{1}{k}} \right)^{\frac{1}{h}-2}.$$

(9.44)

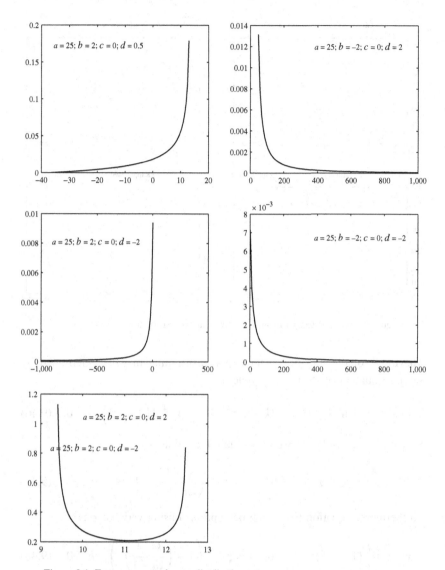

Figure 9.1 Four-parameter kappa distribution.

In Equation (9.44), let $z = 1 - \frac{k}{\alpha}(x - \xi) \Rightarrow \frac{df}{dx} = \frac{df}{dz}\frac{dz}{dx} \Rightarrow \frac{df}{dx} = -\frac{k}{\alpha}\frac{df}{dz}$. Equation (9.44) can be rewritten as

$$\frac{df}{dz} = -\frac{k-1}{\alpha k} z^{\frac{1}{k}-2}\left(1 - hz^{\frac{1}{k}}\right)^{\frac{1}{h}-1} + \frac{h-1}{\alpha k} z^{\frac{2}{k}-2}\left(1 - hz^{\frac{1}{k}}\right)^{\frac{1}{k}-2}$$

$$\Rightarrow \frac{df}{dz} = -\frac{1}{\alpha k} z^{\frac{1}{k}-1}\left(1 - hz^{\frac{1}{k}}\right)^{\frac{1}{h}-1}\left((k-1)z^{-1} - (r-1)z^{\frac{1}{k}-1}\left(1 - hz^{\frac{1}{k}}\right)^{-1}\right).$$

$$(9.45)$$

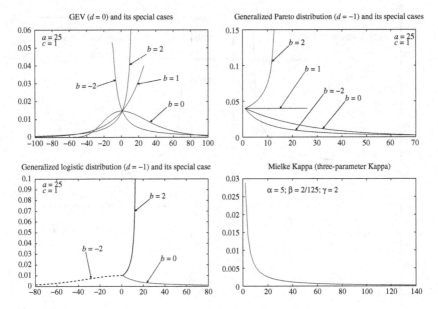

Figure 9.2 Special cases of four-parameter kappa distribution.

The mode of the four-parameter kappa distribution may be determined by setting Equation (9.45) to zero, yielding

$$1 - hz^{\frac{1}{k}} = 0 \text{ if } h \neq 0 \text{ or } (k-1)z^{-1} - (r-1)z^{\frac{1}{k}-1}\left(1 - hz^{\frac{1}{k}}\right)^{-1} = 0. \quad (9.46)$$

Solving Equation (9.46), we may obtain the mode as

$$z = r^{-k}, r \neq 0; \quad \text{or} \quad z = \left(\frac{k-1}{kh-1}\right)^{k}, \; kh \neq 1. \quad (9.47)$$

Furthermore, Equation (9.47) can be expressed using variable x as

$$x = \xi + \frac{k}{\alpha}(1 - r^{-k}); r \neq 0 \quad \text{or} \quad x = \xi + \frac{k}{\alpha}\left(1 - \left(\frac{k-1}{kh-1}\right)^{k}\right), \; kh \neq 0. \quad (9.48)$$

9.4 Derivation of Four-Parameter Kappa Distribution Using Entropy Theory

The Shannon entropy of the four-parameter kappa distribution can be written as

$$H(f) = -\int f(x) \ln f(x) dx. \quad (9.49)$$

From Equation (9.1), the density function in the logarithm may be expressed as

$$\ln f(x) = -\ln \alpha + \left(\frac{1}{k} - 1\right)\ln\left(1 - \frac{k}{\alpha}(x - \xi)\right) + \ln\left(1 - h\left(1 - \frac{k}{\alpha}(x - \xi)\right)^{\frac{1}{k}}\right)^{\frac{1}{k} - 1}.$$

(9.50)

If $h > 0$, substituting Equation (9.50) in Equation (9.49), we have

$$H(f) = -\int\left[-\ln \alpha + \left(\frac{1}{k} - 1\right)\ln\left(1 - \frac{k}{\alpha}(x - \xi)\right)\right.$$

$$\left. + \left(\frac{1}{h} - 1\right)\ln\left(1 - h\left(1 - \frac{k}{\alpha}(x - \xi)\right)^{\frac{1}{k}}\right)\right]f(x)dx.$$

(9.51)

If $h < 0$, the entropy may be expressed as

$$H(f) = -\int\left[-\ln \alpha + \left(\frac{1}{k} - 1\right)\ln\left(1 - \frac{k}{\alpha}(x - \xi)\right)\right.$$

$$\left. + \left(1 - \frac{1}{h}\right)\ln\left(1 - h\left(1 - \frac{k}{\alpha}(x - \xi)\right)^{\frac{1}{k}}\right)^{-1}\right]f(x)dx.$$

(9.52)

The constraints needed for maximizing the entropy and deriving the maximum entropy-based four-parameter kappa distribution are as follows:

$$C_0 = \int f(x)dx = 1$$

(9.53)

$$C_1 = \int \ln\left(1 - \frac{k}{\alpha}(x - \xi)\right)f(x)dx = E\left[\ln\left(1 - \frac{k}{\alpha}(x - \xi)\right)\right]$$

(9.54)

$$C_2 = \int \ln\left(1 - h\left(1 - \frac{k}{\alpha}(x - \xi)\right)^{\frac{1}{k}}\right)f(x)dx = E\left[\ln\left(1 - h\left(1 - \frac{k}{\alpha}(x - \xi)\right)^{\frac{1}{k}}\right)\right], \quad \text{if } h > 0$$

(9.55)

or

$$C_2 = \int \ln\left(1 - h\left(1 - \frac{k}{a}(x - \xi)\right)^{\frac{1}{k}}\right)^{-1} f(x)dx$$

$$= E\left[\ln\left(1 - h\left(1 - \frac{k}{a}(x - \xi)\right)^{\frac{1}{k}}\right)^{-1}\right], \quad \text{if } h < 0.$$

(9.56)

Using the constraints expressed through Equations (9.53)–(9.56), the Lagrangian function L may then be constructed as

$$
h > 0 : L = -\int f(x)\ln f(x)dx - (\lambda_0 - 1)\left[\int f(x)dx - 1\right]
$$
$$
-\lambda_1\left\{\int \ln\left(1 - \frac{k}{\alpha}(x - \xi)\right)f(x)dx - E\left[\ln\left(1 - \frac{k}{\alpha}(x - \xi)\right)\right]\right\}
$$
$$
-\lambda_2\left\{\int \ln\left(1 - h\left(1 - \frac{k}{\alpha}(x - \xi)\right)^{\frac{1}{k}}\right)f(x)dx - E\left[\ln\left(1 - h\left(1 - \frac{k}{\alpha}(x - \xi)\right)^{\frac{1}{k}}\right)\right]\right\}
$$
(9.57)

$$
h < 0 : L = -\int f(x)\ln f(x)dx - (\lambda_0 - 1)\left[\int f(x)dx - 1\right]
$$
$$
-\lambda_1\left\{\int \ln\left(1 - \frac{k}{\alpha}(x - \xi)\right)f(x)dx - E\left[\ln\left(1 - \frac{k}{\alpha}(x - \xi)\right)\right]\right\}
$$
$$
-\lambda_2\left\{\int \ln\left(1 - h\left(1 - \frac{k}{\alpha}(x - \xi)\right)^{\frac{1}{k}}\right)f(x)dx - E\left[\ln\left(1 - h\left(1 - \frac{k}{\alpha}(x - \xi)\right)^{\frac{1}{k}}\right)^{-1}\right]\right\}.
$$
(9.58)

In Equations (9.57) and (9.58), λ_0, λ_1, and λ_2 are the Lagrange multipliers. Differentiating Equations (9.57) and (9.58) with respect to the density function $f(x)$, we obtain the maximum entropy-based PDF for the four-parameter kappa distribution as

$$
f(x) = \exp\left(-\lambda_0 - \lambda_1\ln\left(1 - \frac{k}{\alpha}(x - \xi)\right) - \lambda_2\ln\left(1 - h\left(1 - \frac{k}{\alpha}(x - \xi)\right)^{\frac{1}{k}}\right)\right); \quad h > 0
$$
(9.59)

$$
f(x) = \exp\left(-\lambda_0 - \lambda_1\ln\left(1 - \frac{k}{\alpha}(x - \xi)\right) - \lambda_2\ln\left(1 - h\left(1 - \frac{k}{\alpha}(x - \xi)\right)^{\frac{1}{k}}\right)^{-1}\right); \quad h < 0.
$$
(9.60)

(1) $h > 0$
Substituting Equation (9.59) in Equation (9.53), we have

$$
\exp(\lambda_0) = \int \exp\left(-\lambda_1\ln\left(1 - \frac{k}{\alpha}(x - \xi)\right) - \lambda_2\ln\left(1 - h\left(1 - \frac{k}{\alpha}(x - \xi)\right)^{\frac{1}{k}}\right)\right)dx
$$
$$
\Rightarrow \exp(\lambda_0) = \int \left(1 - \frac{k}{\alpha}(x - \xi)\right)^{-\lambda_1}\left(1 - h\left(1 - \frac{k}{\alpha}(x - \xi)\right)^{\frac{1}{k}}\right)^{-\lambda_2}dx
$$
(9.61)

or

$$\lambda_0 = \ln \int \left(1 - \frac{k}{a}(x - \xi)\right)^{-\lambda_1} \left(1 - h\left(1 - \frac{k}{a}(x - \xi)\right)^{\frac{1}{k}}\right)^{-\lambda_2} dx. \quad (9.62)$$

In Equation (9.61), let $z = h\left(1 - \frac{k}{a}(x - \xi)\right)^{\frac{1}{k}} \Rightarrow dx = -\frac{a}{h^k} z^{k-1} dz$. Equation (9.61) can then be reformulated as

$$\exp(\lambda_0) = \int \left(\frac{z}{h}\right)^{-k\lambda_1} (1 - z)^{-\lambda_2} \left(-\frac{a}{h^k}\right) z^{k-1} dz = -ah^{k\lambda_1 - k} \int z^{-k\lambda_1 + k - 1}(1 - z)^{-\lambda_2} dz. \quad (9.63)$$

Based on the range of variable x given in Equations (9.5)–(9.10), the lower limit for the transformed variable z is one and the upper limit is zero. Equation (9.63) can then be evaluated as

$$\exp(\lambda_0) = ah^{k\lambda_1 - k} \int_0^1 z^{k(1-\lambda_1)-1}(1 - z)^{-\lambda_2} dz$$

$$= ah^{-k(1-\lambda_1)} B\big(k(1 - \lambda_1), (1 - \lambda_2)\big) = \frac{ah^{-k(1-\lambda_1)}\Gamma\big(k(1 - \lambda_1)\big)\Gamma(1 - \lambda_2)}{\Gamma\big(k(1 - \lambda_1) + (1 - \lambda_2)\big)}, \quad (9.64)$$

and λ_0 is then expressed as

$$\lambda_0 = \ln \frac{a}{h^{k(1-\lambda_1)}} - \ln \Gamma\big(k(1 - \lambda_1) + (1 - \lambda_2)\big) + \ln \Gamma\big(k(1 - \lambda_1)\big) + \ln \Gamma(1 - \lambda_2). \quad (9.65)$$

Substituting Equation (9.65) back in Equation (9.59), the entropy-based PDF is expressed as

$$f(x) = \frac{h^{k(1-\lambda_1)}}{aB\big(k(1 - \lambda_1), (1 - \lambda_2)\big)} \left(1 - \frac{k}{a}(x - \xi)\right)^{-\lambda_1} \left(1 - h\left(1 - \frac{k}{a}(x - \xi)\right)^{\frac{1}{k}}\right)^{-\lambda_2}. \quad (9.66)$$

Comparing Equation (9.66) with Equation (9.1), we have

$$\lambda_1 = 1 - \frac{1}{k} \Rightarrow k = (1 - \lambda_1)^{-1}; \quad \lambda_2 = 1 - \frac{1}{h} \Rightarrow h = (1 - \lambda_2)^{-1}. \quad (9.67)$$

To this end, the parameter constraints are $h > 0$; $a > 0$; $\xi \in R$; $k(1 - \lambda_1) > 0$, $k \in R$; $\lambda_2 < 1$.

(2) $h < 0$

Substituting Equation (9.60) in Equation (9.53), we have

$$\exp(\lambda_0) = \int \exp\left(-\lambda_1 \ln\left(1 - \frac{k}{\alpha}(x-\xi)\right) - \lambda_2 \ln\left(1 - h\left(1 - \frac{k}{\alpha}(x-\xi)\right)^{\frac{1}{k}}\right)^{-1}\right) dx$$

$$\Rightarrow \exp(\lambda_0) = \int \left(1 - \frac{k}{\alpha}(x-\xi)\right)^{-\lambda_1} \left(1 - h\left(1 - \frac{k}{\alpha}(x-\xi)\right)^{\frac{1}{k}}\right)^{\lambda_2} dx \tag{9.68}$$

or

$$\lambda_0 = \ln \int \left(1 - \frac{k}{\alpha}(x-\xi)\right)^{-\lambda_1} \left(1 - h\left(1 - \frac{k}{\alpha}(x-\xi)\right)^{\frac{1}{k}}\right)^{\lambda_2} dx. \tag{9.69}$$

In Equation (9.69), let $z = -h\left(1 - \frac{k}{\alpha}(x-\xi)\right)^{\frac{1}{k}} \Rightarrow dx = -\frac{\alpha}{(-h)^k} z^{k-1} dz$. Equation (9.68) can then be reformulated as

$$\exp(\lambda_0) = -\alpha(-h)^{k\lambda_1 - k} \int z^{-k\lambda_1 + k - 1}(1 + z)^{\lambda_2} dz. \tag{9.70}$$

Based on the range of variable x given in Equations (9.5)–(9.10), the lower limit for the transformed variable z is ∞ and the upper limit is $-\infty$. Equation (9.70) can then be evaluated as

$$\exp(\lambda_0) = \alpha(-h)^{k\lambda_1 - k} \int_{-\infty}^{\infty} z^{k(1-\lambda_1)-1}(1 + z)^{\lambda_2} dz$$

$$= \alpha(-h)^{-k(1-\lambda_1)} B\big(k(1 - \lambda_1), -\lambda_2 - k(1 - \lambda_1)\big). \tag{9.71}$$

λ_0 is then expressed as

$$\lambda_0 = \ln\frac{\alpha}{(-h)^{k(1-\lambda_1)}} - \ln\Gamma(-\lambda_2) + \ln\Gamma\big(k(1-\lambda_1)\big) + \ln\Gamma\big(-\lambda_2 - k(1-\lambda_1)\big). \tag{9.72}$$

Substituting Equation (9.71) back in Equation (9.60), the entropy-based PDF is expressed as

$$f(x) = \frac{(-h)^{k(1-\lambda_1)}}{\alpha B\big(k(1-\lambda_1), -\lambda_2 - k(1-\lambda_1)\big)}\left(1 - \frac{k}{\alpha}(x-\xi)\right)^{-\lambda_1}\left(1 - h\left(1 - \frac{k}{\alpha}(x-\xi)\right)^{\frac{1}{k}}\right)^{\lambda_2}. \tag{9.73}$$

Comparing Equation (9.73) with Equation (9.1), we have

$$\lambda_1 = 1 - \frac{1}{k} \Rightarrow k = (1 - \lambda_1)^{-1}; \quad \lambda_2 = \frac{1}{h} - 1 \Rightarrow h = (1 + \lambda_2)^{-1}. \tag{9.74}$$

To this end, the corresponding parameter constraints are $a > 0$; $h < 0$; $\xi \in R$; $k(1 - \lambda_1) > 0$, $k, \lambda_1 \in R$; $\lambda_2 < -k(1 - \lambda_1)$.

9.5 Parameter Estimation

9.5.1 Entropy Method

From the entropy-based PDF, that is, Equation (9.66), there are six parameters that need to be estimated, including four parameters (of four-parameter kappa distribution) and two Lagrange multipliers (λ_1, λ_2). In what follows, we discuss how to set up the equations to estimate the parameters by two approaches.

9.5.1.1 Using the Relation of λ_0 and (λ_1, λ_2), that is, $\lambda_0 = f(\lambda_1, \lambda_2)$

(1) $h > 0$

Taking the partial derivative of Equation (9.65) with respect to λ_1 and λ_2, we have

$$\frac{\partial \lambda_0}{\partial \lambda_1} = k \ln h + k \psi \big(k(1 - \lambda_1) + (1 - \lambda_2)\big) - k \psi \big(k(1 - \lambda_1)\big) \tag{9.75}$$

$$\frac{\partial \lambda_0}{\partial \lambda_2} = \psi \big(k(1 - \lambda_1) + (1 - \lambda_2)\big) - \psi(1 - \lambda_2). \tag{9.76}$$

Taking the partial derivative of Equation (9.62) with respect to λ_1 and λ_2, we have

$$\frac{\partial \lambda_0}{\partial \lambda_1} = -\frac{\int \left[\ln\left(1 - \frac{k}{\alpha}(x - \xi)\right)\right]\left(1 - \frac{k}{\alpha}(x - \xi)\right)^{-\lambda_1}\left(1 - h\left(1 - \frac{k}{\alpha}(x - \xi)\right)^{\frac{1}{k}}\right)^{-\lambda_2} dx}{\int \left(1 - \frac{k}{\alpha}(x - \xi)\right)^{-\lambda_1}\left(1 - h\left(1 - \frac{k}{\alpha}(x - \xi)\right)^{\frac{1}{k}}\right)^{-\lambda_2} dx}$$

$$= -E\left[\ln\left(1 - \frac{k}{\alpha}(x - \xi)\right)\right]$$

$$\tag{9.77}$$

$$\frac{\partial \lambda_0}{\partial \lambda_2} = -\frac{\int \left[\ln\left(1 - h\left(1 - \frac{k}{\alpha}(x - \xi)\right)^{\frac{1}{k}}\right)\right]\xi^{-\lambda_1}\left(1 - h\left(1 - \frac{k}{\alpha}(x - \xi)\right)^{\frac{1}{k}}\right)^{-\lambda_2} dx}{\int \left(1 - \frac{k}{\alpha}(x - \xi)\right)^{-\lambda_1}\left(1 - h\left(1 - \frac{k}{\alpha}(x - \xi)\right)^{\frac{1}{k}}\right)^{-\lambda_2} dx}$$

$$= -E\left[\ln\left(1 - h\left(1 - \frac{k}{\alpha}(x - \xi)\right)^{\frac{1}{k}}\right)\right].$$

$$\tag{9.78}$$

It is seen that we have two equations but we have six parameters for the entropy-based kappa distribution. Thus, four more equations are necessary for parameter estimation. These equations are the second derivatives of $(\partial^2 \lambda_0 / \partial \lambda_1^2)$, $(\partial^2 \lambda_0 / \partial \lambda_2^2)$, $(\partial^2 \lambda_0 / \partial \lambda_1 \partial \lambda_2)$.

Based on Equations (9.75) and (9.76), we have

$$\frac{\partial^2 \lambda_0}{\partial \lambda_1^2} = -k^2 \left(\psi^{(1)} \big(k(1 - \lambda_1) + (1 - \lambda_2) \big) - \psi^{(1)} \big(k(1 - \lambda_1) \big) \right) \quad (9.79)$$

$$\frac{\partial^2 \lambda_0}{\partial \lambda_2^2} = \psi^{(1)} (1 - \lambda_2) - \psi^{(1)} \big(k(1 - \lambda_1) + (1 - \lambda_2) \big) \quad (9.80)$$

$$\frac{\partial^2 \lambda_0}{\partial \lambda_1 \partial \lambda_2} = -k \psi^{(1)} \big(k(1 - \lambda_1) + (1 - \lambda_2) \big). \quad (9.81)$$

Based on Equations (9.77) and (9.78), we have

$$\frac{\partial^2 \lambda_0}{\partial \lambda_1^2} = \frac{\int \left[\ln \left(1 - \frac{k}{\alpha} \xi \right) \right]^2 \left(1 - \frac{k}{\alpha}(x - \xi) \right)^{-\lambda_1} \left(1 - h \left(1 - \frac{k}{\alpha}(x - \xi) \right)^{\frac{1}{k}} \right)^{-\lambda_2} dx}{\int \left(1 - \frac{k}{\alpha}(x - \xi) \right)^{-\lambda_1} \left(1 - h \left(1 - \frac{k}{\alpha}(x - \xi) \right)^{\frac{1}{k}} \right)^{-\lambda_2} dx}$$

$$= E \left[\left(\ln \left(1 - \frac{k}{\alpha}(x - \xi) \right) \right)^2 \right] \quad (9.82)$$

$$\frac{\partial^2 \lambda_0}{\partial \lambda_2^2} = \frac{\int \left[\ln \left(1 - h \left(1 - \frac{k}{\alpha}(x - \xi) \right)^{\frac{1}{k}} \right) \right]^2 \left(1 - \frac{k}{\alpha}(x - \xi) \right)^{-\lambda_1} \left(1 - h \left(1 - \frac{k}{\alpha}(x - \xi) \right)^{\frac{1}{k}} \right)^{-\lambda_2} dx}{\int \left(1 - \frac{k}{\alpha}(x - \xi) \right)^{-\lambda_1} \left(1 - h \left(1 - \frac{k}{\alpha}(x - \xi) \right)^{\frac{1}{k}} \right)^{-\lambda_2} dx}$$

$$= E \left[\left(\ln \left(1 - h \left(1 - \frac{k}{\alpha}(x - \xi) \right)^{\frac{1}{k}} \right) \right)^2 \right] \quad (9.83)$$

$$\frac{\partial^2 \lambda_0}{\partial \lambda_1 \partial \lambda_2} = \frac{\int \left[\ln \left(1 - \frac{k}{\alpha}(x - \xi) \right) \right] \left[\ln \left(1 - h \left(1 - \frac{k}{\alpha}(x - \xi) \right)^{\frac{1}{k}} \right) \right] \left(1 - \frac{k}{\alpha}(x - \xi) \right)^{-\lambda_1} \left(1 - h \left(1 - \frac{k}{\alpha}(x - \xi) \right)^{\frac{1}{k}} \right)^{-\lambda_2} dx}{\int \left(1 - \frac{k}{\alpha}(x - \xi) \right)^{-\lambda_1} \left(1 - h \left(1 - \frac{k}{\alpha}(x - \xi) \right)^{\frac{1}{k}} \right)^{-\lambda_2} dx}$$

$$= E \left[\left[\ln \left(1 - \frac{k}{\alpha}(x - \xi) \right) \right] \left[\ln \left(1 - h \left(1 - \frac{k}{\alpha}(x - \xi) \right)^{\frac{1}{k}} \right) \right] \right]. \quad (9.84)$$

Combining the above equations, we have

$$\ln h^k + k\left(\psi\big(k(1-\lambda_1)+(1-\lambda_2)\big) - \psi\big(k(1-\lambda_1)\big)\right) = -E\left[\ln(1-\frac{k}{\alpha}(x-\xi)\right] \tag{9.85}$$

$$\psi\big(k(1-\lambda_1)+(1-\lambda_2)\big) - \psi(1-\lambda_2) = -E\left[\ln\left(1-h\left(1-\frac{k}{\alpha}(x-\xi)\right)^{\frac{1}{k}}\right)\right] \tag{9.86}$$

$$-k^2\left(\psi^{(1)}\big(k(1-\lambda_1)+(1-\lambda_2)\big) - \psi^{(1)}\big(k(1-\lambda_1)\big)\right) = E\left[\left(\ln\left(1-\frac{k}{\alpha}(x-\xi)\right)\right)^2\right] \tag{9.87}$$

$$\psi^{(1)}(1-\lambda_2) - \psi^{(1)}\big(k(1-\lambda_1)+(1-\lambda_2)\big) = E\left[\left(\ln\left(1-h\left(1-\frac{k}{\alpha}(x-\xi)\right)^{\frac{1}{k}}\right)\right)^2\right] \tag{9.88}$$

$$-k\psi^{(1)}\big(k(1-\lambda_1)+(1-\lambda_2)\big) = E\left[\left[\ln\left(1-\frac{k}{\alpha}(x-\xi)\right)\right]\left[\ln\left(1-h\left(1-\frac{k}{\alpha}(x-\xi)\right)^{\frac{1}{k}}\right)\right]\right], \tag{9.89}$$

where $\psi(x) = \frac{d\ln\Gamma(x)}{dx}$, $\psi^{(1)}(x) = \frac{d^2\ln\Gamma(x)}{dx^2}$.

To this end, we need one more equation for parameter estimation. Comparing Equations (9.1) and (9.66), we can obtain the last equation as

$$B\big(k(1-\lambda_1), 1-\lambda_2\big) = h^{k(1-\lambda_1)}. \tag{9.90}$$

(2) $h < 0$

The first partial derivative of Equation (9.72) with respect to λ_1 and λ_2 yields

$$\frac{\partial\lambda_0}{\partial\lambda_1} = k\left(\ln(-h) - \psi\big(k(1-\lambda_1)\big) + \psi\big(-\lambda_2 - k(1-\lambda_1)\big)\right) \tag{9.91}$$

$$\frac{\partial\lambda_0}{\partial\lambda_2} = \psi(-\lambda_2) - \psi\big(-\lambda_2 - k(1-\lambda_1)\big). \tag{9.92}$$

The first partial derivative of Equation (9.69) with respect to λ_1 and λ_2 yields

$$\frac{\partial\lambda_0}{\partial\lambda_1} = -\frac{\int\left[\ln\left(1-\frac{k}{a}(x-\xi)\right)\right]\left(1-\frac{k}{a}(x-\xi)\right)^{-\lambda_1}\left(1-h\left(1-\frac{k}{a}(x-\xi)\right)^{\frac{1}{k}}\right)^{\lambda_2}dx}{\int\left(1-\frac{k}{a}(x-\xi)\right)^{-\lambda_1}\left(1-h\left(1-\frac{k}{a}(x-\xi)\right)^{\frac{1}{k}}\right)^{\lambda_2}dx}$$

$$= -E\left[\ln\left(1-\frac{k}{a}(x-\xi)\right)\right] \tag{9.93}$$

$$\frac{\partial \lambda_0}{\partial \lambda_2} = \frac{\int \left[\ln\left(1 - h\left(1 - \frac{k}{a}(x - \xi)\right)^{\frac{1}{k}}\right)^{-1}\right]\left(1 - \frac{k}{a}(x - \xi)\right)^{-\lambda_1}\left[\left(1 - h\left(1 - \frac{k}{a}(x - \xi)\right)^{\frac{1}{k}}\right)^{-1}\right]^{-\lambda_2} dx}{\int \left(1 - \frac{k}{a}(x - \xi)\right)^{-\lambda_1}\left(1 - h\left(1 - \frac{k}{a}(x - \xi)\right)^{\frac{1}{k}}\right)^{\lambda_2} dx}$$

$$= E\left[\ln\left(1 - h\left(1 - \frac{k}{a}(x - \xi)\right)^{\frac{1}{k}}\right)^{-1}\right].$$

(9.94)

The second derivative of Equation (9.72) with respect to λ_1 and λ_2 yields

$$\frac{\partial^2 \lambda_0}{\partial \lambda_1^2} = k^2 \left(\psi^{(1)}\left(k(1 - \lambda_1)\right) + \psi^{(1)}\left(-\lambda_2 - k(1 - \lambda_1)\right)\right) \qquad (9.95)$$

$$\frac{\partial^2 \lambda_0}{\partial \lambda_2^2} = \psi^{(1)}\left(-\lambda_2 - k(1 - \lambda_1)\right) - \psi^{(1)}(-\lambda_2) \qquad (9.96)$$

$$\frac{\partial^2 \lambda_0}{\partial \lambda_1 \partial \lambda_2} = -k\psi^{(1)}\left(-\lambda_2 - k(1 - \lambda_1)\right). \qquad (9.97)$$

The second derivative of Equation (9.69) yields

$$\frac{\partial^2 \lambda_0}{\partial \lambda_1^2} = E\left[\left(\ln\left(1 - \frac{k}{a}(x - \xi)\right)\right)^2\right] \qquad (9.98)$$

$$\frac{\partial^2 \lambda_0}{\partial \lambda_2^2} = E\left[\left(\ln\left(1 - h\left(1 - \frac{k}{a}(x - \xi)\right)^{\frac{1}{k}}\right)^{-1}\right)^2\right] \qquad (9.99)$$

$$\frac{\partial^2 \lambda_0}{\partial \lambda_1 \partial \lambda_2} = E\left[\left(\ln\left(1 - \frac{k}{a}(x - \xi)\right)\right)\left(\ln\left(1 - h\left(1 - \frac{k}{a}(x - \xi)\right)^{\frac{1}{k}}\right)^{-1}\right)\right].$$

(9.100)

Combining the above equations yields

$$k\left(\ln(-h) - \psi\left(k(1 - \lambda_1)\right) + \psi\left(-\lambda_2 - k(1 - \lambda_1)\right)\right) = -E\left[\ln\left(1 - \frac{k}{a}(x - \xi)\right)\right]$$

(9.101)

$$\psi(-\lambda_2) - \psi\left(-\lambda_2 - k(1 - \lambda_1)\right) = E\left[\ln\left(1 - h\left(1 - \frac{k}{a}(x - \xi)\right)^{\frac{1}{k}}\right)^{-1}\right] \qquad (9.102)$$

$$k^2\left(\psi^{(1)}\left(k(1 - \lambda_1)\right) + \psi^{(1)}\left(-\lambda_2 - k(1 - \lambda_1)\right)\right) = E\left[\left(\ln\left(1 - \frac{k}{a}(x - \xi)\right)\right)^2\right]$$

(9.103)

$$\psi^{(1)}\left(-\lambda_2 - k(1-\lambda_1)\right) - \psi^{(1)}(-\lambda_2) = E\left[\left(\ln\left(1 - h\left(1 - \frac{k}{a}(x-\xi)\right)^{\frac{1}{k}}\right)^{-1}\right)^2\right]$$

(9.104)

$$-k\psi^{(1)}\left(-\lambda_2 - k(1-\lambda_1)\right) = E\left[\left(\ln\left(1 - \frac{k}{a}(x-\xi)\right)\right)\left(\ln\left(1 - h\left(1 - \frac{k}{a}(x-\xi)\right)^{\frac{1}{k}}\right)^{-1}\right)\right].$$

(9.105)

Comparing Equation (9.73) with Equation (9.1) yields the last equation for parameter estimation:

$$(-h)^{k(1-\lambda_1)} = B\left(k(1-\lambda_1), -\lambda_2 - k(1-\lambda_1)\right).$$

(9.106)

9.5.1.2 Entropy Function of the Kappa Distribution

(1) $h > 0$

In the case of the four-parameter kappa distribution, this method may also be considered as the method of expanded parameter space method. Using Equation (9.65), the entropy $H(f)$ is expressed as

$$H(f) = -\int f(x)\ln f(x)dx$$

$$= -\int f(x) \times \left\{ \ln\left[\frac{h^{k(1-\lambda_1)}}{aB\left(k(1-\lambda_1), 1-\lambda_2\right)}\right] - \lambda_1 \ln\left(1 - \frac{k}{\alpha}(x-\xi)\right)\right.$$

$$\left. - \lambda_2 \ln\left(1 - h\left(1 - \frac{k}{\alpha}(x-\xi)\right)^{\frac{1}{k}}\right)\right\}dx$$

$$= -\ln\left[\frac{h^{k(1-\lambda_1)}}{aB\left(k(1-\lambda_1), 1-\lambda_2\right)}\right] + \lambda_1 E\left[\ln\left(1 - \frac{k}{\alpha}(x-\xi)\right)\right]$$

$$+ \lambda_2 E\left[\ln\left(1 - h\left(1 - \frac{k}{\alpha}(x-\xi)\right)^{\frac{1}{k}}\right)\right]$$

$$= \ln\alpha - k(1-\lambda_1)\ln h + \ln B\left(k(1-\lambda_1), 1-\lambda_2\right) + \lambda_1 E\left[\ln\left(1 - \frac{k}{\alpha}(x-\xi)\right)\right]$$

$$+ \lambda_2 E\left[\ln\left(1 - h\left(1 - \frac{k}{\alpha}(x-\xi)\right)^{\frac{1}{k}}\right)\right].$$

(9.107)

Then, entropy is maximized and parameters are estimated by taking the derivative of $H(f)$ with respect to the parameters and then they are set equal to zero.

$$\frac{\partial H(f)}{\partial \xi} = \frac{k\lambda_1}{\alpha} E\left(\frac{1}{1-\frac{k}{\alpha}(x-\xi)}\right) - \frac{h\lambda_2}{\alpha} E\left[\frac{\left(1-\frac{k}{\alpha}(x-\xi)\right)^{\frac{1}{k}-1}}{1-h\left(1-\frac{k}{\alpha}(x-\xi)\right)^{\frac{1}{k}}}\right] = 0$$

(9.108)

$$\frac{\partial H(f)}{\partial \alpha} = \frac{1}{\alpha} + \frac{\lambda_1}{\alpha} E\left[\frac{\frac{k}{\alpha}(x-\xi)}{1-\frac{k}{\alpha}(x-\xi)}\right] - \frac{\lambda_2 h}{\alpha^2} E\left[\frac{(x-\xi)\left(1-\frac{k}{\alpha}(x-\xi)\right)^{\frac{1}{k}-1}}{1-h\left(1-\frac{k}{\alpha}(x-\xi)\right)^{\frac{1}{k}}}\right] = 0$$

(9.109)

$$\frac{\partial H(f)}{\partial k} = -(1-\lambda_1)\ln h + (1-\lambda_1)(\psi(k(1-\lambda_1)) - \psi(k(1-\lambda_1)+(1-\lambda_2)))$$

$$-\frac{\lambda_1}{\alpha} E\left[\frac{x-\xi}{1-\frac{k}{\alpha}(x-\xi)}\right]$$

$$+\lambda_2 h E\left[\frac{\frac{\ln\left(1-\frac{k}{\alpha}(x-\xi)\right)\left[1-\frac{k}{\alpha}(x-\xi)\right]^{\frac{1}{k}}}{k^2} + \frac{x-\xi}{\alpha k\left(1-\frac{k}{\alpha}(x-\xi)\right)^{\frac{k-1}{k}}}}{1-h\left(1-\frac{k}{\alpha}(x-\xi)\right)^{\frac{1}{k}}}\right]$$

(9.110)

$$\frac{\partial H(f)}{\partial h} = -\frac{k(1-\lambda_1)}{h} - \lambda_2 E\left[\frac{\left(1-\frac{k}{\alpha}(x-\xi)\right)^{\frac{1}{k}}}{1-h\left(1-\frac{k}{\alpha}(x-\xi)\right)^{\frac{1}{k}}}\right] = 0 \qquad (9.111)$$

$$\frac{\partial H(f)}{\partial \lambda_1} = k\ln h - k\left(\psi(k(1-\lambda_1)) - \psi(k(1-\lambda_1)+(1-\lambda_2))\right)$$
$$+ E\left[\ln\left(1-\frac{k}{\alpha}(x-\xi)\right)\right]$$

(9.112)

$$\frac{\partial H(f)}{\partial \lambda_2} = -\psi(1-\lambda_2) + \psi(k(1-\lambda_1) + (1-\lambda_2)) + E\left[\ln\left(1 - h\left(1 - \frac{k}{\alpha}(x-\xi)\right)^{\frac{1}{k}}\right)\right].$$

(9.113)

From Equation (9.67) we can reduce six to four equations as follows.
    From Equation (9.67), we have:

$$\frac{\partial H(f)}{\partial \lambda_1} = \frac{\partial H(f)}{\partial k}\frac{\partial k}{\partial \lambda_1} = \frac{1}{(1-\lambda_1)^2}\frac{\partial H(f)}{\partial k}.$$

(9.114)

Substituting Equation (9.110) in Equation (9.114), we have

$$\frac{\partial H(f)}{\partial \lambda_1} = \frac{1}{(1-\lambda_1)}\Bigg(-\ln h + \psi(k(1-\lambda_1)) - \psi(k(1-\lambda_1) + 1 - \lambda_2)\Bigg)$$
$$+ \frac{1}{(1-\lambda_1)^2}\Bigg(-\frac{\lambda_1}{\alpha}E\left[\frac{x-\xi}{1-\frac{k}{\alpha}(x-\xi)}\right] + \lambda_2 h E\left[\frac{\ln\left(1 - \frac{k}{\alpha}(x-\xi)\right)\left[1 - \frac{k}{\alpha}(x-\xi)\right]^{\frac{1}{k}}}{k^2} + \frac{x-\xi}{\alpha k\left(1 - \frac{k}{\alpha}(x-\xi)\right)^{\frac{k-1}{k}}}}{1 - h\left(1 - \frac{k}{\alpha}(x-\xi)\right)^{\frac{1}{k}}}\right]\Bigg).$$

(9.115)

Substituting Equation (9.67) in Equations (9.115) and (9.112), Equations (9.115) and (9.112) may be expressed with parameters ξ, α, k, and h as

$$\frac{\partial H(f)}{\partial \lambda_1} = -k(\ln h - \psi(1)) + \psi\left(1 + \frac{1}{h}\right) - \frac{k(k-1)}{a}E\left(\frac{x-\xi}{1 - \frac{k}{a}(x-\xi)}\right)$$
$$+ k^2(h-1)E\left[\frac{\ln\left(1 - \frac{k}{\alpha}(x-\xi)\right)\left[1 - \frac{k}{\alpha}(x-\xi)\right]^{\frac{1}{k}}}{k^2} + \frac{x-\xi}{\alpha k\left(1 - \frac{k}{\alpha}(x-\xi)\right)^{\frac{k-1}{k}}}}{1 - h\left(1 - \frac{k}{\alpha}(x-\xi)\right)^{\frac{1}{k}}}\right] = 0$$

(9.116)

$$\frac{\partial H(f)}{\partial \lambda_1} = k\ln h - k\left(\psi(1) - \psi\left(1 + \frac{1}{h}\right)\right) + E\left[\ln\left(1 - \frac{k}{\alpha}(x-\xi)\right)\right]. \quad (9.117)$$

Equating Equations (9.116) and (9.117), we have

$$E\left[\ln\left(1-\frac{k}{a}(x-\xi)\right)\right] - \frac{k(k-1)}{a}E\left(\frac{x-\xi}{1-\frac{k}{a}(x-\xi)}\right) + k^2(h-1)E$$

$$\left[\frac{\dfrac{\ln\left(1-\dfrac{k}{\alpha}(x-\xi)\right)\left[1-\dfrac{k}{\alpha}(x-\xi)\right]^{\frac{1}{k}}}{k^2} + \dfrac{x-\xi}{\alpha k\left(1-\dfrac{k}{\alpha}(x-\xi)\right)^{\frac{k-1}{k}}}}{1-h\left(1-\dfrac{k}{\alpha}(x-\xi)\right)^{\frac{1}{k}}}\right] = 0.$$

$$(9.118)$$

From Equation (9.67), we also have

$$\frac{\partial H(f)}{\partial \lambda_2} = \frac{\partial H(f)}{\partial h}\frac{\partial h}{\partial \lambda_2} = \frac{1}{(1-\lambda_2)^2}\frac{\partial H(f)}{\partial h}. \tag{9.119}$$

Substituting Equation (9.113) in Equation (9.119), we have

$$\frac{\partial H(f)}{\partial \lambda_2} = -\frac{1}{(1-\lambda_2)^2}\left(\frac{k(1-\lambda_1)}{h} + \lambda_2 E\left[\frac{\left(1-\frac{k}{\alpha}(x-\xi)\right)^{\frac{1}{k}}}{1-h\left(1-\frac{k}{\alpha}(x-\xi)\right)^{\frac{1}{k}}}\right]\right). \tag{9.120}$$

Substituting Equation (9.67) in Equations (9.113) and (9.120), we have

$$\frac{\partial H(f)}{\partial \lambda_2} = -\psi\left(\frac{1}{h}\right) + \psi\left(1+\frac{1}{h}\right) + E\left[\ln\left(1-h\left(1-\frac{k}{\alpha}(x-\xi)\right)^{\frac{1}{k}}\right)\right]$$

$$= h + E\left[\ln\left(1-h\left(1-\frac{k}{\alpha}(x-\xi)\right)^{\frac{1}{k}}\right)\right] \Rightarrow E\left[\ln\left(1-h\left(1-\frac{k}{\alpha}(x-\xi)\right)^{\frac{1}{k}}\right)\right]$$

$$= -h$$

$$(9.121)$$

and

$$\frac{\partial H(f)}{\partial \lambda_2} = -h - h(h-1)E\left[\frac{\left(1-\frac{k}{\alpha}(x-\xi)\right)^{\frac{1}{k}}}{1-h\left(1-\frac{k}{\alpha}(x-\xi)\right)^{\frac{1}{k}}}\right]. \tag{9.122}$$

Substituting Equation (9.121) in Equation (9.122), we have

$$E\left[\ln\left(1-h\left(1-\frac{k}{\alpha}(x-\xi)\right)^{\frac{1}{k}}\right)\right]+h(1-h)E\left[\frac{\left(1-\frac{k}{\alpha}(x-\xi)\right)^{\frac{1}{k}}}{1-h\left(1-\frac{k}{\alpha}(x-\xi)\right)^{\frac{1}{k}}}\right]=0.$$

(9.123)

Furthermore, substituting Equation (9.67) in Equations (9.108) and (9.109), we have

$$\frac{\partial H(f)}{\partial \xi}=\frac{k-1}{\alpha}E\left(\frac{1}{1-\frac{k}{\alpha}(x-\xi)}\right)-\frac{h-1}{\alpha}E\left[\frac{\left(1-\frac{k}{\alpha}(x-\xi)\right)^{\frac{1}{k}-1}}{1-h\left(1-\frac{k}{\alpha}(x-\xi)\right)^{\frac{1}{k}}}\right]=0$$

(9.124)

$$\frac{\partial H(f)}{\partial \alpha}=\frac{1}{\alpha}+\frac{k-1}{\alpha k}E\left[\frac{\frac{k}{\alpha}(x-\xi)}{1-\frac{k}{\alpha}(x-\xi)}\right]-\frac{h-1}{\alpha^2}E\left[\frac{(x-\xi)\left(1-\frac{k}{\alpha}(x-\xi)\right)^{\frac{1}{k}-1}}{1-h\left(1-\frac{k}{\alpha}(x-\xi)\right)^{\frac{1}{k}}}\right]=0.$$

(9.125)

To this end, the four equations needed for parameter estimation are Equations (9.118), (9.123)–(9.125).

(2) $h < 0$

In the case of $h < 0$, the entropy function is expressed as

$$H(f)=-\ln\left(\frac{(-h)^{k(1-\lambda_1)}}{\alpha B\left(k(1-\lambda_1),-\lambda_2-k(1-\lambda_1)\right)}\right)+\lambda_1 E\left(\ln\left(1-\frac{k}{\alpha}(x-\xi)\right)\right)$$

$$-\lambda_2 E\left[\ln\left(1-h\left(1-\frac{k}{\alpha}(x-\xi)\right)^{\frac{1}{k}}\right)\right].$$

(9.126)

Then, entropy, that is, Equation (9.126), may be optimized by taking the partial derivatives with respect to parameters $\xi, \alpha, k, h, \lambda_1$, and λ_2 as in the following:

$$\frac{\partial H(f)}{\partial \xi} = \frac{\lambda_1 k}{a} E\left[\left(1 - \frac{k}{a}(x-\xi)\right)^{-1}\right] + \frac{h\lambda_2}{\alpha} E\left[\frac{\left(1 - \frac{k}{\alpha}(x-\xi)\right)^{\frac{1}{k}-1}}{1 - h\left(1 - \frac{k}{\alpha}(x-\xi)\right)^{\frac{1}{k}}}\right] = 0$$

(9.127)

$$\frac{\partial H(f)}{\partial \alpha} = \frac{1}{\alpha} + \frac{\lambda_1}{\alpha} E\left[\frac{\frac{k}{\alpha}(x-\xi)}{1 - \frac{k}{\alpha}(x-\xi)}\right] + \frac{\lambda_2 h}{\alpha^2} E\left[\frac{(x-\xi)\left(1 - \frac{k}{\alpha}(x-\xi)\right)^{\frac{1}{k}-1}}{1 - h\left(1 - \frac{k}{\alpha}(x-\xi)\right)^{\frac{1}{k}}}\right] = 0$$

(9.128)

$$\frac{\partial H(f)}{\partial k} = -(1-\lambda_1)\Big(\ln(-h) - \psi\big(k(1-\lambda_1)\big) + \psi\big(-\lambda_2 - k(1-\lambda_1)\big)\Big)$$

$$-\frac{\lambda_1}{\alpha} E\left[\frac{x-\xi}{1 - \frac{k}{\alpha}(x-\xi)}\right] - \lambda_2 h E\left[\frac{\frac{\ln\left(1 - \frac{k}{\alpha}(x-\xi)\right)\left(1 - \frac{k}{\alpha}(x-\xi)\right)^{\frac{1}{k}}}{k^2} + \frac{x-\xi}{\alpha k\left(1 - \frac{k}{\alpha}(x-\xi)\right)^{1-\frac{1}{k}}}}{1 - h\left(1 - \frac{k}{a}(x-\xi)\right)^{\frac{1}{k}}}\right] = 0$$

(9.129)

$$\frac{\partial H(f)}{\partial h} = -\frac{k(1-\lambda_1)}{h} + \lambda_2 E\left[\frac{\left(1 - \frac{k}{a}(x-\xi)\right)^{\frac{1}{k}}}{1 - h\left(1 - \frac{k}{a}(x-\xi)\right)^{\frac{1}{k}}}\right] = 0 \qquad (9.130)$$

$$\frac{\partial H(f)}{\partial \lambda_1} = k\Big(\ln(-h) - \psi\big(k(1-\lambda_1)\big) + \psi\big(-\lambda_2 - k(1-\lambda_1)\big)\Big) + E\left[\ln\left(1 - \frac{k}{a}(x-\xi)\right)\right] = 0$$

(9.131)

$$\frac{\partial H(f)}{\partial \lambda_2} = \psi(-\lambda_2) - \psi\big(-\lambda_2 - k(1-\lambda_1)\big) - E\left[\ln\left(1 - h\left(1 - \frac{k}{\alpha}(x-\xi)\right)^{\frac{1}{k}}\right)\right].$$

(9.132)

Similar to the case of $h > 0$, six equations may be reduced to four using Equation (9.74). From Equation (9.74), we have $dk/d\lambda_1 = 1/(1 - \lambda_1)^2$. Then, Equation (9.131) may be expressed as

$$\frac{\partial H(f)}{\partial \lambda_1} = \frac{\partial H(f)}{\partial k} \frac{dk}{d\lambda_1}$$

$$= \frac{1}{(1 - \lambda_1)^2} \left\{ -(1 - \lambda_1)\Big(\ln(-h) - \psi(k(1 - \lambda_1)) + \psi(-\lambda_2 - k(1 - \lambda_1)) \Big) \right.$$

$$- \frac{\lambda_1}{a} E \left[\frac{x - \xi}{1 - \frac{k}{a}(x - \xi)} \right]$$

$$\left. - \lambda_2 h E \left[\frac{\dfrac{\ln\left(1 - \frac{k}{a}(x - \xi)\right)\left(1 - \frac{k}{a}(x - \xi)\right)^{\frac{1}{k}}}{k^2} + \dfrac{x - \xi}{ak\left(1 - \frac{k}{a}(x - \xi)\right)^{1 - \frac{1}{k}}}}{1 - h\left(1 - \frac{k}{a}(x - \xi)\right)^{\frac{1}{k}}} \right] \right\} = 0.$$

$$(9.133)$$

Now substituting Equation (9.74) in Equation (9.133), we have

$$-k\left(\ln(-h) - \psi(1) + \psi\left(-\frac{1}{h}\right) \right) - \frac{k^2 - k}{ak} E \left[\frac{x - \xi}{1 - \frac{k}{a}(x - \xi)} \right]$$

$$- k^2(1 - h)E \left[\frac{\dfrac{\ln\left(1 - \frac{k}{a}(x - \xi)\right)\left(1 - \frac{k}{a}(x - \xi)\right)^{\frac{1}{k}}}{k^2} + \dfrac{x - \xi}{ak\left(1 - \frac{k}{a}(x - \xi)\right)^{1 - \frac{1}{k}}}}{1 - h\left(1 - \frac{k}{a}(x - \xi)\right)^{\frac{1}{k}}} \right] = 0.$$

$$(9.134)$$

Similarly, Equation (9.132) can be rewritten as

$$k\left(\ln(-h) - \psi(1) + \psi\left(-\frac{1}{h}\right) \right) + E\left[\ln\left(1 - \frac{k}{a}(x - \xi)\right) \right] = 0.   (9.135)$$

Combining Equations (9.134) and (9.135), we have

$$E\left[\ln\left(1-\frac{k}{\alpha}(x-\xi)\right)\right]-\frac{k^2-k}{\alpha}E\left[\frac{x-\xi}{1-\frac{k}{\alpha}(x-\xi)}\right]$$

$$-k^2(1-h)E\left[\frac{\frac{\ln\left(1-\frac{k}{\alpha}(x-\xi)\right)\left(1-\frac{k}{\alpha}(x-\xi)\right)^{\frac{1}{k}}}{k^2}+\frac{x-\xi}{\alpha k\left(1-\frac{k}{\alpha}(x-\xi)\right)^{1-\frac{1}{k}}}}{1-h\left(1-\frac{k}{a}(x-\xi)\right)^{\frac{1}{k}}}\right]=0.$$

$$(9.136)$$

From Equation (9.74), we have $dh/d\lambda_2 = -1/(1+\lambda_2)^2$ and Equation (9.132) can be rewritten as

$$\frac{\partial H(f)}{\partial\lambda_2}=\frac{\partial H(f)}{\partial h}\frac{dh}{d\lambda_2}$$

$$=-\frac{1}{(1+\lambda_2)^2}\left[-\frac{k(1-\lambda_1)}{h}+\lambda_2 E\left[\frac{\left(1-\frac{k}{a}(x-\xi)\right)^{\frac{1}{k}}}{1-h\left(1-\frac{k}{a}(x-\xi)\right)^{\frac{1}{k}}}\right]\right].$$

$$(9.137)$$

Substituting Equation (9.74) in Equation (9.137), we have

$$\frac{\partial H(f)}{\partial\lambda_2}=-h^2\left[-\frac{1}{h}+\frac{1-h}{h}E\left[\frac{\left(1-\frac{k}{a}(x-\xi)\right)^{\frac{1}{k}}}{1-h\left(1-\frac{k}{a}(x-\xi)\right)^{\frac{1}{k}}}\right]\right]=0\Rightarrow h$$

$$(9.138)$$

$$=h(1-h)E\left[\frac{\left(1-\frac{k}{a}(x-\xi)\right)^{\frac{1}{k}}}{1-h\left(1-\frac{k}{a}(x-\xi)\right)^{\frac{1}{k}}}\right].$$

Similarly, Equation (9.132) may be rewritten as

$$\frac{\partial H(f)}{\partial \lambda_2} = \psi\left(-\frac{1}{h}+1\right) - \psi\left(-\frac{1}{h}\right) - E\left(\ln\left(1 - h\left(1 - \frac{k}{a}(x-\xi)\right)^{\frac{1}{k}}\right)\right)$$
$$= -h - E\left(\ln\left(1 - h\left(1 - \frac{k}{a}(x-\xi)\right)^{\frac{1}{k}}\right)\right).$$

(9.139)

Combining Equations (9.138) and (9.139), we have

$$h(1-h)E\left[\frac{\left(1 - \frac{k}{a}(x-\xi)\right)^{\frac{1}{k}}}{1 - h\left(1 - \frac{k}{a}(x-\xi)\right)^{\frac{1}{k}}}\right] + E\left(\ln\left(1 - h\left(1 - \frac{k}{a}(x-\xi)\right)^{\frac{1}{k}}\right)\right) = 0.$$

(9.140)

Furthermore, Equations (9.127) and (9.128) can be rewritten as

$$\frac{\partial H(f)}{\partial \xi} = \frac{k-1}{a}E\left[\left(1 - \frac{k}{a}(x-\xi)\right)^{-1}\right] + \frac{1-h}{\alpha}E\left[\frac{\left(1 - \frac{k}{a}(x-\xi)\right)^{\frac{1}{k}-1}}{1 - h\left(1 - \frac{k}{a}(x-\xi)\right)^{\frac{1}{k}}}\right] = 0$$

(9.141)

$$\frac{\partial H(f)}{\partial a} = \frac{1}{\alpha} + \frac{k-1}{\alpha k}E\left[\frac{\frac{k}{a}(x-\xi)}{1 - \frac{k}{a}(x-\xi)}\right] + \frac{1-h}{\alpha^2}E\left[\frac{(x-\xi)\left(1 - \frac{k}{a}(x-\xi)\right)^{\frac{1}{k}-1}}{1 - h\left(1 - \frac{k}{a}(x-\xi)\right)^{\frac{1}{k}}}\right] = 0.$$

(9.142)

We can use Equations (9.136) and (9.140)–(9.142) to estimate the parameters for the entropy-based four-parameter kappa distributions for $h < 0$.

9.5.2 MLE Method

From Equation (9.4), the log-likelihood function of kappa-distributed random variable $x = (x_1, x_2, \ldots, x_n)$ is expressed as

$$\ln L = -n\ln\alpha + \left(\frac{1}{k} - 1\right)\sum_{i=1}^{n}\ln\left(1 - \frac{k}{\alpha}(x_i - \xi)\right) + (1-h)\ln F(x_i).$$

(9.143)

The parameters may then be estimated by maximizing Equation (9.143) by taking the partial derivatives with respect to parameters and setting the derivatives equal to zero as follows:

$$\frac{\partial \ln L}{\partial a} = \frac{n}{\alpha} - \frac{k-1}{\alpha^2}\sum_{i=1}^{n}\frac{x_i - \xi}{S_i} + \frac{h-1}{\alpha^2}\sum_{i=1}^{n}\frac{(x_i - \xi)}{W_i} = 0 \qquad (9.144)$$

$$\frac{\partial \ln L}{\partial h} = (h-1)\sum_{i=1}^{n}\left(\frac{S_i^{\frac{1}{k}}}{hW_i} + \frac{\ln W_i}{h^2}\right) - \sum_{i=1}^{n}\ln F_i = 0 \qquad (9.145)$$

$$\frac{\partial \ln L}{\partial k} = -\frac{1-k}{\alpha k}\sum_{i=1}^{n}\frac{x_i - \xi}{S_i} - (h-1)\sum_{i=1}^{n}\frac{\frac{S_i^{\frac{1}{k}}\ln(S_i)}{k^2} + \frac{(x_i-\xi)S_i^{\frac{1}{k}-1}}{\alpha k}}{W_i} - \frac{1}{k^2}\sum_{i=1}^{n}\ln S_i = 0 \qquad (9.146)$$

$$\frac{\partial \ln L}{\partial \xi} = \frac{1-k}{\alpha k}\sum_{i=1}^{n}\frac{1}{S_i} + \frac{h-1}{\alpha}\sum_{i=1}^{n}\frac{S_i^{\frac{1}{k}-1}}{W_i} = 0. \qquad (9.147)$$

In Equations (9.144)–(9.147) $S = 1 - \frac{k}{\alpha}(x - \xi)$, $W = 1 - h\left(1 - \frac{k}{\alpha}(x - \xi)\right)^{1/k}$, and F represents the cumulative distribution.

9.5.3 MOM

The rth moment about the origin (also called the rth noncentral moment) is expressed as

$$E(X^r) = \int x^r f(x)dx. \qquad (9.148)$$

Previously, we have shown that variable x falls into different ranges with different sets of parameters. Additionally, we have also shown that the quantile function may be expressed explicitly through the cumulative probability distribution, that is, Equation (9.3). Thus, Equation (9.148) may be reformulated as

$$E(X^r) = \int x^r f(x)dx = \int_0^1 x(F)^r dF = \int_0^1 \left(\xi + \frac{\alpha}{k}\left(1 - \left(\frac{1-F^h}{h}\right)^k\right)\right)^h dF. \qquad (9.149)$$

Following Hosking (1994), the rth moment exists if any of the following conditions is fulfilled:

(1) All moments exist if $h, k \geq 0$;
(2) if $k \geq 0, h < 0$, the rth moment exists iff $r < -1/kh$; and
(3) if $k < 0$, the rth moment exists iff $r < -1/k$.

Furthermore, let $y = 1 - \frac{b}{a}(x - c)$; Equation (9.149) may be rewritten for the transformed random variable Y as

$$E(Y^r) = E\left(\left(1 - \frac{k}{\alpha}(x - \xi)\right)^r\right) = \int_0^1 \left(\frac{1 - F^h}{h}\right)^{rk} dF. \qquad (9.150)$$

If $h > 0$ and let $w = F^h \Rightarrow F = w^{1/h}$, we have

$$dw = hF^{h-1}dF \Rightarrow dF = h^{-1}F^{1-h}dw = h^{-1}w^{\frac{1-h}{h}}dw.$$

Equation (9.150) may be solved as

$$E\left(\left(1 - \frac{k}{\alpha}(x - \xi)\right)^r\right) = h^{-1-rk} \int_0^1 (1 - w)^{rk} w^{\frac{1-h}{h}} dw = h^{-(1+rk)} B\left(\frac{1}{h}, 1 + rk\right).$$

$$(9.151)$$

If $h < 0$ and let $w = F^h - 1 \Rightarrow F = (1 + w)^{1/h}$, we have

$$dw = hF^{h-1}dF \Rightarrow dF = h^{-1}F^{1-h}dw = h^{-1}(1 + w)^{\frac{1-h}{h}}dw.$$

Equation (9.150) may be solved as

$$E\left(\left(1 - \frac{k}{\alpha}(x - \xi)\right)^r\right) = \int_0^1 \left(\frac{F^h - 1}{-h}\right)^{rk} dF$$

$$= (-h)^{-rk} \int_\infty^0 w^{rk} h^{-1}(1 + w)^{\frac{1-h}{h}} dw \qquad (9.152)$$

$$= (-h)^{-(1+rk)} \int_0^\infty w^{rk}(1 + w)^{\frac{1-h}{h}} dw$$

$$= (-h)^{-(1+rk)} B\left(rk + 1, -rk - \frac{1}{h}\right).$$

To this end, if the first four moments exist, the MOM may be applied for parameter estimation.

9.5.4 Method of L-Moments

There is an expression for cumulative distribution function for kappa distribution. As a result, we will introduce the parameter estimation using L-moments. Following Greenwood et al. (1979), the probability weighted moments (PWMs) are expressed as

$$M_{p,r,s} = E\big(X^p F^r (1-F)^s\big) = \int_0^1 x(F)^p F^r (1-F)^s dF. \tag{9.153}$$

In Equation (9.153), let $p = 1$, $s = 0$, and $r \in [0, 1, 2, \ldots]$; we have

$$\beta_r = M_{1,r,0} = E(XF^r) = \int_0^1 X(F)F^r dF. \tag{9.154}$$

Equation (9.154) represents the population PWMs; its sample estimate may be computed from the ordered sample $x_1 \leq x_2 \cdots \leq x_n$ as

$$b_r = \frac{1}{n}\sum_{i=1}^{n} \frac{(i-1)(i-2)\cdots(i-r)}{(n-1)(n-2)\cdots(n-r)} x_i, \qquad r = 0, 1, 2, \ldots, n-1. \tag{9.155}$$

Hosking (1990) showed that the L-moments are linear combinations of PWMs for the order statistics of variable x as

$$\lambda_{r+1} = \sum_{j=0}^{r} p_{r,j}^* \beta_j, \quad r = 0, 1, 2, \ldots; \quad p_{r,j}^* = (-1)^{r-j}\binom{r}{j}\binom{r+j}{j} = \frac{(-1)^{r-j}(r+j)!}{(j!)^2(r-j)!}. \tag{9.156}$$

Furthermore, the first four L-moments may be expressed through the PWMs as follows:

Measure of location:

$$\lambda_1 = \beta_0 = E(X) \tag{9.157}$$

Measure of scale:

$$\lambda_2 = 2\beta_1 - \beta_0 \tag{9.158}$$

$$\lambda_3 = 6\beta_2 - 6\beta_1 + \beta_0 \tag{9.159}$$

$$\lambda_4 = 20\beta_3 - 30\beta_2 + 12\beta_1 - \beta_0. \tag{9.160}$$

The L-skewness (τ_3) and L-kurtosis (τ_4) are expressed as

$$\tau_3 = \frac{\lambda_3}{\lambda_2}; \quad \tau_4 = \frac{\lambda_4}{\lambda_2}. \tag{9.161}$$

As discussed in Hosking (1994), β_r exists if the mean of distribution exists. Thus, in the case of the four-parameter kappa distribution, β_r exists under the following conditions: (1) $k > -1$ if $h \geq 0$ or (2) $-1 < k < -1/h$ if $h < 0$. Substituting Equation (9.3) in Equation (9.154), we have

$$\beta_r = \int_0^1 \left(\xi + \frac{\alpha}{k} \left(1 - \left(\frac{1 - F^h}{h} \right)^k \right) \right) F^r dF = \frac{1}{r+1} \left(\xi + \frac{\alpha}{k} \right) - \frac{\alpha}{k} \int_0^1 \left(\frac{1 - F^h}{h} \right)^k F^r dF.$$

$$(9.162)$$

Depending on different parameter ranges in regard to h and k, the integral of Equation (9.162) can be evaluated as follows:

(1) $h > 0, k > -1$

Let $w = F^h \Rightarrow F = w^{1/h}$, $dw = hF^{h-1}dF \Rightarrow dF = h^{-1}w^{(1-h)/h}dw$; the integral may be evaluated as

$$\int_0^1 \left(\frac{1 - F^h}{h} \right)^k F^r dF = h^{-(k+1)} \int_0^1 (1 - w)^k w^{\frac{r+1-h}{h}} dw = h^{-(1+k)} B \left(k + 1, \frac{r+1}{h} \right).$$

$$(9.163)$$

Substituting Equation (9.163) in Equation (9.162), we have

$$\beta_r = \frac{1}{r+1} \left(\xi + \frac{\alpha}{k} \right) - \frac{\alpha}{k} h^{-(k+1)} B \left(k + 1, \frac{r+1}{h} \right). \qquad (9.164)$$

Equation (9.164) may be rearranged as

$$(r+1)\beta_r = \xi + \frac{\alpha}{k} \left(1 - (r+1)h^{-(k+1)} B \left(k + 1, \frac{r+1}{h} \right) \right). \qquad (9.165)$$

Equation (9.165) can be rewritten as

$$r\beta_{r-1} = \xi + \frac{\alpha}{k} \left(1 - rh^{-(k+1)} B \left(k + 1, \frac{r}{h} \right) \right). \qquad (9.166)$$

(2) $h = 0, k > -1$

Previously, we have shown that the GEV distribution is obtained under this condition. Substituting Equation (9.16) in Equation (9.154), we have

$$\beta_r = \int_0^1 \left(\xi + \frac{\alpha}{k} \left(1 - (-\ln F)^k \right) \right) F^r dF = \frac{1}{r+1} \left(\xi + \frac{\alpha}{k} \right) - \frac{\alpha}{k} \int_0^1 (-\ln F)^k F^r dF.$$

$$(9.167)$$

Let $w = -\ln F \Rightarrow F = \exp(-w)$, $dw = -\frac{1}{F}dF \Rightarrow dF = -\exp(-w)dw$; the integral in Equation (9.167) may be evaluated as

$$\int_0^1 (-\ln F)^k F^r dF = \int_0^\infty w^k \exp(-(r+1)w)dw = (r+1)^{-(k+1)}\Gamma(k+1).$$

$$(9.168)$$

Substituting Equation (9.168) in Equation (9.167), we have

$$\beta_r = (r+1)^{-1}\left(\xi + \frac{\alpha}{k}\right) - \frac{\alpha}{k}(r+1)^{-(k+1)}\Gamma(k+1). \qquad (9.169)$$

Similarly, Equation (9.169) may be rewritten as

$$(r+1)\beta_r = \xi + \frac{\alpha}{k}\left(1 - (r+1)^{-k}\Gamma(k+1)\right) \qquad (9.170)$$

$$r\beta_{r-1} = \xi + \frac{\alpha}{k}\left(1 - r^{-k}\Gamma(k+1)\right). \qquad (9.171)$$

(3) $h < 0$, $k \in \left(-1, -\frac{1}{h}\right)$

In Equation (9.154), let $w = F^h - 1$; we have $F = (1+w)^{1/h}$, $dF = \frac{1}{h}(1+w)^{(1/h)-1}dw$. The integral of Equation (9.154) can then be evaluated as

$$\int_0^1 \left(\frac{1-F^h}{h}\right)^k F^r dF = (-h)^{-k}\int_\infty^0 w^k\left(\frac{1}{h}\right)(1+w)^{\frac{r}{h}-1+\frac{r}{h}}dw$$

$$= (-h)^{-(k+1)}\int_0^\infty w^k(1+w)^{-\left(\left(-\frac{r+1}{h}+1\right)\right)}dw \qquad (9.172)$$

$$= (-h)^{-(k+1)}B\left(k+1, -k-\frac{r+1}{h}\right).$$

Substituting Equation (9.172) in Equation (9.154), we have

$$\beta_r = \frac{1}{r+1}\left(\xi + \frac{\alpha}{k}\right) - \frac{\alpha}{k}(-h)^{-(k+1)}B\left(k+1, -k-\frac{r+1}{h}\right). \qquad (9.173)$$

Equation (9.173) can be rearranged as

$$(r+1)\beta_r = \xi + \frac{\alpha}{k}\left(1 - (-h)^{-(k+1)}(r+1)B\left(k+1, -k-\frac{r+1}{h}\right)\right)$$

$$(9.174)$$

$$rB_{r-1} = \xi + \frac{\alpha}{k}\left(1 - (-h)^{-(k+1)}rB\left(k+1, -k-\frac{r}{h}\right)\right). \quad (9.175)$$

Till now we have discussed the PWMs with respect to the range of parameter h. We also need to pay attention to the general three-parameter kappa distribution such that $k = 0$. Substituting Equation (9.13) in Equation (9.154), we have

$$\beta_r = \int_0^1 \left[\xi - \alpha \ln\left(\frac{1-F^h}{h}\right)\right] F^r dF = \frac{\xi}{r+1} - \alpha \int_0^1 F^r \ln\left(\frac{1-F^h}{h}\right) dF. \quad (9.176)$$

β_r may then be evaluated under the condition of $h > 0$, $h = 0$, and $h > 0$, as

$$(r+1)\beta_r = \xi + \alpha\left(\gamma + \ln h + \psi\left(1 + \frac{r+1}{h}\right)\right); \quad h > 0 \quad (9.177)$$

$$(r+1)\beta_r = \xi + \alpha(\gamma + \ln(r+1)); \quad h = 0 \quad (9.178)$$

$$(r+1)\beta_r = \xi + \alpha\left(\gamma + \ln(-h) + \psi\left(-\frac{r+1}{h}\right)\right). \quad (9.179)$$

Similarly, Equations (9.177)–(9.179) may also be rearranged as

$$rB_{r-1} = \xi + \alpha\left(\gamma + \ln h + \psi\left(1 + \frac{r}{h}\right)\right); \quad h > 0 \quad (9.180)$$

$$rB_{r-1} = \xi + \alpha(\gamma + \ln(r)); \quad h = 0 \quad (9.181)$$

$$rB_{r-1} = \xi + \alpha\left(\gamma + \ln(-h) + \psi\left(-\frac{r}{h}\right)\right); \quad h < 0. \quad (9.182)$$

In Equations (9.177)–(9.182), $\gamma = 0.5772$ (Euler's constant). Following Hosking (1994), the LM method proposes the constraints on the parameters as

(1) $k > -1$; (2) $hk > -1$, if $h < 0$; (3) $h > -1$; and (4) $k + 0.725h > -1$. Based on Equations (9.177)–(9.182) and Equations (9.157)–(9.161), the L-skewness (τ_3) and L-kurtosis (τ_4) can be expressed for the parameters fulfilling the above constraints as follows:

For $h > 0$, $k > -1$:

$$\tau_3 = \frac{6B\left(k+1, \frac{3}{h}\right) - 6B\left(k+1, \frac{2}{h}\right) + B\left(k+1, \frac{1}{h}\right)}{2B\left(k+1, \frac{2}{h}\right) - B\left(k+1, \frac{1}{h}\right)};$$

$$\tau_4 = \frac{12B\left(k+1, \frac{2}{h}\right) - 30B\left(k+1, \frac{3}{h}\right) + 20B\left(k+1, \frac{4}{h}\right) - B\left(k+1, \frac{1}{h}\right)}{2B\left(k+1, \frac{2}{h}\right) - B\left(k+1, \frac{1}{h}\right)}.$$

$$(9.183)$$

For $h = 0$, $k > -1$:

$$\tau_3 = -\frac{2^k\left(\frac{2}{3^k} - \frac{3}{2^k} + 1\right)}{2^k - 1};$$

$$(9.184)$$

$$\tau_4 = -\frac{2^k\left(\frac{5}{4^k} - \frac{10}{3^k} + 3(2^{1-k}) - 1\right)}{2^k - 1}.$$

For $h < 0$, $k \in \left(-1, -\frac{1}{h}\right)$:

$$\tau_3 = -\frac{S_1 - 6S_2 + 6S_3}{S_1 - 2S_2};$$

$$(9.185)$$

$$\tau_4 = \frac{S_1 - 12S_2 + 30S_3 - 20S_4}{S_1 - 2S_2},$$

where

$$S_1 = B\left(k+1, -\frac{1+hk}{h}\right), \quad S_2 = B\left(k+1, -\frac{hk+2}{h}\right),$$

$$S_3 = B\left(k+1, -\frac{hk+3}{h}\right), \quad S_4 = B\left(k+1, -\frac{hk+4}{h}\right).$$

For $k = 0$, $h > 0$:

$$\tau_3 = -\frac{\psi\left(\frac{1+h}{h}\right) - 3\psi\left(\frac{h+2}{h}\right) + 2\psi\left(\frac{h+3}{h}\right)}{\psi\left(\frac{1+h}{h}\right) - \psi\left(\frac{h+2}{h}\right)};$$

$$(9.186)$$

$$\tau_4 = \frac{\psi\left(\frac{1+h}{h}\right) - 6\psi\left(\frac{2+h}{h}\right) + 10\psi\left(\frac{3+h}{h}\right) - 5\psi\left(\frac{4+h}{h}\right)}{\psi\left(\frac{1+h}{h}\right) - \psi\left(\frac{h+2}{h}\right)}.$$

For $k = 0$, $h = 0$: $\tau_3 \approx 0.17$, $\tau_4 \approx 0.15$.

For $k = 0$, $h < 0$:

$$\tau_3 = -\frac{\psi\left(\dfrac{h-1}{h}\right) - 3\psi\left(\dfrac{h-2}{h}\right) + 2\psi\left(\dfrac{h-3}{h}\right)}{\psi\left(\dfrac{h-1}{h}\right) - \psi\left(\dfrac{h-2}{h}\right)};$$

$$\tau_4 = \frac{\psi\left(\dfrac{h-1}{h}\right) - 6\psi\left(\dfrac{h-2}{h}\right) + 10\psi\left(\dfrac{h-3}{h}\right) - 5\psi\left(\dfrac{h-4}{h}\right)}{\psi\left(\dfrac{h-1}{h}\right) - \psi\left(\dfrac{h-2}{h}\right)}.$$

(9.187)

Figure 9.3 depicts the L-moment ratio diagram by fixing parameter h ($h \in [-4, 4]$). The lower bound of L-kurtosis is also illustrated. The upper bound of the L-kurtosis is in line with that of the generalized logistic distribution (i.e., $h = -1$). From the view point of L-moments, the necessary conditions to apply the L-moment method are

(1) $L_2 > 0$; (2) $\tau_3, \tau_4 \in [-1, 1]$; and (3) $\tau_4 \in \left[\dfrac{5\tau_3^2 - 1}{4}, \dfrac{5\tau_3^2 + 1}{6}\right]$.

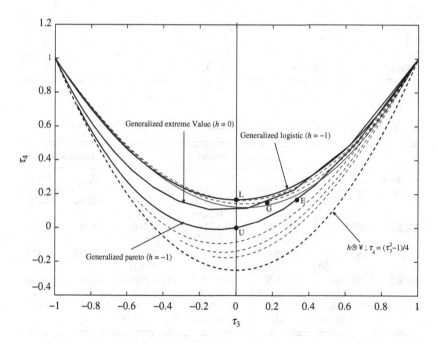

Figure 9.3 L-moment ratio diagram.

9.6 Application

9.6.1 Peak Flow

As in the previous chapters, the same annual peak flow dataset (USGS09239500) is applied to evaluate the appropriateness of the four-parameter kappa distribution. The sample L-moments are computed as $L_1 = 102.99$, $L_2 = 17.64$, $\tau_3 = 0.05$, and $\tau_4 = 0.12 \in [-0.247, 0.169]$. Thus, the L-moment method may be applied with the use of the generalized Pareto distribution to initiate parameter estimation, that is, $h = 1.001$ and $k = (1 - 3\tau_3)/(1 + \tau_3) = 0.808$. From the viewpoint of the entropy method, the reduced four-parameter entropy approach is applied for parameter estimation to reduce the computational burden. Table 9.1 lists the parameters estimated with the use of LM, MLE, MOM, and entropy methods. Table 9.1 shows that the parameters estimated using MLE, MOM, and entropy methods yield similar estimates. Table 9.2 lists the confidence bound estimated using the parametric bootstrap method with 1,000 repetitions, as well as the

Table 9.1. *Parameters estimated with different estimation methods: peak flow.*

Method	ξ	α	k	h
LM	93.047	26.612	0.140	−0.145
MLE	91.837	28.509	0.187	−0.0498
MOM	91.897	28.285	0.180	−0.058
Entropy	91.839	28.507	0.187	−0.0499

Table 9.2. *Confidence bound and goodness-of-fit results: peak flow.*

| Method | Confidence bound | | | | Goodness-of-fit | |
	ξ	α	k	h	D_n	P-value
LM	[84.566, 107.607]	[7.351, 34.916]	[−0.166, 0.326]	[−0.734, 0.486]	0.052	0.267
MLE	[85.995, 97.679][a]	[25.523, 31.495]	[0.134, 0.239]	[−0.216, 0.116]	N/A	N/A
	[82.623, 106.635]	[5.291, 38.026]	[−0.132, 0.371]	[−0.635, 0.561]	0.053	0.485
MOM	[80.362, 99.696]	[254.504, 38.368]	[0.110, 0.356]	[−0.204, 0.638]	0.053	0.668
Entropy	[87.138, 100.188]	[23.165, 31.716]	[0.08, 0.374]	[−0.996, −0.0499]	0.0533	0.63

[a] The confidence bound is estimated using observed Fisher's information.

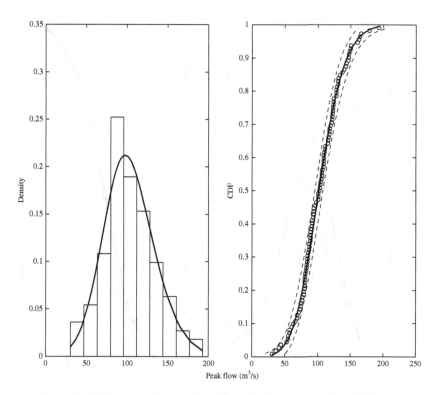

Figure 9.4 Comparison of empirical and fitted distribution: peak flow (LM).

goodness-of-fit results using the Kolmogorov–Smirnov (KS) test. The detailed procedures were discussed previously in Chapter 2. The goodness-of-fit results indicate all methods yield similar KS statistic values and all methods may be applied for parameter estimation. Figures 9.4–9.7 compare the empirical and parametric PDFs and CDFs for different methods. Using these figures, it is shown that all methods yield similar results visually, which are in agreement with the results listed in Tables 9.1 and 9.2. Furthermore, it is worth noting that the sample moments of the peak flow dataset fulfill the conditions for the MOM and LM methods.

9.6.2 Maximum Daily Precipitation

The maximum daily precipitation at Brenham, Texas (GHCND: USC0411048), is applied for the analysis. Applying MLE and entropy methods, Table 9.3 lists the parameters estimated, 95% confidence bound, and goodness-of-fit results. $h \to 0$ for the maximum entropy-based kappa

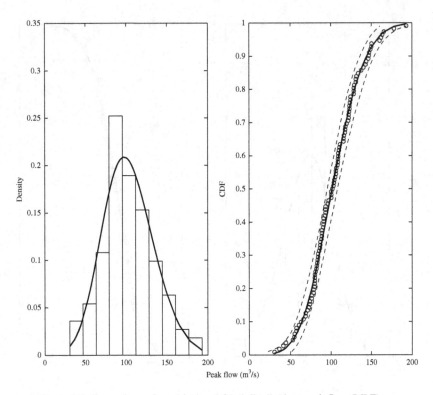

Figure 9.5 Comparison of empirical and fitted distribution: peak flow (MLE).

distribution indicates that this distribution converges to the GEV distribution for the maximum daily precipitation dataset. The goodness-of-fit test results indicate the KS statistic value computed from the maximum entropy-based kappa distribution is slightly larger than that computed from the kappa distribution fitted with MLE. Figures 9.8 and 9.9 compare the empirical distribution and fitted distribution. The comparison visually indicates that the fitted distribution can properly model the maximum daily precipitation.

9.6.3 Total Flow Deficit

The total flow deficit at Tilden, Texas, is applied for the analysis. Similar to the initial analysis, it shows the GEV distribution (i.e., the special case of kappa distribution with $h = 0$). Applying MLE and entropy methods, Table 9.4 lists the parameters estimated, 95% confidence bound, and goodness-of-fit results. The estimated parameters listed in the table show that the entropy estimation converges to the MLE estimation. The goodness-of-fit test results indicate that

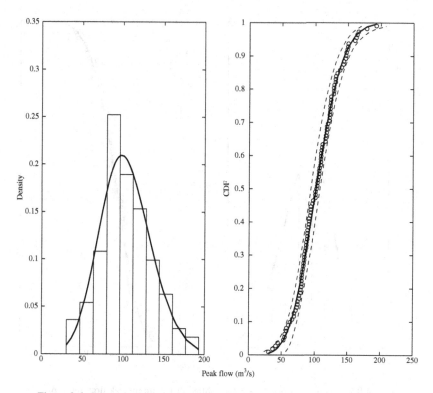

Figure 9.6 Comparison of empirical and fitted distribution: peak flow (MOM).

there is minimal difference for KS test statistic values. The GEV distribution, in other words, kappa distribution with $h = 0$, may be applied to model the maximum daily precipitation. Figures 9.10 and 9.11 compare the empirical distribution and fitted distribution. The comparison visually indicates that the fitted distribution tends to overestimate the flow deficit at high quantiles.

9.7 Conclusion

This chapter revisited the four-parameter kappa distribution. As a generalized four-parameter distribution, several popular distribution families may be considered as the special cases of the four-parameter kappa distribution, including generalized extreme value, generalized Pareto, and generalized logistic distributions. Certain conditions need to be fulfilled to apply the parameter estimation

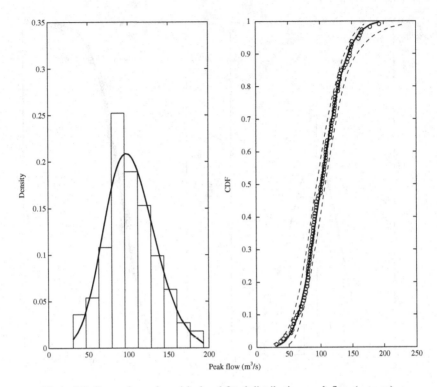

Figure 9.7 Comparison of empirical and fitted distribution: peak flow (entropy).

Table 9.3. *Parameters estimated, 95% confidence bound, goodness-of-fit test results: maximum daily precipitation.*

Method	Estimated parameter and confidence bound				Goodness-of-fit	
	k	α	ζ	h	D_n	P-Value
MLE	0.162 [0.005, 0.319]	31.524 [26.51, 36.448]	76.472 [69.812, 82.496]	0.102 [−0.487, 0.819]	0.0469	0.632
Entropy	0.162 [0.015, 0.459]	31.524 [26.966, 38.157]	76.473 [70.523, 84.604]	7.09E−13 [−3.32E−05, 1.42E−12]	0.057	0.697

with LM and MOM methods; however, there is no such constraint for parameter estimation with MLE and entropy methods. In the case of the entropy method, besides the two Lagrange multipliers for the corresponding two constraints, all four population parameters are embedded in the entropy function. Thus, the

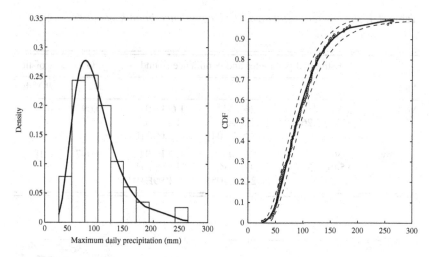

Figure 9.8 Comparison of empirical and fitted distribution: maximum daily precipitation (MLE).

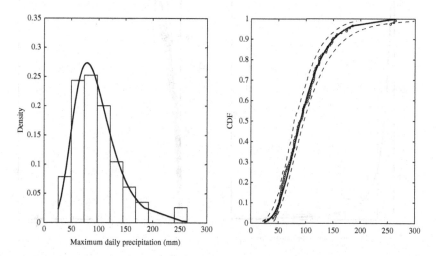

Figure 9.9 Comparison of empirical and fitted distribution: maximum daily precipitation (entropy).

relation of population parameters and Lagrange multipliers can be applied to reduce the parameter space. The application example with annual peak flow indicates all four methods yield similar results and can be applied for frequency analysis. In the case of the random variables that cannot be estimated with the

Table 9.4. *Parameters estimated, 95% confidence bound, goodness-of-fit test results: total flow deficit.*

| Method | Estimated parameter and confidence bound | | | Goodness-of-fit | |
	k	a	c	D_n	P-Value
MLE	1.036 [0.775, 1.264]	2.06E+04 [1.42E+04, 2.59E+04]	1.60E+04 [1.15E+04, 2.00E+04]	0.097	0.21
Entropy	1.036 [0.546, 1.382]	2.06E+04 [1.92E+04, 2.20E+05]	1.60E+04 [1.55E+04, 1.99E+04]	0.097	0.16

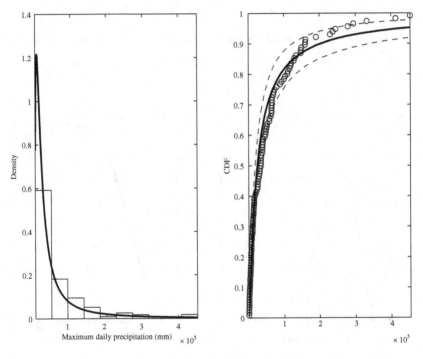

Figure 9.10 Comparison of empirical and fitted distribution: maximum daily precipitation (MLE).

LM or MOM method, one can always apply the MLE and entropy methods for frequency study. The application of maximum daily precipitation shows that the kappa distribution (or its special case GEV: $h \to 0$) may be applied to model the maximum daily precipitation dataset. The application of the total flow deficit

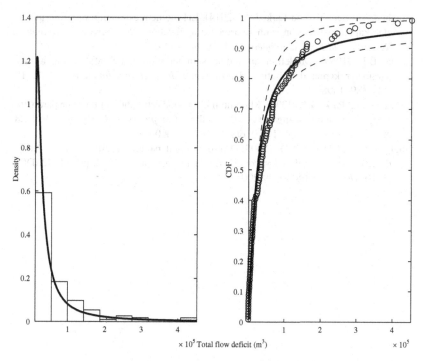

Figure 9.11 Comparison of empirical and fitted distribution: maximum daily precipitation (entropy).

indicates that the special case, GEV distribution, may be applied to model the total flow deficit. In the case of the total flow deficit, the GEV distribution tends to overestimate the flow deficit for high quantiles.

Reference

Dupuis, D.J. and Winchester, C. (2001). More on the four-parameter kappa distribution. *Journal of Statistical Computation and Simulation*, Vol. 71, No. 2, pp. 99–113, DOI: 10.1080/00949650108812137.

Greenwood, J.A., Landwehr, J.M., and Matalas, N.C. (1979). Probability weighted moments: Definition and relation to parameters of several distributions expressible in inverse form. *Water Resources Research*, Vol. 15, pp. 1049–1054.

Hosking, J.R.M. (1990). L-moments: Analysis and estimation of distributions using linear combination of order statistics. *Journal of the Royal Statistical Society, Series B* Vol. 52, No. 1, pp. 105–124.

Hosking, J.R.M. (1994). The four-parameter kappa distribution. *IBM Journal of Research and Development*, Vol. 38, No. 3, pp. 251–258.

Murshed, M., Seo, Y., and Park, J.-S. (2014). LH-moment estimation of a four parameter kappa distribution with hydrologic applications. *Stochastic Environmental
Research and Risk Assessment*, Vol. 28, pp. 253–262.

Parida, B.P. (1999). Modelling of Indian summer monsoon rainfall using a four-
parameter kappa distribution. *International Journal of Climatology*, Vol. 19,
pp. 1389–1398.

Park, J.-S. and Park, B.J. (2002). Maximum likelihood estimation of the four-parameter
Kappa distribution using the penalty method. *Computers & Geosciences*, Vol. 28,
No. 1, pp. 65–68. DOI: 10.1016/S0098-3004(1)00069-3.

Singh, V.P. and Deng, Z.Q. (2003). Entropy-based parameter estimation for kappa
distribution. *Journal of Hydrologic Engineering*, Vol. 8, No. 2, pp. 81–92. DOI:
10.1061/(ASCE)1084-0699(2003)8:2(81).

10

Four-Parameter Exponential Gamma Distribution

10.1 Introduction

The four-parameter exponential gamma (FPEG) distribution is one of the several generalized distributions that may be used in hydrology and hydro-meteorology due to its kinship to the gamma distribution. The probability density function (PDF) of the FPEG distribution can be expressed as

$$f(x) = \frac{c^a}{b\Gamma(a)}(x-d)^{\frac{a}{b}-1} \exp\left(-c(x-d)^{\frac{1}{b}}\right); \quad a,b,c > 0, d \in \mathbb{R}, x \geq d,$$

(10.1)

where a and b are the shape parameters, c is the scale parameter, and d is the location parameter. Figure 10.1 shows the distribution shapes for different values of the parameters.

Some commonly applied frequency distributions may be deduced from the FPEG distribution, by fixing some parameter values in Equation (10.1). These special cases include gamma, Pearson type III, Weibull, three-parameter Weibull, Kritsky and Menkel, chi-square, Poisson, half-normal normal, and exponential distributions. These special cases are discussed here.

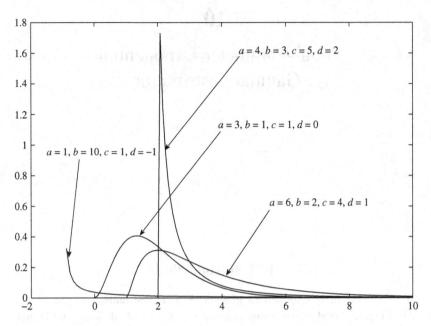

Figure 10.1 Shape of FPEG distributions for different parameters.

Case I: $b = 1$

By setting $b = 1$, Pearson III, gamma, Chi-squared, and exponential distributions may be derived.

• **Pearson III Distribution**

The Pearson III distribution is obtained by setting $b = 1$ as

$$f(x; a, c, d) = \frac{c}{\Gamma(a)} [c(x - d)]^{a-1} \exp(-c(x - d)). \qquad (10.2)$$

• **Gamma Distribution**

Setting $d = 0$ in Equation (10.2), the gamma distribution is obtained as

$$f(x; a, c) = \frac{c^a}{\Gamma(a)} x^{a-1} \exp(-cx). \qquad (10.3)$$

• **Chi-Squared Distribution**

The chi-squared distribution is obtained if $c = 1/2$, $a = k/2$, $k \in \mathbb{N}^+$ in Equation (10.3) as

$$f(x; a, c) = \frac{x^{\frac{k}{2}-1}}{2^{\left(\frac{k}{2}\right)}\Gamma\left(\frac{k}{2}\right)} \exp\left(-\frac{x}{2}\right). \tag{10.4}$$

• Exponential Distribution

Furthermore, the exponential distribution is obtained if $a = 1$ in Equation (10.3) as

$$f(x; c) = c\exp(-cx). \tag{10.5}$$

Case II: $a = 1$

Setting $a = 1$, the three-parameter Weibull distribution (Weibull distribution with the location parameter), two-parameter Weibull distribution, and exponential distribution may be derived.

• Three-Parameter Weibull Distribution

The three-parameter Weibull distribution (i.e., Weibull distribution with the location parameter) is obtained if $a = 1$ as

$$f(x; b, c, d) = \frac{c}{b}(x-d)^{\frac{1}{b}-1} \exp\left(-c(x-d)^{\frac{1}{b}}\right). \tag{10.6}$$

• Two-Parameter Weibull Distribution

Setting $d = 0$ in Equation (10.6), the two-parameter Weibull distribution is obtained as

$$f(x; b, c) = \frac{c}{|b|}x^{\frac{1}{b}-1} \exp\left(-cx^{\frac{1}{b}}\right). \tag{10.7}$$

Setting $b = 1/k$ and $c = \lambda^{-1/b}$, Equation (10.7) may be rewritten as

$$f(x; b, c) = \frac{k}{\lambda}\left(\frac{x}{\lambda}\right)^{k-1} \exp\left(-\left(\frac{x}{\lambda}\right)^{k}\right). \tag{10.8}$$

It is seen from Equation (10.7) that the exponential distribution is again derived if $b = 1$.

Case III: $d = 0$

• **Three-parameter generalized gamma distribution (Kritsky and Menkel) Distribution**

The three-parameter generalized gamma distribution or Kritsky and Menkel distribution is obtained by setting $d = 0$ as

$$f(x; a, b, c) = \frac{c^a}{b\Gamma(a)} x^{\frac{a}{b}-1} \exp\left(-cx^{\frac{1}{b}}\right), \quad b > 0. \tag{10.9a}$$

In Equation (10.9a) setting $a = \lambda$, $b = 1/s$, and $c = \theta^b$, Equation (10.9a) may be expressed in the exact form in Ashkar and Ouarda (1998) as

$$f(x; a, b, c) = \frac{s\theta^{\lambda s}}{\Gamma(\lambda)} x^{\lambda s - 1} \exp\left(-(\theta x)^s\right) = \frac{s\theta}{\Gamma(\lambda)} (\theta x)^{\lambda s - 1} \exp\left(-(\theta x)^s\right). \tag{10.9b}$$

• **Poisson Distribution**

Setting $b = 1$, $c = 1$, and $a = k + 1$, $k \in \mathbb{N}^+$ in Equation (10.9a), the Poisson distribution is obtained as

$$f(x; a) = \frac{x^k}{k!} \exp(-x). \tag{10.10}$$

Case IV: $a = b = 1/2$

Setting $a = b = 1/2$, the half-normal distribution is obtained as

$$f(x; c, d) = \frac{c^{\frac{1}{2}}}{\frac{1}{2}\Gamma\left(\frac{1}{2}\right)} \exp\left(-c(x-d)^2\right) = \frac{\sqrt{2}}{\sqrt{\pi}\left(\frac{1}{\sqrt{2c}}\right)} \exp\left(-\frac{(x-d)^2}{2\left(\frac{1}{\sqrt{2c}}\right)^2}\right). \tag{10.11}$$

Figure 10.2 graphs the chart of FPEG and its special cases.

From Equation (10.1), the cumulative distribution function (CDF) of the FPEG distribution can be expressed as

$$F(x) = \int_d^x f(x)dx = \int_d^x \frac{c^a}{b\Gamma(a)} (x-d)^{\frac{a}{b}-1} \exp\left(-c(x-d)^{\frac{1}{b}}\right)dx. \tag{10.12}$$

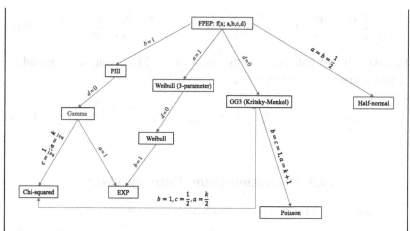

Figure 10.2 Chart of FPEG and its special cases.

In Equation (10.12), let

$$z = c(x-d)^{\frac{1}{b}} \Rightarrow x = \left(\frac{z}{c}\right)^b + d \Rightarrow dx = \frac{b}{c^b} z^{b-1} dz. \tag{10.13}$$

Substituting Equation (10.13) in Equation (10.12), we have

$$F(x) = \int_0^z \frac{c^a}{b\Gamma(a)} \left(\frac{z}{c}\right)^{a-b} \exp(-z) \frac{b}{c^b} z^{b-1} dz = \frac{1}{\Gamma(a)} \int_0^z z^{a-1} \exp(-z) dz = \gamma(a,z). \tag{10.14}$$

The integral result of Equation (10.14) shows that the CDF of the FPEG may be expressed as the lower incomplete gamma function after variable transformation.

10.2 Generating Differential Equation

Taking the logarithm of Equation (10.1), we obtain

$$\ln f(x) = \ln\left(\frac{c^a}{b\Gamma(a)}\right) + \left(\frac{a}{b} - 1\right) \ln(x-d) - c(x-d)^{\frac{1}{b}}. \tag{10.15}$$

Differentiation of Equation (10.15) yields

$$\frac{1}{f(x)}\frac{df(x)}{dx} = \left(\frac{a}{b}-1\right)\frac{1}{x-d}-\frac{c}{b}(x-d)^{\frac{1}{b}-1} = \frac{(a-b)-c(x-d)^{\frac{1}{b}}}{b(x-d)}. \quad (10.16)$$

Equation (10.16) indicates that the mode of the FPEG can be computed by setting Equation (10.16) to zero as

$$(a-b)-c(x-d)^{\frac{1}{b}} = 0 \Rightarrow x_{\text{mode}} = d + \left(\frac{a-b}{c}\right)^{b}. \quad (10.17)$$

10.3 Derivation Using Entropy Theory

The Shannon entropy of a given distribution can be written as

$$H(X) = -\int f(x)\ln f(x)dx. \quad (10.18)$$

Substitution of Equation (10.15) in Equation (10.18) results in

$$H(X) = -\int_{d}^{\infty} f(x)\left[\ln\left(\frac{c^{a}}{b\Gamma(a)}\right) + \left(\frac{a}{b}-1\right)\ln(x-d) - c(x-d)^{\frac{1}{b}}\right]dx. \quad (10.19)$$

Equation (10.19) yields the following constraints for the FPEG distribution:

$$\int_{d}^{\infty} f(x)dx = 1 \quad (10.20)$$

$$\int_{d}^{\infty} \ln(x-d)f(x)dx = E\big(\ln(x-d)\big) \quad (10.21)$$

$$\int_{d}^{\infty} (x-d)^{\frac{1}{b}}f(x)dx = E\left((x-d)^{\frac{1}{b}}\right). \quad (10.22)$$

Based on these constraints, the corresponding Lagrangian function is given as

$$L = -\int_{d}^{\infty} f(x)\ln f(x)dx - (\lambda_0 - 1)\left(\int_{d}^{\infty} f(x)dx - 1\right)$$
$$-\lambda_1\left(\ln(x-d)f(x)dx - E[\ln(x-d)]\right) \quad (10.23)$$
$$-\lambda_2\left(\int_{d}^{\infty} (x-d)^{\frac{1}{b}}f(x)dx - E\left[(x-d)^{\frac{1}{b}}\right]\right)$$

And the entropy-based FPEG distribution is then obtained by maximizing Equation (10.23); that is, differentiating Equation (10.23) with respect to $f(x)$ and equating the derivative to zero, we then obtain the entropy-based PDF of FPEG distribution as

$$f(x) = \exp\left(-\lambda_0 - \lambda_1 \ln(x-d) - \lambda_2(x-d)^{\frac{1}{b}}\right). \tag{10.24}$$

In Equations (10.23) and (10.24), λ_0, λ_1, and λ_2 are the Lagrange multipliers. Substituting Equation (10.24) in Equation (10.20), we have

$$\int_d^\infty \exp\left(-\lambda_0 - \lambda_1 \ln(x-d) - \lambda_2(x-d)^{\frac{1}{b}}\right) dx = 1. \tag{10.25}$$

Equation (10.25) may be rewritten as

$$\exp(\lambda_0) = \int_d^\infty \exp\left(-\lambda_1 \ln(x-d) - \lambda_2(x-d)^{\frac{1}{b}}\right) dx. \tag{10.26}$$

Equation (10.26) shows that λ_0 is a function of λ_1, and λ_2. Let $z = \lambda_2(x-d)^{1/b}$ $\Rightarrow x = (z/\lambda_2)^b + d \Rightarrow dx = (b/\lambda_2^b)z^{b-1}dz$. Equation (10.26) may be solved as

$$\exp(\lambda_0) = \int_0^\infty \left(\frac{z}{\lambda_2}\right)^{-b\lambda_1} \exp(-z)\frac{b}{\lambda_2^b}z^{b-1}dz$$

$$= \frac{b}{\lambda_2^{b-b\lambda_1}}\int_0^z z^{-b\lambda_1+b-1}\exp(-z)dz = \frac{b}{\lambda_2^{b(1-\lambda_1)}}\Gamma(b(1-\lambda_1)); \quad \lambda_2 > 0, \lambda_1 < 1.$$

$$\tag{10.27}$$

Substituting Equation (10.27) in Equation (10.24), the entropy-based PDF of the FPEG may be expressed as

$$f(x) = \frac{\lambda_2^{b(1-\lambda_1)}}{b\Gamma(b(1-\lambda_1))}(x-d)^{-\lambda_1}\exp\left(-\lambda_2(x-d)^{\frac{1}{b}}\right). \tag{10.28}$$

Comparing Equation (10.28) with Equation (10.1), we can find the relation between the Lagrange multipliers and the distribution parameters as

$$\lambda_1 = 1 - \frac{a}{b}, \quad \lambda_2 = c. \tag{10.29}$$

10.4 Parameter Estimation

The FPEG distribution parameters are estimated using the regular entropy method, parameter space expansion method, method of moments (MOM), probability weighted moments, L-moments, cumulative moments, and maximum likelihood estimation (MLE).

10.4.1 Regular Entropy Method

From Equation (10.27), the zeroth Lagrange multiplier can be written as

$$\lambda_0 = \ln b - b(1 - \lambda_1)\ln \lambda_2 + \ln\left(\Gamma\left(b(1 - \lambda_1)\right)\right). \tag{10.30}$$

Differentiating λ_0 with respect to λ_1 and λ_2 we get

$$\frac{\partial \lambda_0}{\partial \lambda_1} = b\ln \lambda_2 - b\psi\left(b(1 - \lambda_1)\right) \tag{10.31}$$

$$\frac{\partial \lambda_0}{\partial \lambda_2} = -\frac{b(1 - \lambda_1)}{\lambda_2}. \tag{10.32}$$

From Equation (10.26), the zeroth Lagrange multiplier can be written as

$$\lambda_0 = \ln\left(\int\limits_d^\infty \exp\left(-\lambda_1\ln(x - d) - \lambda_2(x - d)^{\frac{1}{b}}\right)dx\right). \tag{10.33}$$

Differentiating Equation (10.33) with respect to λ_1 and λ_2, we have

$$\frac{\partial \lambda_0}{\partial \lambda_1} = -\frac{\int\limits_d^\infty \ln(x - d) \exp\left(-\lambda_1\ln(x - d) - \lambda_2(x - d)^{\frac{1}{b}}\right)dx}{\int\limits_d^\infty \exp\left(-\lambda_1\ln(x - d) - \lambda_2(x - d)^{\frac{1}{b}}\right)dx} = -E[\ln(x - d)] \tag{10.34}$$

$$\frac{\partial \lambda_0}{\partial \lambda_2} = -\frac{\int\limits_d^\infty (x - d)^{\frac{1}{b}} \exp\left(-\lambda_1\ln(x - d) - \lambda_2(x - d)^{\frac{1}{b}}\right)dx}{\int\limits_d^\infty \exp\left(-\lambda_1\ln(x - d) - \lambda_2(x - d)^{\frac{1}{b}}\right)dx} = -E\left[(x - d)^{\frac{1}{b}}\right]. \tag{10.35}$$

Equating Equation (10.31) to Equation (10.34) and Equation (10.32) to Equation (10.35), we have

$$b\ln \lambda_2 - b\psi\left(b(1 - \lambda_1)\right) = -E[\ln(x - d)] \tag{10.36}$$

$$\frac{b(1-\lambda_1)}{\lambda_2} = E\left[(x-d)^{\frac{1}{b}}\right]. \tag{10.37}$$

Equation (10.28) indicates there are four parameters that need to be estimated (i.e., two Lagrange multipliers and two embedded distribution parameters); thus, we need two more equations that can be accomplished by taking the second derivative of Equations (10.32)–(10.35) as follows:

$$\frac{\partial^2 \lambda_0}{\partial \lambda_1^2} = b^2 \psi^{(1)}\left(b(1-\lambda_1)\right) \tag{10.38a}$$

or

$$\frac{\partial^2 \lambda_0}{\partial \lambda_1^2} = \frac{\int\limits_d^\infty \left(\ln(x-d)\right)^2 \exp\left(-\lambda_1 \ln(x-d) - \lambda_2(x-d)^{\frac{1}{b}}\right) dx}{\left(\int\limits_d^\infty \exp\left(-\lambda_1 \ln(x-d) - \lambda_2(x-d)^{\frac{1}{b}}\right) dx\right)^2}$$

$$- \left(\frac{\int\limits_d^\infty \ln(x-d) \exp\left(-\lambda_1 \ln(x-d) - \lambda_2(x-d)^{\frac{1}{b}}\right) dx}{\int\limits_d^\infty \exp\left(-\lambda_1 \ln(x-d) - \lambda_2(x-d)^{\frac{1}{b}}\right) dx}\right)^2 = \mathrm{var}\left(\ln(x-d)\right)$$

$$\tag{10.38b}$$

$$\frac{\partial^2 \lambda_0}{\partial \lambda_2^2} = \frac{b(1-\lambda_1)}{\lambda_2^2} \tag{10.39a}$$

or

$$\frac{\partial^2 \lambda_0}{\partial \lambda_2^2} = \frac{\int\limits_d^\infty (x-d)^{\frac{2}{b}} \exp\left(-\lambda_1 \ln(x-d) - \lambda_2(x-d)^{\frac{1}{b}}\right) dx}{\left(\int\limits_d^\infty \exp\left(-\lambda_1 \ln(x-d) - \lambda_2(x-d)^{\frac{1}{b}}\right) dx\right)^2}$$

$$- \left(\frac{\int\limits_d^\infty (x-d)^{\frac{1}{b}} \exp\left(-\lambda_1 \ln(x-d) - \lambda_2(x-d)^{\frac{1}{b}}\right) dx}{\int\limits_d^\infty \exp\left(-\lambda_1 \ln(x-d) - \lambda_2(x-d)^{\frac{1}{b}}\right) dx}\right)^2 = \mathrm{var}\left((x-d)^{\frac{1}{b}}\right).$$

$$\tag{10.39b}$$

Now, the other two equations are

$$b^2 \psi^{(1)}\left(b(1-\lambda_1)\right) = \mathrm{var}\left(\ln(x-d)\right) \tag{10.40}$$

$$\frac{b(1-\lambda_1)}{\lambda_2^2} = \text{var}\left((x-d)^{\frac{1}{b}}\right). \tag{10.41}$$

To this end, the parameters $\{\lambda_1, \lambda_2, b, d\}$ may be estimated by solving Equations (10.36), (10.37), (10.40), and (10.41) simultaneously.

10.4.2 Parameter Space Expansion Method

Substituting Equation (10.28) in Equation (10.18), we can obtain the corresponding Shannon entropy as

$$H(X) = -\ln\left(\frac{\lambda_2^{b(1-\lambda_1)}}{b\Gamma(b(1-\lambda_1))}\right) + \lambda_1 E[\ln(x-d)] + \lambda_2 E\left[(x-d)^{\frac{1}{b}}\right]$$

$$= -b(1-\lambda_1)\ln\lambda_2 + \ln b + \ln\Gamma(b(1-\lambda_1)) + \lambda_1 E[\ln(x-d)] + \lambda_2 E\left[(x-d)^{\frac{1}{b}}\right]. \tag{10.42}$$

Differentiating Equation (10.42) with respect to $\{\lambda_1, \lambda_2, b, d\}$ and setting them to zero, we have

$$\frac{\partial H(X)}{\partial \lambda_1} = b\ln\lambda_2 - b\psi(b(1-\lambda_1)) + E[\ln(x-d)] = 0 \tag{10.43}$$

$$\frac{\partial H(X)}{\partial \lambda_2} = -\frac{b(1-\lambda_1)}{\lambda_2} + E\left[(x-d)^{\frac{1}{b}}\right] = 0 \tag{10.44}$$

$$\frac{\partial H(X)}{\partial b} = -(1-\lambda_1)\ln\lambda_2 + \frac{1}{b} + (1-\lambda_1)\psi(b(1-\lambda_1)) - \frac{\lambda_2}{b^2}E\left[(x-d)^{\frac{1}{b}}\ln(x-d)\right] = 0 \tag{10.45}$$

$$\frac{\partial H(X)}{\partial d} = -\lambda_1 E\left[\frac{1}{x-d}\right] - \frac{\lambda_2}{b}E\left[(x-d)^{\frac{1}{b}-1}\right] = 0. \tag{10.46}$$

Now the parameters may be estimated by solving Equations (10.43)–(10.46) simultaneously.

10.4.3 MLE Method

For the random variables $\mathbf{x} = \{x_1, \ldots, x_n\}$ following the FPEG distribution, its likelihood and log-likelihood functions are given as

$$L = \prod_{i=1}^{n} \frac{c^a}{b\Gamma(a)} (x_i - d)^{\frac{a}{b}-1} \exp\left(-c(x_i - d)^{\frac{1}{b}}\right) \quad (10.47)$$

$$LL = n\left(a \ln c - \ln b - \ln \Gamma(a)\right) + \left(\frac{a}{b} - 1\right) \sum_{i=1}^{n} \ln(x_i - d) - c \sum_{i=1}^{n} (x_i - d)^{\frac{1}{b}}. \quad (10.48)$$

Differentiating Equation (10.48) with respect to the distribution parameters and setting them to zero, we have

$$\frac{\partial LL}{\partial a} = n \ln c - n\psi(a) + \frac{1}{b} \sum_{i=1}^{n} \ln(x_i - d) = 0 \quad (10.49)$$

$$\frac{\partial LL}{\partial b} = -\frac{n}{b} - \frac{a}{b^2} \sum_{i=1}^{n} \ln(x_i - d) + \frac{c}{b^2} \sum_{i=1}^{n} (x_i - d)^{\frac{1}{b}} \ln(x_i - d) = 0 \quad (10.50)$$

$$\frac{\partial LL}{\partial c} = \frac{na}{c} - \sum_{i=1}^{n} (x_i - d)^{\frac{1}{b}} = 0 \quad (10.51)$$

$$\frac{\partial LL}{\partial d} = -\left(\frac{a}{b} - 1\right) \sum_{i=1}^{n} \frac{1}{x_i - d} + \frac{c}{b} \sum_{i=1}^{n} (x_i - d)^{\frac{1}{b}-1} = 0. \quad (10.52)$$

To this end, the parameters may be estimated by solving Equations (10.49)–(10.52) simultaneously. For the parameters estimated with the MLE method, the corresponding confidence interval may be estimated using the observed Fisher information matrix that evaluated as $(-H)^{-1}$, that is, the inverse of the negative Hessian matrix. The elements of the Hessian matrix are obtained by taking the second derivatives for Equations (10.49)–(10.52) as follows:

$$\frac{\partial^2 LL}{\partial a^2} = -n\psi^{(1)}(a) \quad (10.53)$$

$$\frac{\partial^2 LL}{\partial a \partial b} = -\frac{1}{b^2} \sum_{i=1}^{n} \ln(x_i - d) \quad (10.54)$$

$$\frac{\partial^2 LL}{\partial a \partial c} = \frac{n}{c} \quad (10.55)$$

$$\frac{\partial^2 LL}{\partial a \partial d} = -\frac{1}{b} \sum_{i=1}^{n} \frac{1}{x_i - d} \quad (10.56)$$

$$\frac{\partial^2 LL}{\partial b^2} = \frac{n}{b^2} + \frac{2a}{b^3} \sum_{i=1}^{n} \ln(x_i - d) - \frac{2c}{b^3} \sum_{i=1}^{n} (x_i - d)^{\frac{1}{b}} \ln(x_i - d)$$
$$- \frac{c}{b^4} \sum_{i=1}^{n} (x_i - d)^{\frac{1}{b}} \left(\ln(x_i - d)\right)^2 \quad (10.57)$$

$$\frac{\partial^2 LL}{\partial b \partial c} = \frac{1}{b^2} \sum_{i=1}^{n} (x_i - d)^{\frac{1}{b}} \ln(x_i - d) \qquad (10.58)$$

$$\frac{\partial^2 LL}{\partial b \partial d} = \frac{a}{b^2} \sum_{i=1}^{n} \frac{1}{x_i - d} - \frac{c}{b^3} \sum_{i=1}^{n} \ln(x_i - d)(x_i - d)^{\frac{1}{b}-1} - \frac{c}{b^2} \sum_{i=1}^{n} (x_i - d)^{\frac{1}{b}-1}$$

$$(10.59)$$

$$\frac{\partial^2 LL}{\partial c^2} = -\frac{na}{c^2} \qquad (10.60)$$

$$\frac{\partial^2 LL}{\partial c \partial d} = \frac{1}{b} \sum_{i=1}^{n} (x_i - d)^{\frac{1}{b}-1} \qquad (10.61)$$

$$\frac{\partial^2 LL}{\partial d^2} = -\left(\frac{a}{b} - 1\right) \sum_{i=1}^{n} (x_i - d)^{-2} - \frac{c}{b}\left(\frac{1}{b} - 1\right) \sum_{i=1}^{n} (x_i - d)^{\frac{1}{b}-2}. \qquad (10.62)$$

Now we can evaluate the 95% confidence interval as

$$CI = \begin{bmatrix} \hat{a} \\ \hat{b} \\ \hat{c} \\ \hat{d} \end{bmatrix} \pm 1.96\left(\text{diag}[-H]^{-1}\right)^{0.5}. \qquad (10.63)$$

In Equation (10.63), the constant 1.96 is obtained from the standard normal distribution.

10.4.4 MOM

The rth noncentral moments of FPEG distribution may be expressed as

$$E(X^r) = \int_d^{\infty} \frac{x^r c^a}{b\Gamma(a)} (x - d)^{\frac{a-b}{b}} \exp\left(-c(x - d)^{\frac{1}{b}}\right) dx. \qquad (10.64)$$

Applying variable transformation, that is, Equation (10.13), Equation (10.64) may be rewritten as

$$E(X^r) = \frac{1}{\Gamma(a)} \int_0^{\infty} \left[\left(\frac{z}{c}\right)^b + d\right]^r z^{a-1} \exp(-z) dz. \qquad (10.65)$$

From Equation (10.65), the first four moments about the origin are given as

$$E(X) = \frac{1}{\Gamma(a)} \int_0^\infty \left[\left(\frac{z}{c}\right)^b + d \right] z^{a-1} \exp(-z)dz$$

$$= \frac{1}{\Gamma(a)} \left(\int_0^\infty (c^{-b}z^{a+b-1} + d \ z^{a-1}) \exp(-z)dz \right) = \frac{\Gamma(a+b)}{c^b\Gamma(a)} + d$$

$$(10.66)$$

$$E(X^2) = \frac{1}{\Gamma(a)} \int_0^\infty \left[\left(\frac{z}{c}\right)^b + d \right]^2 z^{a-1}\exp(-z)dz = \frac{\Gamma(a+2b)}{c^{2b}\Gamma(a)} + \frac{2d\Gamma(a+b)}{c^b\Gamma(a)} + d^2$$

$$(10.67)$$

$$E(X^3) = \frac{1}{\Gamma(a)} \int_0^\infty \left[\left(\frac{z}{c}\right)^b + d \right]^3 z^{a-1}\exp(-z)dz$$

$$(10.68)$$

$$= \frac{\Gamma(a+3b)}{c^{3b}\Gamma(a)} + \frac{3d\Gamma(a+2b)}{c^{2b}\Gamma(a)} + \frac{3d^2\Gamma(a+b)}{c^b\Gamma(a)} + d^3$$

$$E(X^4) = \frac{1}{\Gamma(a)} \int_0^\infty \left[\left(\frac{z}{c}\right)^b + d \right]^4 z^{a-1} \exp(-z)dz$$

$$= \frac{\Gamma(a+4b)}{c^{4b}\Gamma(a)} + \frac{4d\Gamma(a+3b)}{c^{3b}\Gamma(a)} + \frac{6d^2\Gamma(a+2b)}{c^{2b}\Gamma(a)} + \frac{4d^3\Gamma(a+b)}{c^b\Gamma(a)} + d^4.$$

$$(10.69)$$

From Equations (10.66)–(10.69), the coefficient of variation, coefficient of skewness, and coefficient of excess kurtosis are given as

$$C_v = \frac{\sqrt{\Gamma(a+2b)\Gamma(a) - \Gamma^2(a+b)}}{\Gamma(a+b) + dc^b\Gamma(a)}$$

$$(10.70)$$

$$C_s = \frac{\Gamma^2(a)\Gamma(a+3b) - 3\Gamma(a)\Gamma(a+b)\Gamma(a+2b) + 2\Gamma^3(a+b)}{\left(\Gamma(a)\Gamma(a+2b) - \Gamma^2(a+b) \right)^{\frac{3}{2}}}$$

$$(10.71)$$

$$C_k = \frac{\Gamma^3(a)\Gamma(a+4b) - 4\Gamma^2(a)\Gamma(a+b)\Gamma(a+3b) + 12\Gamma(a)\Gamma^2(a+b)\Gamma(a+2b) - 3\Gamma^2(a)\Gamma^2(a+2b) - 6\Gamma^4(a+b)}{\left(\Gamma(a)\Gamma(a+2b) - \Gamma^2(a+b) \right)^2}.$$

$$(10.72)$$

Based on the structure of the PDF of the FPEG distribution, Song et al. (2017) applied the first two moments and $E((x - d)^{-1})$ and $E((x - d)^{1/b})$ for parameter estimation with MOM as

$$E\left((X-d)^{-1}\right) = \int_d^\infty \frac{c^a}{b\Gamma(a)}(x-d)^{\frac{a-2b}{b}}\exp\left(-c(x-d)^{\frac{1}{b}}\right)dx \qquad (10.73)$$

$$E\left((X-d)^{\frac{1}{b}}\right) = \int_d^\infty \frac{c^a}{b\Gamma(a)}(x-d)^{\frac{a+1}{b}-1}\exp\left(-c(x-d)^{\frac{1}{b}}\right)dx. \qquad (10.74)$$

Applying variable transformation, that is, Equation (10.13), Equations (10.73) and (10.74) may be evaluated as

$$E\left((X-d)^{-1}\right) = \frac{c^b}{\Gamma(a)}\int_0^\infty z^{a-b-1}\exp(-z)dz = \frac{c^b\Gamma(a-b)}{\Gamma(a)}; \quad \exists a > b \quad (10.75)$$

$$E\left((X-d)^{\frac{1}{b}}\right) = \frac{1}{\Gamma(a)}\int_0^\infty \frac{1}{c}z^a\exp(-z)dz = \frac{\Gamma(a+1)}{c\Gamma(a)} = \frac{a}{c}. \qquad (10.76)$$

With the constraints of Equation (10.75), Equations (10.66)–(10.68) and (10.76) are applied for parameter estimation using MOM by equating these equations to the corresponding sample moments.

Unlike the MLE method, the confidence bound for the estimated parameters estimated using MOM and entropy are approximated with the parametric bootstrap method outlined in the Appendix. Furthermore, the confidence bound may also be approximated with the parametric bootstrap method for the parameters estimated using MLE.

10.5 Application

In what follows, the real-world data are applied to investigate the applicability of the FPEG distribution.

10.5.1 Peak Flow

Peak flow data at USGS09239500 is applied for a case study example. MLE, MOM, and entropy methods are applied for parameter estimation. The parameter values estimated from the Pearson III distribution (i.e., $b = 1$ for FPEG) are used as the initial parameter values, which are $a = 62.04$, $b = 1$, $c = 0.25$, and $d = -142.94$. MOM is applied for the estimation of initial parameters.

Using the initial parameters listed above, Table 10.1 lists the parameters estimated using MLE, MOM, and entropy, their confidence interval, and

Table 10.1. *Parameters estimated, confidence intervals, and GoF results: peak flow.*

	Parameter estimated and confidence interval				GoF	
	a	b	c	d	D_n	*P*-value
MLE	47.78	0.94	0.15	−124.59	0.056	0.54
Confidence interval	[37.11, 53.86]	[0.92, 0.97]	[0.13, 0.17]	[−152.69, −99.93]		
MOM	62.04	1	0.25	−140.98	0.057	0.55
Confidence interval	[51.28, 74.64]	[0.96, 1.02]	[0.21, 0.28]	[−180.86, −117.54]		
	λ_1	λ_2	b	d		
Entropy	−49.60	0.15	0.94	−124.58	0.056	0.575
Confidence interval	[−56.99, −37.18]	[0.13, 0.17]	[0.92, 0.97]	[−150.41, −100.37]		
Equation (10.29):	$a = 47.78, b = 0.94, c = 0.15, d = -124.58$					

Table 10.2. *Quantiles (m^3/s) computed from empirical and fitted FPEG distributions: peak flow.*

Probability	Empirical	MLE	MOM	Entropy
0.5	101.88	101.55	101.67	101.55
0.8	128.20	128.61	128.59	128.61
0.9	146.76	143.54	143.42	143.54
0.95	161.25	156.28	156.06	156.28
0.99	191.24	181.21	180.76	181.20

goodness-of-fit (GoF) results. The results indicate that (1) the parameters estimated using MOM for FPEG converge to the initial parameter values, that is, Pearson III distribution; (2) the parameters estimated using MLE and entropy for FPEG yield very similar results if Equation (10.29) is applied to compute the equivalent population parameters for the entropy method; (3) all three methods yield very similar GoF results; and (4) FPEG may be properly applied to study the frequency for the peak flow dataset. Additionally, it is worth noting that the confidence intervals of the parameters are approximated with the parametric bootstrap method for all three methods.

Table 10.2 lists the quantiles computed from empirical distribution and the fitted FPEG distribution. From the table, it is seen that there is minimal difference between the quantiles computed from the empirical distribution and the

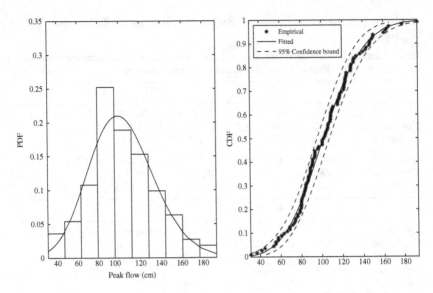

Figure 10.3 Comparison of empirical distribution with fitted FPEG distribution (MLE): peak flow.

fitted distribution for $Q_{0.5}$, $Q_{0.8}$, $Q_{0.9}$, and $Q_{0.95}$. Even though the largest absolute difference is found at $Q_{0.99}$ between the empirical distribution and the fitted FPEG with parameters estimated using MOM (i.e., $10.48\,\mathrm{m^3/s}$), the relative difference is about 5.5%. However, the results of the largest difference shown at $Q_{0.99}$ indicate that FPEG may underestimate the magnitude of peak flow for large events (in other words, events of higher return periods).

Figures 10.3–10.5 compare the empirical distribution with the fitted FPEG distribution. The comparison shows that all three methods yield very similar performances and confirms the application of FPEG to study the frequency of peak flow.

10.5.2 Maximum Daily Precipitation

The maximum daily precipitation at Brenham, Texas (GHCND: USC0411048), is applied for analysis. MLE and entropy methods are applied for parameter estimation. The parameter values estimated from the Pearson III distribution [i.e., $b = 1$ for FPEG] are used as the initial parameter values. Table 10.3 lists the parameters estimated, the corresponding confidence bound, and GoF test results. The results indicate that (1) the parameters estimated using MLE and entropy for FPEG yield very similar results if Equation (10.29) is applied to compute the equivalent population parameters for the entropy method; (2) the

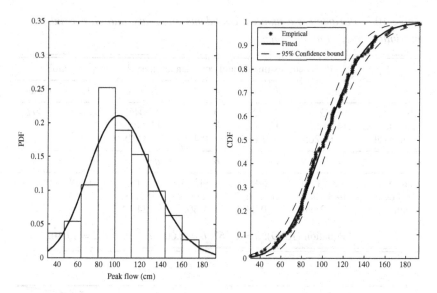

Figure 10.4 Comparison of empirical distribution with fitted FPEG distribution (MOM): peak flow.

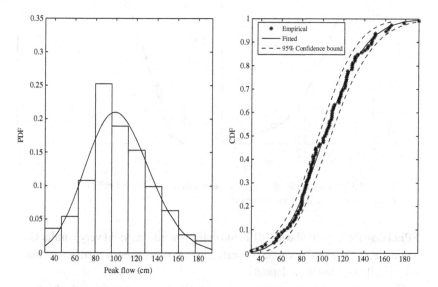

Figure 10.5 Comparison of empirical distribution with fitted FPEG distribution (entropy): peak flow.

Table 10.3. *Parameters estimated, confidence intervals, and GoF results: maximum daily precipitation.*

	Parameter estimated and confidence interval				GoF	
	a	b	c	d	D_n	P-value
MLE	6.383	1.342	0.249	17.045	0.04	0.892
Confidence interval	[4.952, 10.265]	[1.241, 1.655]	[0.245, 0.488]	[3.849, 31.289]		
	λ_1	λ_2	b	d		
Entropy	−3.741	0.249	1.343	17.102	0.039	0.945
Confidence interval	[−4.758, −3.044]	[0.239, 0.280]	[1.293, 1.406]	[10.218, 24.123]		

Equation (10.29): $a = 6.367$, $b = 2.607$, $c = 0.249$, $d = 17.102$

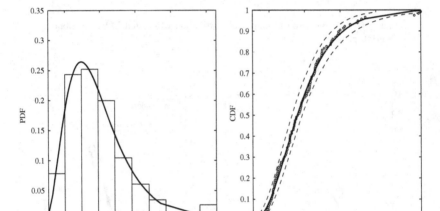

Figure 10.6 Comparison of empirical distribution with fitted FPEG distribution (MLE): maximum daily precipitation.

FPEG distribution fitted with MLE and entropy methods yield very similar GoF results; and (3) FPEG may be properly applied to frequency analysis of maximum daily precipitation datasets.

Figures 10.6 and 10.7 compare the empirical distribution with the fitted FPEG distribution using MLE and entropy methods. The comparison shows that the FPEG distributions fitted with MLE and entropy methods yield very similar performances and confirms the application of FPEG to do frequency analysis of maximum daily precipitation.

10.5.3 Total Flow Deficit

The total flow deficit at Tilden, Texas, is applied for the analysis. The MLE and entropy methods are applied for parameter estimation. The parameter values estimated from the Pearson III distribution [i.e., $b = 1$ for FPEG] are used as initial parameter estimates, which are $a = 1.013$, $b = 1$, $c = 1.63 \times 10^{-5}$, and $d = 585.16$. Using the initial parameter estimates listed above, Table 10.4 lists the parameters estimated, the corresponding confidence

Table 10.4. *Parameters estimated, confidence intervals, and GoF results: total flow deficit.*

	Parameter estimated and confidence interval				GoF	
	a	b	c	d	D_n	P-value
MLE	0.681	1.035	1.612E−05	631.813	0.074	0.259
Confidence	[0.525,	[1.010,	[1.168E−5,	[538.784,		
interval	0.769]	1.049]	1.612E−5]	921.813]		
	λ_1	λ_2	b	d		
Entropy	0.342	1.612E−05	1.035	631.879	0.074	0.219
Confidence	[0241,	[1.161E−05,	[1.101,	[542.04,		
interval	0.487]	1.614E−05]	1.048]	935.004]		
Equation (10.29)	$a = 0.681, b = 1.35, c = 1.612E - 05, d = 631.879$					

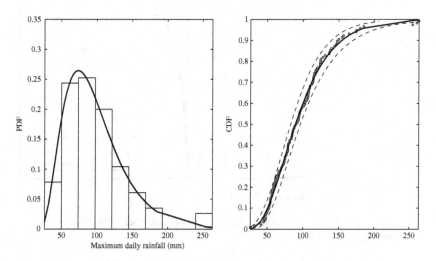

Figure 10.7 Comparison of empirical distribution with fitted FPEG distribution (entropy): maximum daily precipitation.

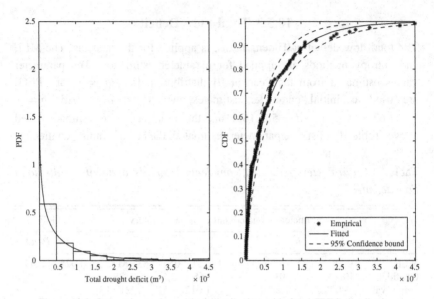

Figure 10.8 Comparison of empirical distribution with fitted FPEG distribution (MLE): total flow deficit.

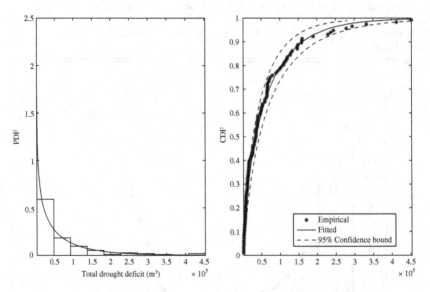

Figure 10.9 Comparison of empirical distribution with fitted FPEG distribution (entropy): total flow deficit.

bound, and GoF test results. The results indicate that (1) the parameters estimated using MLE and entropy for FPEG yield very similar results if Equation (10.29) is applied to compute the equivalent population parameters for the entropy method; (2) the FPEG distribution fitted with the MLE and entropy methods yield very similar GoF results; and (3) FPEG may be properly applied for frequency analysis of the total flow deficit.

Figures 10.8 and 10.9 compare the empirical distribution with fitted FPEG distribution using MLE and entropy methods. The comparison shows that the FPEG distributions fitted with MLE and entropy methods yield very similar performances and confirms the application of FPEG to do frequency analysis of the total flow deficit.

10.6 Conclusion

This chapter revisited the FPEG distribution. As a generalized four-parameter distribution, several popular distribution families may be derived as the special cases, including the Pearson type III (PIII) distribution and its special cases, three-parameter Weibull distribution and its special cases, and three-parameter gamma distribution and its special cases. The MLE, MOM, and entropy methods may be applied for the parameter estimation. In the case of the entropy method, besides the two Lagrange multipliers for the corresponding two constraints, two population parameters are embedded in the entropy function. The application example with the annual peak flow indicates all three methods yield very similar results and can be applied for frequency analysis. Additionally, we also evaluated the application of the FPEG distribution to maximum daily precipitation and total flow deficit with MLE and entropy methods. The investigation of maximum daily precipitation and total flow deficit indicates that the FPEG distribution may be applied to model these extreme events.

References

Ashkar, F. and Ouarda, T.B.M.J. (1998). Approximate confidence intervals for quantiles of gamma and generalized gamma distributions. *Journal of Hydrologic Engineering*, Vol. 3, pp. 43–51.

Song, S.B., Song, X., and Kang, Y. (2017). Entropy-based parameter estimation for the four parameter exponential gamma distribution. *Entropy*, Vol. 19, p. 189. DOI: 10.3390/e19050189.

11

Summary

Comparison of Real-World Applications

11.1 Introduction

In this final chapter, we summarize and compare the applications of the generalized distributions to real-world datasets, including peak flow, maximum daily precipitation, and total flow deficit, which have been investigated throughout the book. Besides the formal goodness-of-fit study using the Kolmogorov–Smirnov (KS) test statistics, the following two performance measures are added in this chapter:

Mean absolute bias (MAB):

$$\text{MAB} = \sum_{i=1}^{n} \frac{|X_{obs} - X_{pred}|}{X_{obs}} \Big/ n \tag{11.1}$$

and

Root mean square error (RMSE):

$$\text{RMSE} = \sqrt{\frac{\sum_{i=1}^{n} (X_{obs} - X_{pred})^2}{n}}. \tag{11.2}$$

In Equations (11.1) and (11.2), n is the sample size.

Table 11.1 lists the first four sample moments for the three real-world datasets applied throughout the book. It is shown that total flow deficit is most skewed to the right with a heavy tail followed by maximum daily precipitation, and peak flow is least skewed and has almost no tail. Thus, the real-world datasets investigated include less skewed and tailed, moderate skewed and tailed, and skewed and heavy tailed. Table 11.2 lists the empirical quantiles for each dataset. Additionally, we use BSM, Hal-A, Hal-B, Hal-IB, GG3, GBL,

Table 11.1. *Sample moments for three real-world datasets.*

Variable	Mean	Standard deviation	Skewness	Kurtosis	Sample size
Peak flow (m^3/s)	102.99	31.23	0.25	2.97	111
Maximum daily precipitation (mm)	97.32	43.76	1.42	6.04	115
Total flow deficit (m^3)	6.32E+04	8.38E+04	2.39	9.39	115

Table 11.2. *Quantiles obtained from empirical distribution.*

Probability	Peak flow (m^3/s)	Maximum daily precipitation (mm)	Total flow deficit (m^3)
0.5	101.880	89.916	3.510E+04
0.9	146.764	148.488	1.595E+05
0.95	161.253	175.768	2.513E+05
0.99	191.240	263.124	4.464E+05

FP, and EG to represent Burr–Singh–Maddala, Halphen type A, Halphen type B, Halphen type inverse B, generalized gamma, generalized beta Lomax, Feller–Pareto, and exponential gamma distributions, respectively.

11.2 Summary on Peak Flow Dataset

The peak flow data at USGS09239500 is applied as a case study throughout the book. In what follows we summarize the performances of fitted distributions to the peak flow dataset. As pointed out, the peak flow is slightly skewed to the right with a minimal tail (skewness and kurtosis for normal distribution is zero and three, respectively). Table 11.3 summarizes the distributions that may be applied, the KS test statistics, and the performance measures. It shows that except for the Hal-IB distribution, all other distributions may be applied to model the peak flow dataset.

For the distributions fitted with the maximum likelihood estimation (MLE) method, we found that (1) according to the KS statistics, the goodness-of-fit for the distributions may be ranked as: BSM > kappa > GG3 > EG > Hal-A > Hal-B > FP with the range of [0.046, 0.086]; (2) according to the performance measure MAB, fitted Hal-A and GBL distributions yield the smallest and the largest MAB of 1.862% and 2.896%, respectively; and (3) according to the

Table 11.3. *Distributions, KS statistics, and performance measures: peak flow.*

Distribution	Fit (Y/N)		KS statistic	MAB (%)	RMSE (m³/s)
BSM	Y	MLE	0.046	2.124	2.413
		Entropy	0.057	2.153	2.423
Hal-A (3-parameter)	Y	MLE	0.059	1.862	2.385
		Entropy	0.048	1.555	2.026
Hal-B (3-parameter)	Y	MLE	0.065	2.126	2.789
		Entropy	0.064	2.126	2.789
Hal-IB (3-parameter)	N	MLE	–	–	–
		Entropy	–	–	–
GG3 (3-parameter)	Y	MLE	0.054	2.206	2.563
		Entropy	0.054	2.206	2.563
GBL (4-parameter)	Y	MLE	0.059	2.896	3.003
		Entropy	0.064	2.897	3.003
FP (5-parameter)	Y	MLE	0.086	2.348	2.429
		Entropy	0.057	2.38	2.47
Kappa (4-parameter)	Y	MLE	0.053	2.15	2.47
		Entropy	0.053	2.15	2.47
EG (4-parameter)	Y	MLE	0.056	2.348	2.429
		Entropy	0.056	2.38	2.47

performance measure RMSE, the fitted Hal-A and GBL distributions again yield the smallest and the largest RMSE of 2.385 m³/s and 3.003 m³/s, respectively.

For the fitted maximum entropy-based distributions, we found that (1) according to the KS statistics, the goodness-of-fit for the distributions may be ranked as: Hal-A > FP > kappa > GBL > EG = BSM > Hal-B = GG3 with the range of [0.048, 0.064]; (2) according to the performance measure MAB, fitted Hal-A and GBL distributions again yield the smallest and the largest MAB of 1.555% and 2.897%, respectively; and (3) according to the performance measure RMSE, the fitted Hal-A and GBL distributions yield the smallest and the largest RMSE of 2.026 m³/s and 3.003 m³/s, respectively.

Overall, we found that (1) maximum entropy-based GG3, kappa, and four-parameter EG distributions converge to the fitted distribution using MLE; (2) maximum entropy-based BSM, Hal-B, GBL, and kappa distributions yield slightly larger KS statistics than those fitted with MLE; (3) maximum entropy-based Hal-A and FP distributions yield slightly smaller KS statistics than those fitted with MLE; and (4) statistically, the Hal-A distribution may be the preferred distribution to model the peak flow dataset followed by the BSM distribution.

Table 11.4. *Computed quantiles for the fitted distributions (MLE):*
peak flow (m³/s)

Quantile	BSM	Hal-A	Hal-B	GG3
0.5	101.99	99.04	103.06	101.72
0.9	143.08	143.44	142.97	143.93
0.95	155.81	158.38	154.25	156.29
0.99	182.14	189.36	175.40	179.74
	GBL	FP	Kappa	EG
0.5	99.77	100.87	101.47	100.87
0.9	148.30	144.13	144.15	144.13
0.95	164.43	157.61	156.79	157.61
0.99	197.65	184.34	179.79	184.34

Table 11.5. *Computed quantiles for the fitted distributions (entropy):*
peak flow (m³/s)

Quantile	BSM	Hal-A	Hal-B	GG
0.5	101.93	99.85	102.54	101.72
0.9	143.15	144.48	143.27	143.94
0.95	155.91	159.15	154.94	156.29
0.99	182.11	189.17	176.94	179.74
	GBL	FP	Kappa	EG
0.5	99.59	100.87	101.47	100.88
0.9	146.10	144.13	144.15	143.99
0.95	161.49	157.61	156.79	157.42
0.99	193.10	184.34	179.79	184.06

Tables 11.4 and 11.5 list the computed quantiles from the fitted distributions. Figures 11.1 and 11.2 compare the observed peak flow with the predicted peak flow. Comparisons show that (1) Hal-A and GBL distributions seem to overestimate peak flow at low quantiles; (2) Hal-B and kappa distributions seem to underestimate peak flow at high quantiles; and (3) in general, all eight fitted distributions may properly model the peak flow dataset visually.

11.3 Summary on Maximum Daily Precipitation Dataset

The maximum daily precipitation at Brenham, Texas (GHCND: USC0041148), is applied as the second case study throughout the book. As pointed out in

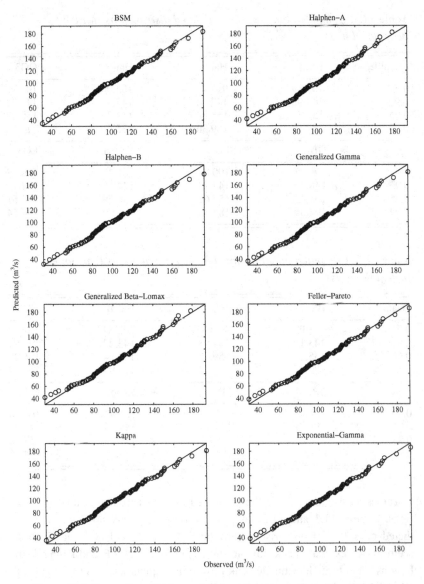

Figure 11.1 Comparison of observed and predicted peak flow from the fitted distributions (MLE).

Table 11.1, the maximum daily precipitation is moderately skewed to the right with tail. Table 11.6 summarizes the distributions that may be applied, the KS test statistics, and the performance measures. It shows that except for Hal-IB distribution, all other distributions may be applied to model the maximum daily precipitation dataset.

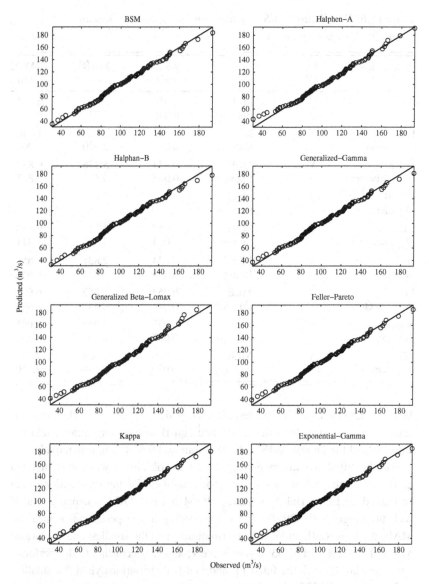

Figure 11.2 Comparison of observed and predicted peak flow from the fitted distributions (entropy).

For the distributions fitted with the MLE method, we found that (1) according to the KS statistics, the goodness-of-fit for the distributions may be ranked as: EG > Hal-A > GBL > GG3 = kappa > FP > BSM > Hal-B with the range of [0.046, 0.074]; (2) according to the performance measure MAB, the fitted EG and Hal-B distributions yield the smallest and the largest

Table 11.6. *Distributions, KS statistics, and performance measures:*
maximum daily precipitation.

Distribution	Fit (Y/N)		KS statistic	MAB (%)	RMSE (mm)
BSM	Y	MLE	0.058	3.064	6.053
		Entropy	0.046	2.832	6.611
Hal-A (3-parameter)	Y	MLE	0.042	2.880	7.040
		Entropy	0.039	2.940	6.900
Hal-B (3-parameter)	Y	MLE	0.074	5.376	9.838
		Entropy	0.075	5.383	9.837
Hal-IB (3-parameter)	N	MLE	–	–	–
		Entropy	–	–	–
GG3 (3-parameter)	Y	MLE	0.047	3.71	9.334
		Entropy	0.054	3.358	9.421
GBL (4-parameter)	Y	MLE	0.044	2.831	6.656
		Entropy	0.044	2.831	6.656
FP (5-parameter)	Y	MLE	0.053	3.853	6.608
		Entropy	0.047	2.840	6.498
Kappa (4-parameter)	Y	MLE	0.047	2.855	6.685
		Entropy	0.057	3.03	8.59
EG (4-parameter)	Y	MLE	0.040	2.831	7.437
		Entropy	0.039	2.840	7.410

MAB of 2.831% and 5.376%, respectively; and (3) according to the performance measure RMSE the fitted BSM and Hal-B distributions again yield the
smallest and the largest RMSE of 6.053 mm and 9.838 mm, respectively.

For the fitted maximum entropy-based distributions, we found that (1)
according to the KS statistics, the goodness-of-fit for the distributions may
be ranked as: EG = Hal-A > GBL > BSM > FP > GG3 > kappa > Hal-B
with the range of [0.039, 0.075]; (2) according to the performance measure
MAB, the fitted GBL and Hal-B distributions yield the smallest and the largest
MAB of 2.831% and 5.383%, respectively; and (3) according to the performance measure RMSE, the fitted FP and Hal-B distributions yield the smallest
and the largest RMSE of 6.498 mm and 9.837 mm, respectively.

Overall, we found that (1) the maximum entropy-based GBL and Hal-B distributions converge to the fitted GBL and Hal-B distributions using MLE; (2) the
maximum entropy-based BSM, Hal-A, and FP distributions yield slightly smaller
KS statistics than those fitted with MLE; and (3) the maximum entropy-based GG3
and kappa distributions yield slightly larger KS statistics than those fitted with MLE.

Tables 11.7 and 11.8 list the computed quantiles from the fitted distributions. Figures 11.3 and 11.4 compare the observed maximum daily

Table 11.7. *Computed quantiles for the fitted distributions (MLE): maximum daily precipitation (mm).*

Quantile	BSM	Hal-A	Hal-B	GG3
0.5	89.492	88.817	95.409	91.936
0.9	150.430	154.849	152.623	152.791
0.95	178.110	180.404	170.514	173.643
0.99	256.713	237.929	205.616	217.202
	GBL	FP	Kappa	EG
0.5	88.780	86.923	88.736	89.260
0.9	153.049	159.580	153.382	154.102
0.95	179.158	193.134	179.549	178.481
0.99	242.182	283.507	241.488	232.517

Table 11.8. *Computed quantiles for the fitted distributions (entropy): maximum daily precipitation (mm).*

Quantile	BSM	Hal-A	Hal-B	GG3
0.5	88.787	88.837	92.254	91.456
0.9	153.507	155.208	156.199	151.061
0.95	179.693	180.885	176.780	171.912
0.99	242.509	238.658	217.725	216.171
	GBL	FP	Kappa	EG
0.5	88.780	88.727	90.117	89.239
0.9	153.049	153.307	149.903	154.191
0.95	179.158	179.767	172.610	178.627
0.99	242.182	244.196	223.750	232.807

precipitation with the predicted maximum daily precipitation. Comparisons of all distributions showed similar trends: (1) for low quantiles, the predicted values match the observed values very well, except for the Hal-B distribution; (2) for medium quantiles, the fitted distributions again properly predict the maximum daily precipitation; and (3) for high quantiles, the fitted distributions seem to underestimate the maximum daily precipitation.

11.4 Summary on Total Flow Deficit Dataset

The total flow deficit at Tilden, Texas, is applied as the last case study throughout the book. As pointed out in Table 11.1, the total flow deficit is

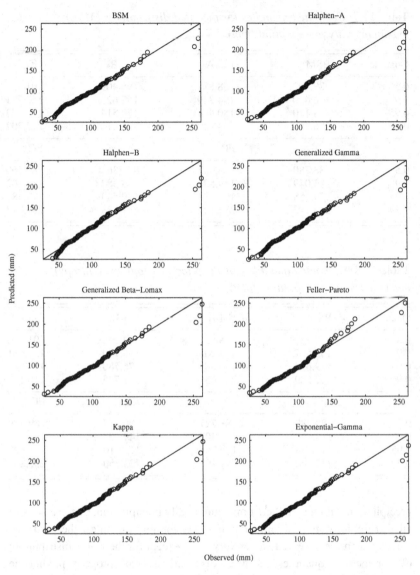

Figure 11.3 Comparison of observed and predicted maximum daily precipitation from the fitted distributions (MLE).

skewed to the right with tail. Table 11.9 summarizes the distributions that may be applied, the KS test statistics, and the performance measures. Table 11.9 shows that (1) the Hal-IB and maximum entropy-based FP distributions may not be applied to model total flow deficit because they failed the goodness-of-

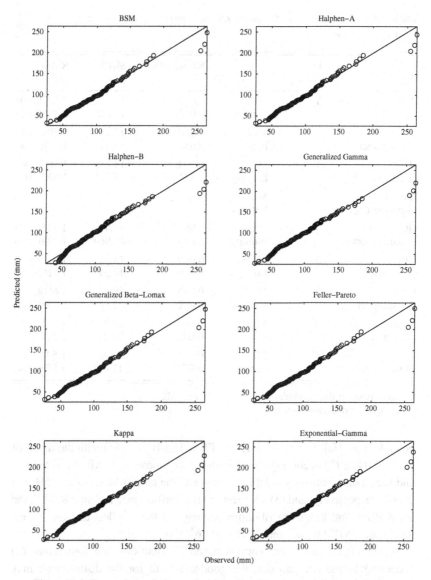

Figure 11.4 Comparison of observed and predicted maximum daily precipitation from the fitted distributions (entropy).

fit test and (2) the Hal-A and GG3 distributions fitted with MLE and maximum entropy-based Hal-B distribution barely pass the goodness-of-fit test.

For the distributions fitted with the MLE method, we found that (1) according to the KS statistics, the goodness-of-fit for the distributions may

Table 11.9. *Distributions, KS statistics, and performance measures: total flow deficit.*

Distribution	Fit (Y/N)		KS statistic	MAB (%)	RMSE (m^3)
BSM	Y	MLE	0.072^a	12.14	9.506E+03
		Entropy	0.053	14.875	1.486E+04
Hal-A	Y	MLE	0.050	14.96	1.58E+04
(3-parameter)		Entropy	0.0512	17.44	1.33E+04
Hal-B	Y	MLE	0.094	18.319	1.88E+04
(3-parameter)		Entropy	0.111^a	25.735	1.63E+04
Hal-IB	N	MLE	–	–	–
(3-parameter)		Entropy	–	–	–
GG3	Y	MLE	0.086^a	17.448	1.331E+04
(3-parameter)		Entropy	0.078	14.964	1.578E+04
GBL	Y	MLE	0.069	14.51	1.720E+04
(4-parameter)		Entropy	0.069	14.51	1.720E+04
FP	Y	MLE	0.085	19.22	1.680E+05
(5-parameter)		Entropy	0.092^b	23.16	5.998E+04
Kappa ($h = 0$)	Y	MLE	0.097	29.4	2.37E+05
(4-parameter)		Entropy	0.097	29.4	2.37E+05
EG	Y	MLE	0.074	11.451	1.52E+04
(4-parameter)		Entropy	0.074	11.451	1.52E+04

^{a} Barely passing the goodness-of-fit test with $\alpha = 0.05$.
^{b} Failing the goodness-of-fit test.

be ranked as: Hal-A > GBL > EG > FP > Hal-B > kappa with the range of [0.05, 0.097]; (2) according to the performance measure MAB, the fitted EG and kappa distributions yield the smallest and the largest MAB of 11.45% and 29.4%, respectively; and (3) according to the performance measure RMSE, the fitted BSM and kappa distributions again yield the smallest and the largest RMSE of 9.51E+03 m^3 and 2.37E+05 m^3, respectively.

For the fitted maximum entropy-based distributions, we found that (1) according to the KS statistics, the goodness-of-fit for the distributions may be ranked as: Hal-A > BSM > GBL > EG > GG3 > kappa with the range of [0.051, 0.097]; (2) according to the performance measure MAB, the fitted EG and kappa distributions yield the smallest and the largest MAB of 11.45% and 29.4%, respectively; and (3) according to the performance measure RMSE, the fitted Hal-A and kappa distributions yield the smallest and the largest RMSE of 1.33E+04 m^3 and 2.37E+05 m^3, respectively.

Overall, we found that (1) the maximum entropy-based GBL, EG, and kappa distributions converge to those fitted with the MLE method and (2)

Table 11.10. *Computed quantiles for the fitted distributions (MLE): total flow deficit (m^3).*

Quantile	BSM	Hal-A	Hal-B	GG3
0.5	3.266E+04	2.921E+04	3.79E+04	3.714E+04
0.9	1.555E+05	1.679E+05	1.60E+05	1.670E+05
0.95	2.270E+05	2.444E+05	2.15E+05	2.288E+05
0.99	4.450E+05	4.409E+05	3.41E+05	3.791E+05
	GBL	FP	Kappa	EG
0.5	3.107E+04	2.553E+04	2.518E+04	3.495E+04
0.9	1.549E+05	2.163E+05	2.006E+05	1.602E+05
0.95	2.411E+05	4.213E+05	4.270E+05	2.202E+05
0.99	5.742E+05	1.603E+06	2.327E+06	3.657E+05

Table 11.11. *Computed quantiles for the fitted distributions (entropy): total flow deficit (m^3).*

Quantile	BSM	Hal-A	Hal-B	GG
0.5	2.954E+04	3.145E+04	4.11E+04	3.547E+04
0.9	1.609E+05	1.600E+05	1.71E+05	1.595E+05
0.95	2.495E+05	2.253E+05	2.29E+05	2.186E+05
0.99	5.465E+05	3.871E+05	3.63E+05	3.621E+05
	GBL	FP	Kappa	EG
0.5	3.107E+04	2.541E+04	2.518E+04	3.495E+04
0.9	1.549E+05	1.799E+05	2.006E+05	1.602E+05
0.95	2.411E+05	3.256E+05	4.270E+05	2.202E+05
0.99	5.742E+05	1.044E+06	2.327E+06	3.657E+05

the maximum entropy-based BSM and GG3 distributions yield slightly smaller KS statistics than those fitted with MLE.

Tables 11.10 and 11.11 list the computed quantiles from the fitted distributions. Figure 11.5 and 11.6 compare the observed total flow deficit with the predicted total flow deficit. Comparisons for all distributions showed similar trends: (1) for low and medium quantiles, the predicted values match the observed values reasonably well; (2) the fitted BSM, Hal-A, Hal-B, GG, and EG distributions are found to underestimate the high quantiles of total flow deficit, while the FP and kappa distributions are found to overestimate the total flow deficit; and (3) the GBL distribution is found to be the best fit visually, followed by the BSM and Hal-A distributions.

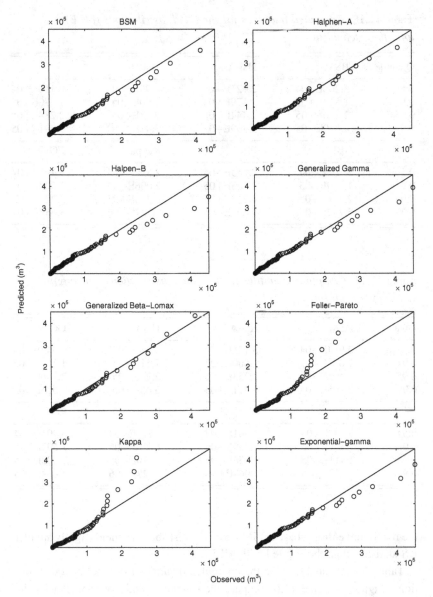

Figure 11.5 Comparison of observed and predicted total flow deficit from the fitted distributions (MLE).

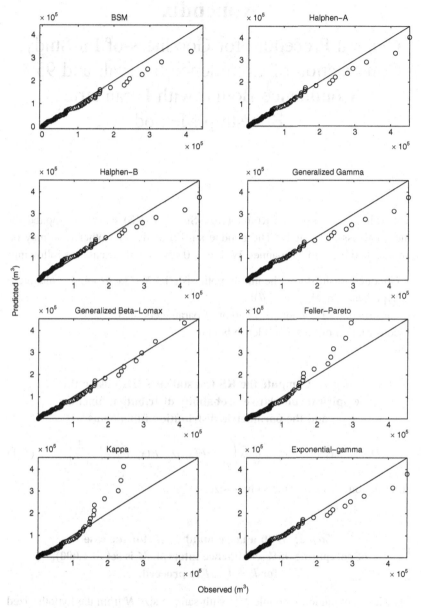

Figure 11.6 Comparison of observed and predicted total flow deficit from the fitted distributions (entropy).

Appendix

General Procedure for Goodness-of-Fit Study, Construction of Confidence Interval, and 95% Confidence Bound with Parametric Bootstrap Method

Throughout this book, the Kolmogorov–Smirnov (KS) statistic is applied for the goodness-of-fit study. The goodness-of-fit study with other tests may be approached in a similar manner. With $\alpha = 0.05$, we will discuss the following:

- Goodness-of-fit study: the null hypothesis (H_0) $X \sim F(x, \hat{\theta})$ and the alterative hypothesis $(H_a)X \neq F(x, \hat{\theta})$;
- construction of confidence interval of parameters; and
- construction of 95% confidence bound.

Step 1: **Compute the KS test statistics (D_N) using the empirical cumulative probability distribution function and the parametric distribution function as**

$$D_N = \max(\hat{\delta}_i); \quad \hat{\delta}_i = \max\left(\frac{i}{n} - F(x_i, \hat{\theta}), F(x_i, \hat{\theta}) - \frac{i-1}{n}\right). \quad \text{(A.1)}$$

In Equation (A.1), n is the sample size.

Step 2: **With a large number M (for the ease of computing 95% confidence interval, M is set to 1,000), for $k = 1 : M$ to proceed:**

(1) Generate random variable $X^{(k)}$ with sample size N from the hypothesized probability distribution $F(x; \hat{\theta})$.
(2) Re-estimate the parameter $\hat{\theta}^{*(k)}$ using the generated random variable $X^{(k)}$.
(3) Using Equation (A.1), compute $D_N^{*(k)}$.

(4) With $P = [0.01, 0.99]$ with step of 0.01, compute $X_1^{*(k)} = F^{-1}\left(P, \hat{\theta}^{*(k)}\right)$.

(5) Take the difference between the re-estimated parameter and parameters for the hypothesized distribution using

$$d = \hat{\theta} - \hat{\theta}^{*(k)}. \tag{A.2}$$

(6) Repeat Steps (1)–(4) for $M = 1,000$.

Step 3: Evaluate P-value using

$$P\text{-value} = \frac{\sum_{i=1}^{M} 1\left(D_N^{*(i)} > D_N\right)}{M}. \tag{A.3}$$

Step 4: Estimate the 95% confidence interval for the parameters.

(1) Sort parameter difference matrix d as the increasing order by column;

(2) For 95% confidence interval, the critical value of $d_{0.025}$ and $d_{0.975}$ are used to representing 2.5 and 97.5 percentile. Now, the 25th and 975th rows will be applied as the critical value for $M = 1,000$. And the confidence interval is given as

$$\theta \in [\hat{\theta} + d_{0.025}, \ \hat{\theta} + d_{0.975}]. \tag{A.4}$$

Step 5: Estimate the 95% bound.

(1) Sort the simulated 100 by M data matrix X_1 as the increasing order by rows.

(2) Similar to the 95% confidence interval for parameters, the 25th column and 975th column will be applied as the confidence bound.

Index